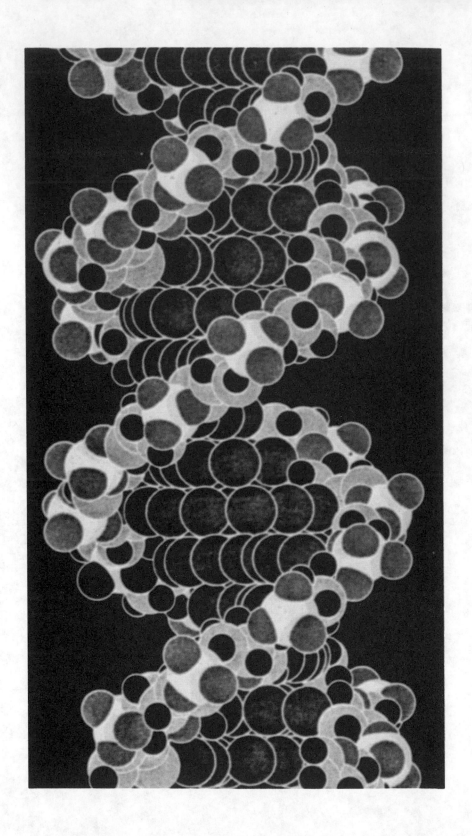

CORRECTING THE CODE

Inventing the Genetic Cure for the Human Body

LARRY THOMPSON

SIMON & SCHUSTER
New York   London   Toronto   Sydney   Tokyo   Singapore

SIMON & SCHUSTER
Rockefeller Center
1230 Avenue of the Americas
New York, New York 10020

Designed by Deirdre C. Amthor
Manufactured in the United States of America

1 3 5 7 9 10 8 6 4 2

Library of Congress Cataloging-in-Publication Data
Thompson, Larry.
Correcting the code : inventing the genetic cure for the human body / Larry
Thompson.
      p.   cm.
Includes bibliographical references and index.
1. Gene therapy. 2. Anderson, W. French, date. I. Title.
RB155.8.T48   1994
616'.042—dc20   93-42299
                            CIP

ISBN: 0-671-77082-9

For my wonderful daughters,
Dana Wynne and Julia Marie,
who gave their father the time to tell this story

Contents

Preface

So often, in the development of a field of thought or science, there is a key moment, an experiment, an event, that is both symbolically important, and yet almost anticlimactic. The first genetic treatment in September 1990 of a human being with an inherited illness was such a moment. In a sense, the injection of genetically altered white blood cells into the child's body was little more than a transfusion. The child sat on a bed, an intravenous tube was hooked to her arm, and her own white blood cells were infused into her circulation.

Yet those white blood cells were like no other cells in the history of the universe. They had been genetically engineered in the laboratory to make an enzyme the child had been born without. The treatment marked the first time that cells of a human had been genetically altered on purpose, and put back into the person's body.

That simple treatment opened up the field of human gene therapy to researchers worldwide who quickly launched dozens of tests in more than 100 people in the next two and a half years. Their work already is creating a revolution in the treatment of human disorders that will one day shift therapies away from catastrophic treatments like the use of poisonous chemotherapies to approaches that manipulate biology itself and heal from the inside out. Medicine in the next century will be profoundly different because of this experiment.

The revolution will not come tomorrow, but the first steps toward this brave new world have been taken. Even now, researchers are struggling to turn gene transfer techniques into treatments for diseases ranging from cancer to Duchenne's muscular dystrophy, from hemophilia to Parkinson's disease, from high cholesterol to AIDS.

The path to this day has taken decades and is littered with scientific careers that foundered along the way. This is the story of those who envisioned a day in their lifetime when it would be possible to put genes in people and change their genetic makeup. It is the story of the blind alleys and false starts, of the incremental progress that emerged from the genetic engineering revolution of the 1970s, and culminated on September 14, 1990, with the treatment of a four-year-old girl from the suburbs of Cleveland, Ohio. And it is especially the story of a handful of scientists around the country who deeply believed in the vision of gene therapy, particularly W. French Anderson, who spent most of his life in a government laboratory at the National Heart, Lung and Blood Institute, and who devoted more than two decades of his life to the pursuit of gene therapy.

It is also a story about competition and conflict among visionaries. Many scientists working on the fundamental techniques of gene transfer considered Anderson irrelevant during the early days. While they pursued the basics, Anderson concentrated on turning their developments into treatments for patients, and forging a path through the political and regulatory jungles of Washington that had to be traversed before the first person could be treated.

While Anderson was the first to cross the finish line with an approved gene therapy experiment, his achievements depended on many who went before him. Science is a collegial process in which researchers rely on each other for support, ideas, and hands to do the work. Many, many scientists who made substantial contributions to the development of gene therapy are not mentioned in this book. This book concentrates on the key players needed to tell the story while avoiding the confusing blizzard of characters who also participated in the field. These individuals, nonetheless, deserve credit for their contributions.

I do want to thank W. French Anderson for the time he gave to make this book possible. R. Michael Blaese and Kenneth Culver of the National Cancer Institute were equally generous with their time and ideas. And I want to specially thank the families of the first two children to receive gene therapy, the DeSilvas and the Cutshalls, whose courage and hope remain inspiring. For the many other scientists who sat for interviews and helped me understand the science, I am grateful. Any mistakes that may remain are mine. I want also to thank my editor, Gary Luke, for his support and encouragement, and my agent, Barbara Lowenstein, who made this all possible.

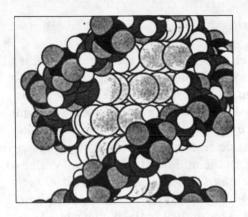

Chapter 1
A Cure for
Ashanthi

*"Diseases desperate grown
By desperate application are relieved,
Or not at all."*
   —*A quotation from* Hamlet,
   *Act IV, Scene III, on the wall
   in French Anderson's office*

Restless reporters, pens and microphones poised, packed a small plain conference room at the National Institutes of Health, the federal government's Mecca of biomedical research. Earlier that day, a chaotic round of urgent phone calls from NIH press officers summoned print and broadcast journalists from the major newspapers, networks, and trade magazines. The briefing would announce the beginning of a new era in medicine. No longer would doctors treat disease only with scalpels and powerful, often toxic, drugs. The time had come for illnesses to be cured from the inside out, by changing the patient's own genetic makeup. The molecules of heredity in a person would be manipulated for the first time. After years of front-page stories, public hearings, ethics debates, congressional questions, academic conferences, squabbles among scientists, lawsuits, general ballyhoo, and hype, human gene therapy was about to make its long-prophesied but often delayed debut.

At 3:30 P.M., Thursday, September 13, 1990, the scientists strode in, smiling and nodding to acquaintances as they wended their way through the throng of reporters. Dr. Kenneth Culver, thirty-five, was the youngest team member and the one who did the hands-on work with the patients. He worried about getting his share of the credit for the work they had done. In science, credit is everything. Without it, no one would know what Culver contributed to the first genetic treatment. Without acknowledgment, his efforts would mean little in his own career. Instead of getting a boost to help him with grants and promotions sometime in the future, he would be a footnote in the history of science—at best, just one author of the scientific paper describing the experiment, buried in the middle of a long list of collaborators. Culver had relaxed a bit earlier that week when his bosses included him in a photo shoot for *Time* magazine. Even if the photo never ran, to Culver it symbolized his arrival in the big leagues. They now accepted him as a member of the research team, not a technical support person. This day's press briefing, with most of the major media present, would inaugurate his Warhol time. Culver leaned against the wall and waited.

Dr. R. Michael Blaese (pronounced *blaze*), deputy chief of the National Cancer Institute's metabolism branch, stood just off to the side of the podium like a giant teddy bear. At 6 foot 3 inches, 250 pounds, he would tower imposingly were it not for his habit of hunching over to ease the pressure on his bad back. As the principal investigator, Blaese was the titular leader of the study and Culver's boss, and together they made a wonderful team. Blaese, by nature, tended to be quiet, thoughtful, cautious. An immunology expert, he had good, sometimes revolutionary, ideas. Culver, a pediatrician and bone marrow transplant expert who was extremely organized, energetic, and skilled with his hands, gave those ideas life. Blaese's more mature and circumspect view balanced Culver's youthful exuberance. Together, they got things done.

As the television lights snapped on, however, it was Dr. W. French Anderson, the slight, 5 foot, 9 inch, 150-pound chief of the heart institute's Molecular Hematology Branch, who first stepped to the microphone. Although Anderson officially took a back seat to Blaese, he had led the way to this experiment. Anderson, more than the other two, more than practically any other scientist in the country, burned with an uncontained fervor to begin human gene therapy. He had dedicated two decades of his life to the struggle of making it more than a dream. Since 1983, his vision and ambition guided the NIH gene therapy research effort that led directly to the treatment they were about to test. Through a combination of persistence in the laboratory and political savvy, Anderson had hurdled endless obstacles that stood between him and his goal. Now Anderson, a former track star, could see the finish line.

"I think Mike and Ken and I are amazed to see the size of the group here, particularly on such short notice," Anderson began, his silver hair shining in the floodlights. He went on to explain that in the morning, if everything went well, they would begin using gene therapy to treat a child with a severe, life-threatening immune disorder. They intended to put a normal copy of a critical gene—one she had been born without—into her white blood cells. The treatment, they hoped, would eventually restore her immune system. The importance of the experiment, however, went well beyond this one child, Anderson explained. "If this works, gene therapy might very well become a major new revolution in medicine. It should provide cures for what are presently incurable diseases."

The girl they were about to treat, whose name would not be released to protect her family's privacy, had been born with the same illness that had confined David, the boy in the bubble, to the germ-free plastic compartment that became part of his name, said Mike Blaese, taking over for Anderson. David had spent twelve years in his spotless plastic box at the Baylor College of Medicine in Houston. The sterile container was a lifeboat, shielding him from the sea of germs that would quickly kill him. And it forever separated him from his mother's kiss, his father's embrace.

Long before AIDS, David taught scientists, and the public, about the importance of the immune system and the dramatic consequence of its loss. He had been born with the most common form of severe combined immune deficiency, or SCID. A defective gene on the X chromosome—not identified until 1993, and not detectable in prenatal tests—causes the disease. Like people at the end stages of AIDS, children with SCID have no immune defenses against even the most benign bacteria and fungi. Unlike AIDS patients, whose disease is caused by a deadly virus that selectively destroys the immune system's white blood cells, SCID arises because of a gene mutation inherited at birth. The effect, however, is identical; the defective gene destroys the same white blood cells lost in AIDS patients. David, the longest known SCID survivor, died after suffering an infection by the Epstein–Barr virus, a member of the herpes family. The infection occurred during a bone marrow transplant intended to cure the SCID and free him from his plastic capsule.

• • •

The girl scheduled to receive the pioneering genetic treatment also had SCID. Like David, she inherited a genetic mutation, but hers was on a different chromosome, number 20, not on the X chromosome. The girl's mutation destroyed her ability to make a single enzyme called adenosine deaminase, or ADA. The ADA enzyme prevents the buildup of a poisonous chemical in the body. As the toxin accumulates, it kills only one thing, a white blood cell called the T cell. But it's a critical loss because T cells coordinate the body's immunological defenses against all infections. Without them, the body lacks immunity to everything. A common cold proves profoundly painful, and chicken pox becomes lethal.

Such immune disorders are deadly but rare. Fewer than 1 in 100,000 newborn babies will have it, and in 1993 only about forty children lived with ADA deficiency worldwide. Until the late 1980s, babies with the disease usually died by age two, but that changed in 1987 when Michael S. Hershfield at Duke University reported in *The New England Journal of Medicine* that a novel formulation of the missing ADA enzyme could be injected into the children's bodies to restore partial immunity. The treatment was not perfect, since the children failed to produce a completely normal immune system, and the long-term effectiveness of the expensive drug was unknown.

To launch the next generation of treatment, Blaese explained to the media, the NIH team would put millions of copies of the normal ADA gene inside the white blood cells of their patients. Treatment would be a demanding process, and the first steps in preparing for it already had taken place. Ten days earlier, at the beginning of September, the child and her family had come to the NIH's gigantic 500-bed hospital, the Warren Grant Magnuson Clinical Center, known on campus as Building 10. In the blood center, Ken Culver had drained several units of the girl's blood. Technicians extracted just the white blood cells, returning the red blood cells and plasma to her body. Then, Culver took the harvested white blood cells to a sterile lab where he intentionally infected them with a specially modified mouse virus, the Moloney murine leukemia virus. The scientists had genetically engineered the virus to be harmless and to carry the normal human ADA gene. Once it infected human cells, it could intertwine the gene into the cell's chromosomes. This procedure would make the ADA gene a permanent part of some of the girl's white blood cells, perhaps as many as 10 percent.

The genetically treated white blood cells would be able to make their own ADA enzyme. These cells, along with the unmodified white blood cells also growing in the laboratory culture, would be injected into the child. Once in the body, the cells would continue making ADA, detoxifying the cells carrying the gene and, if Anderson and Blaese were right, detoxifying the rest of the body so the child would develop normal immunity.

Theoretically, the treatment should work. Laboratory tests and animal studies showed that this normal gene would go into defective cells and safely produce the missing enzyme. The technique genetically corrected defective cells that were growing in the laboratory. They had been successfully and safely introduced into monkeys. The only thing left was the acid test: Try gene therapy in a human patient.

"We feel that gene therapy is potentially a major new therapeutic option," Anderson said at the conclusion of the press briefing. "It should have significant clinical effects in the next century. The technical ability to do gene therapy has been around now for several years. We feel that, as important as anything else, it is time to get started."

After about forty-five minutes, the press conference began to wind down. The reporters ran out of questions, and the camera crews snapped off their lights. Stragglers late for the briefing and reporters too shy to ask questions from the floor collared Anderson, Blaese, and Culver individually for follow-up questions. Paul Van Nevel, a thoughtful, longtime press official for the National Cancer Institute, stood in the back of the room throughout the press briefing. "I've been to a lot of press conferences," he said later, "but that one gave me goose bumps." It announced a sea change in the way doctors would one day treat patients. In the future, medicine would be different because of the experiment announced that day.

Even as Anderson pondered what was about to happen, that he and his colleagues were about to launch the gene therapy era, he worried about the dozens of things that could go wrong. In the morning, the final safety tests of the genetically altered white blood cells could show bacterial or viral contamination. Or hypothetically the inserted gene could convert a normal white blood cell into a leukemia, adding cancer to the child's already substantial list of medical problems. Many of the theoretical and technical questions, Anderson had argued, could be answered only by doing the experiment in people.

Still, on the experiment's eve, there was a major, unresolved legal matter: Was the Food and Drug Administration going to give them permission to proceed in the morning? Although the researchers were ready to go and the proposed experiment had been evaluated and accepted by more than a dozen governmental committees, although the press conference had been held to announce the new treatment, the FDA, the agency with the ultimate say, had yet to give its consent. Though the family, with their sick child, had arrived in Bethesda, the girl's cells growing to their peak in the laboratory, and the scientists fearing delay would damage the experiment, the FDA could still say no.

At the press briefing, Anderson explained that the FDA staff had given a preliminary go-ahead earlier in the week. The paperwork now had to percolate through the bureaucracy, gathering the needed signatures. In Washington, that can take a lifetime. The researchers, however, were optimistic that they would have the approvals within twenty-four hours. Despite their arms-length, and sometimes antagonistic, relationship, the regulatory branch of the federal government that oversees biomedical research, the FDA, had given private and personal assurances to the biomedical research branch of the government, the NIH.

At least, that is what Anderson's boss, Dr. Claude Lenfant, director of the National Heart, Lung and Blood Institute, thought when he ordered Anderson to hold the press briefing before the experiment began. Lenfant feared a media circus on the day of the experiment, similar to what had happened when Barney Clark received the first artificial heart transplant at the University of Utah. At the same time, he worried that Anderson and the others would be too busy taking care of their patient to broadcast the exciting news once the experiment started. Lenfant wanted to ensure the heart institute got its share of the credit. Despite Anderson's reputation as a public scientist, who tended to report his findings in the lay media before publishing them in scientific journals, it was his first press briefing in his twenty-five years at NIH.

Mike Blaese and Ken Culver thought the fix was in, too, as they blinked into the camera lights, dazzled by the media attention. As far as they knew, the FDA approval to start the trial in the morning was on its way and would arrive at any moment.

Anderson, however, had heard differently. Just before he walked into the news briefing, Diane Striar, a longtime heart institute spokeswoman, pulled Anderson aside with bad news. Her office, always cautious about public pronouncements, had called the FDA press office to make sure that the approval was imminent. An FDA press officer called the chief of the FDA division authorized to give final approval.

Its boss said no. Dr. Gerald V. Quinnan, Jr., director of the FDA Center for Biologics Evaluation and Research, said NIH "had a gentleman's agreement with the FDA that [it] would not start until the FDA had reviewed information that needed to be submitted," Quinnan said. As far as he knew, that process was not complete. Key FDA staffers who worked for Quinnan had been reviewing the gene therapy experiment, and they could not be reached. Anderson, Blaese, and Culver, the FDA chief said, could not begin.

Anderson was stunned. Through the doorway, he could see the roomful of waiting reporters, and knew he had to show up for the briefing his boss had called. There was nothing to do but carry on. Anderson decided not to tell anyone this latest news, not his boss, Lenfant, not his partners, Blaese and Culver. He worried that the news would upset his colleagues, that they might get flustered in front of the media and minimize the positive news of NIH scientists marching into the future. "The last thing we wanted to do was present a negative story. If we said we didn't know whether the FDA was going to approve it, and that was in the paper, then they [FDA] would sit on this," Anderson worried. Even if the press briefing had not been his idea, maybe, Anderson reasoned, he could use it to generate pressure on the FDA to approve the experiment. "We needed to build [such] a juggernaut . . . that the FDA had to approve it," he said. "They couldn't say no."

Besides, there already was plenty of pressure. A biological clock—the genetically transformed T cells growing in the laboratory—ticked loudly. In another twenty-four hours, they would begin to deteriorate. The cells either had to be transplanted back into the girl or had to be frozen to preserve them. Freezing always killed some cells, reducing the total available for treatment. Besides, the entire family, already anxious about the advent of an experimental treatment to be used on their baby, had arrived from Ohio and was waiting expectantly a quarter-mile away in the NIH Children's Inn. Everything scientific was in place. The word began to spread through wire stories and radio reports that the experiment could begin tomorrow if only the FDA would give its assent.

Even after the briefing ended, Anderson did not tell his colleagues about the possible delay. "That was wise," Culver later said of Anderson's decision. "I didn't need to be distracted. Everything was my problem, the cells and the patients." When he later learned about the FDA delay, Blaese laughed, saying, "I would not have gone to the press conference if I had known. He [Anderson] probably knew that. I felt pretty strongly that it was a go because they [FDA] had told us to bring the family in." But Blaese must have had some premonition of trouble because during the press conference he told the media, "If they [FDA] change their mind now, they are nuts."

Nuts or not, FDA approval was no longer certain, and Anderson decided that something else needed to be done to apply some pressure. He had worked hard to build a consensus for human gene therapy and didn't want it to fail at the last minute because of bureaucratic intransigence. The consensus building had required as much skill in politics as it did in the laboratory—maybe more. Research often seems like an arcane, high-tech world filled with lockstep logic, inexplicable gadgets, and ideas that only superbrains can comprehend, but in addition scientific progress often requires leadership, vision, and political skills. These attributes are dismissed as unimportant by most scientists, who value only that remarkably rare, lonely spark of innovative creativity that they describe as *elegant*.

Anderson's willingness to engage in the politics of science was critical for gene therapy to advance, but that won him neither friends nor respect among scientists. The biological technology needed to move genes from place to place in human cells was worked out in the mid 1980s, mostly in the labs of other scientists, although Anderson's team developed their own approaches. What remained was politics: convincing research review committees that the time had come to try gene engineering in human patients. Few researchers were willing to spend the time it took to give an idea public life and political persuasion. At this, Anderson had become a practiced master.

In the face of possible FDA rejection, Anderson decided to play one more card—the Rifkin card. Jeremy Rifkin, an antigenetics rabble-rouser and the president of the Foundation for Economic Trends in Washington, D.C., had sent a staffer to the NIH media briefing to hand out a press release. It announced Rifkin's intention to "demand a moratorium on all human gene therapy experiments until an advisory committee on human eugenics [was] established to review the social and ethical implications of the genetic engineering of humans." Rifkin, a short, balding, old-time liberal activist with a trademark mustache and Wharton doctorate in economics, had been a thorn in the scientific community's side for years. He sometimes succeeded in slowing down research with lawsuits or the threat of lawsuits. Rifkin disliked the concept of human gene therapy, which he considered a high-tech approach to eugenics, a modern Nazilike program devoted to genetic improvement of the human race.

After getting back to his office, Anderson began spreading the word through the government grapevine that if the FDA did not approve the pending gene therapy experiment, then it must have caved in to pressure from Rifkin. Since most scientists take a dim view of Rifkin's antiscience activities, for the FDA to acquiesce in Rifkin's demands would be most objectionable. Anderson would never know if his grapevine whispering had any effect; FDA insiders just smiled at the suggestion that such politicking made any difference.

. . .

Just down West Drive from the Warren Grant Magnuson Clinical Center, Raj DeSilva turned his brown, customized, seven-seat Mitsubishi van into the parking lot of the Children's Inn. Traveling with three small children, all girls, all born within a year and a half of one another, can be strenuous for any family. For the DeSilvas it was an ordeal. Two of the girls, Anoushka, the oldest, and Dilani, the youngest, suffered brain damage a few months after birth and were confined to wheelchairs by the resulting physical and mental handicaps.

Ashanthi Vinodani DeSilva, the middle child, appeared to be the healthiest of the three as she bounced through the van's open side door and onto the macadam. Yet Ashanthi was the reason the DeSilvas had driven nearly 400 miles to the National Institutes of Health. Ashanthi (pronounced *ah-shan-tee*; most people call her Ashi) was born September 2, 1986, with ADA deficiency that destroyed her body's defenses. Endless illnesses nearly killed her before her second birthday until a remarkable new drug pulled her back from the abyss. Now, she would be the first human in the history of the universe to undergo gene therapy, a potentially revolutionary treatment. As a medical pioneer, four-year-old Ashi DeSilva would be the first to step into a brave new future where the genes in her body would be intentionally changed to fight her illness.

As the NIH doctors explained the impending experiment to the press, the DeSilvas struggled to settle their family into NIH's Children's Inn, a bright, cheerful residence designed to make sick kids comfortable while they endure the rigors and pain of experimental treatments. Raj, forty-eight, and his wife, Van DeSilva, thirty-seven, had just completed the ten-hour drive from North Olmstead, Ohio, a Cleveland suburb, to Bethesda, Maryland. Raj unpacked the wheelchairs for Anoushka, five, and Dilani, two, and lifted them gently into the carriers. He pushed one wheelchair and Van the other as Ashi trailed behind. They passed through the inn's automatically sliding glass doors, picked up their keys at the front desk, and piled into the elevator that lifted them to the bright, open second floor filled with the smells of cooking and the clatter of children and utensils. Finally, the DeSilvas reached their rooms, one for the girls, one for the adults. The exhaustion of travel and anxiety about the unexpected washed over the parents. Only Ashi's energy remained. She wanted to play, to go to the inn's huge toy room.

The future had rushed in too fast for the DeSilvas, almost too fast to absorb. With three sick children to worry about, the family suddenly had to make sophisticated, informed choices about the risks and benefits of a future technology that promised to shake up medicine. And the doctors who developed this new treatment wanted to try it out on their baby.

Raj DeSilva knew all about new technologies and how their inventors quickly become advocates. He was a chemical engineer by training, and a research-and-development manager at B. F. Goodrich in Avon Lake, Ohio, about thirty miles outside Cleveland. He had seen scientists and engineers lose their objectivity. He had seen ideas that looked good on paper go bad when turned into multimillion-dollar manufacturing plants. Before any new technology would be turned on his child, he had questions—lots of them—about the unknowns in this experiment and the risk being taken with his baby's life.

For Van DeSilva, the issue of this experiment was much simpler. Although she was a trained nurse, Van relied more on a mother's hope and confidence. "They are the nicest people," Van said of the NIH scientists. "They are good, good people. Good doctors. This was not an experiment. I see it as a cure for Ashanthi. It is a cure for my Ashanthi."

For the family, the road to a possible cure had been long and painful. An improbable string of unfortunate events began in 1986 in Sri Lanka, when Van was pregnant with Ashi. The DeSilvas were planning to emigrate to the United States, but by the time the permissions finally arrived, the DeSilvas felt Van was too far along in her pregnancy to endure the transcontinental flight. They decided the second child would be born in Sri Lanka, and then the family would move to Ohio.

Before Ashi was born, disaster struck. Anoushka, the oldest daughter, began to run a fever. The family considered it typical for any child and did not much worry about it. Suddenly, the fever, apparently caused by an influenza infection, turned into encephalitis, a viral attack on the brain itself. As the child's immune system fought the virus, it caused an inflammation in the brain, a swelling that damaged the nerves controlling voluntary movement, thus crippling the child. Anoushka, who at ten months was just beginning to stand and walk independently, lost control of her legs and feet. Although not paralyzed, she could no longer walk. Anoushka entered a Colombo hospital in July 1986, and remained there until September, when Ashanthi was born just down the hall.

Even as the family struggled with Anoushka's problems, there were signs at birth that Ashi might not be completely normal. For example, the stump of Ashi's umbilical cord became infected. With everything else going on—her sister's illness and the upcoming move to America—no one attached much significance to Ashi's unusual infection, not even the doctors. The cord contamination went away, but Ashi's lungs seemed to remain more congested than most newborns'. About a month later, after Anoushka had been released from the hospital, the DeSilvas took their two daughters and moved to the United States. Shortly after arriving in Ohio, Ashi developed an unexplained high fever, but then got over it. It was initially worrisome, but the family dismissed it as a minor childhood illness. It foreshadowed what was to come.

Ashanthi would ultimately be diagnosed as suffering from severe combined immune deficiency. SCID children gradually suffer more and more serious infections by common germs that would not bother a healthy individual. Up to the mid 1980s, a SCID child usually died of an overwhelming ailment before the second birthday.

Although the disease is present at conception, SCID children are actually healthiest at birth, and Ashanthi was no exception. In the womb, her mother's blood provided the missing adenosine deaminase enzyme, or ADA, that Ashi could not make. Her mother's ADA detoxified Ashi's white blood cells, leaving them healthy. At birth, Ashanthi had an essentially normal immune system. After delivery, however, without the supply of her mother's detoxifying enzyme, a metabolic poison that began killing her white blood cells started accumulating in Ashi's body.

In addition, antibodies from Van DeSilva circulated in Ashi's blood at birth, just as they do in every baby's blood for the first six months or so of life. These maternal antibodies provide passive immunity that temporarily protects an infant from infections while its immune system matures. Antibodies are special proteins produced by certain white blood cells whenever a virus or bacteria enters the body. The immune system produces one set of antibodies for every virus and bacteria it encounters, and the antibodies circulating in Van DeSilva reflected her years of exposure to various germs. They persisted in Van's blood for months, even years, to provide long-term immunity that quickly neutralizes any infectious invader.

Many of Van's antibodies had crossed the placenta into Ashi's blood, giving her instant, if transient, protection. As Ashi aged, however, her own immune system should have taken over and begun producing its own antibodies as the immunologic shield she received from her mother faded. That, however, could never happen. Ashi's white blood cells were dying almost as soon as her bone marrow gave them birth. She had inherited defective forms of the ADA gene. Instead of carrying two normal copies of the ADA gene in their own bodies, Van and Raj DeSilva each had a single normal copy of the ADA gene and one mutated version. As a result, their bodies made half as much ADA enzyme as usual, but enough for them to have normal immunity. For Ashi, it was a different story. She had inherited two dysfunctional copies of the ADA gene, one broken gene from each parent (a one-in-four chance). Without any normal copies of the ADA gene, she could not produce any ADA enzyme. Without the enzyme, she could not eliminate the metabolic toxin that began sabotaging her immune system.

By the time she was nine months old, Ashi started to get sick. At first, the illnesses were simple, a runny nose, an upset stomach. Then she began to have a continuous cold because her weakened white blood cells could not kill off the normally innocuous rhinoviruses that cause most runny noses. Then more fierce, flulike symptoms began, causing fevers, intensifying the infant's misery. Ashi worsened. She began to suffer bacterial infections and had a difficult time breathing, presumably because of continuous low-grade lung ailments. She began to vomit frequently and could not hold down food. This led to weight loss and a condition known as "failure to thrive." She was beginning to wither away. Ashi, previously a quiet, happy infant, began to cry all the time from the continual discomfort she suffered.

The DeSilvas began the long odyssey so familiar to families with a child suffering a rare illness. They tried doctor after doctor, searching for an explanation and a solution. By now they were living in

North Olmstead, Ohio, and the local physicians jumped to the first, most likely conclusions: asthma, maybe bronchitis, probably allergies. The doctors said the DeSilvas needed to clean their home of dust and mites and mold. They refurbished their apartment, making clean rooms, bought special sheets and pillowcases made from natural fibers, and got a device to clear her lungs. The doctors put Ashi on Bactrim, a mixture of wide-spectrum antibiotics, to prevent infections, and prescribed Alupent and Ventolin, two bronchodilators used to make breathing easier for asthmatics, even though Ashi lacked the wheezing symptoms of that disorder. None of it helped.

Then, at about age one, Ashi received one of the standard childhood vaccinations made from live, but weakened, strains of virus. It nearly killed her. After returning home from the doctor's office, Ashi spiked a tremendous fever, and then she turned blue. Panicked, Van DeSilva rushed her daughter to the emergency room. The ER doctor said she had an ear infection and sent her home.

Meanwhile, Van became pregnant with a third child early in 1987. Since she was in her midthirties and had two sick children, her obstetrician did an amniocentesis to check for genetic disease. With a long needle, he plucked some of the infant's cells from the fluid filling the placenta. The fetal cells can be examined to detect some chromosome abnormalities. Since chromosomes carry DNA, the genetic storage chemical, damage to the chromosome usually indicates damage to the genes—the way a badly dented box predicts that its contents are damaged. The doctor performing the test on Dilani's cells found a chromosomal translocation, a kind of a mix-up in which a piece from one of the twenty-four types of chromosomes in every cell breaks off and attaches itself to a different chromosome. Usually, this suggests disease is likely, but the Ohio doctor assured the DeSilvas that the chromosome defect was harmless. Van decided not to have an abortion; Dilani appeared healthy at birth.

A few months later, early in 1988, B. F. Goodrich sent Raj DeSilva to Hattiesburg, Mississippi. Raj had been managing, from Ohio, the multimillion-dollar purchase and conversion of a manufacturing plant. Goodrich needed him on site for a few months, so he decided to move his family with him. Ashi was no better; if anything she was worse. The vomiting intensified to the point where the family fed her enriched baby formulas to keep her alive. "We knew the doctors were shooting in the dark," Raj said. "They didn't know what was wrong with her."

So the family took its own precautions. "We said, 'Let's not expose her to any severe infection or disease.' So we started keeping her away from crowds. We protected her," said Raj, who always washed his hands whenever he came in from the outside. He hoped this would minimize the risk of infecting his daughter.

Ashi stayed home. No visits to the playground, preschool, or trips to the mall, and no toddler friends to teach her social skills. By her second birthday, Ashanthi DeSilva had become a recluse. Of course, it meant Van stayed in, too. Raj left for work each day; Van was home-bound with three children, one with brain damage, one miserable from incessant illnesses, and an infant. Their decision to limit Ashi's exposure meant no baby-sitters they did not know to be sniffle-free. For Van, that decision meant little relief from the long hours, days, weeks of caring for her children. When the illnesses struck her children, she took a philosophical view: At least they had access to doctors and medical insurance and a home; they were clean and comfortable. And she had Raj, who worked hard, provided well, and deeply loved his family.

Early in 1988, Raj moved them into a two-bedroom, ground-floor apartment in Hattiesburg, Mississippi. Three days later, Dilani, the infant, began to get sick, probably a cold she picked up from Ashi, the parents reasoned. Although she was running a high temperature, Raj DeSilva went to work. By midmorning, Van called, distressed by how lifeless Dilani appeared. The wife of a Goodrich workmate rushed Van and the girls to a hospital in Jackson, eighty miles away. Dilani arrived perilously ill. Emergency room doctors immediately admitted her, and put her on oxygen. The doctors did not know what was wrong with the child, who seemed to be deteriorating as they watched. By the time Raj reached the hospital, Dilani already had suffered severe brain damage and was so ill that the doctors expected her to die. The DeSilvas were stunned; their third child too had now been struck down. Maybe the chromosomal translocation identified before birth had caused it. None of the doctors who had seen the child, from Mississippi to Maryland, ever figured out what happened.

As suddenly as the malady came, it seemed to go. In less than a week, the Jackson doctors handed over a stable, but brain-damaged, child. Dilani, a mere four months old, suffered such profound nerve injury from the episode that she could not even recognize her mother. The accumulation of tragedy threatened to overwhelm the family. Not only did the DeSilvas have a physically handicapped oldest and a chronically ill middle child, but now the youngest was inexplicably and profoundly impaired.

The only good news in the summer of 1988 came from B. F. Goodrich: Raj was finished with the Mississippi project, and the De-Silvas could move back to Ohio. Anoushka and Dilani's conditions remained stable. Ashi, however, stayed dreadfully sick and unable to eat; she continued losing weight. The family returned to their local Ohio pediatrician, who still didn't know what ailed her. In desperation, Raj called his older brother, Anthony DeSilva, a New York physician specializing in allergies. "Nothing is working," Raj complained. "They can't figure out what is wrong with her."

Anthony did not believe his niece had allergies, and neither did Raj. She was too sick; simple allergies don't cause the vomiting and failure to thrive Ashi suffered. "Ask the pediatrician to order a special test that measures the levels of the different classes of antibodies in her blood," Anthony suggested. "Maybe she is deficient in one of them. Then, have them send me a copy of the test results, and we'll see what they say."

Raj, who can be forcefully persuasive when it comes to his children's health, had the pediatrician run the specialized test. A pediatric allergist was called in to run a full range of allergy tests just to make sure nothing had been missed. Allergy testing, of course, would prove nothing. Ashi's crippled immune system could not mount a reaction to the jabs that injected dust and molds and cat dander under the skin. When the antibody test results came back, the pediatrician called Raj and said everything seemed normal. Raj wanted to see the test report for himself. Next to the measurements from Ashi's blood, the printout listed the normal values for the different classes of antibodies: IgG, IgA, IgE, IgM.

"This is low," Raj said confronting the physician about the antibody values. "They are all low, or below the normal range. How can you say these are normal?" he demanded. In every class of antibodies, Ashanthi had fewer antibodies in her blood, on average, than normal people. Clearly, she was sick all the time because she lacked the immunity to stave off infection. The physician took the paper from Raj's hand and looked at it closely. "Yeah," he said. "You are right. It's not normal. I don't know what to make of it."

Dr. Velma Paschal, the local pediatric allergist who ran the allergy tests, agreed. "These certainly are abnormal," she told the DeSilvas. "I would like to show them to Dr. Ricardo Sorensen," a pediatric immunologist at Rainbow Babies and Children's Hospital, part of University Hospital in Cleveland. She worked with Sorensen in the children's immunology clinic. They scheduled an appointment.

Cleveland is a twenty-five-minute drive from North Olmstead. Raj and Van DeSilva piled Ashi and the others into the car in September 1988 and headed for Cleveland in their endless pursuit of answers. They had been to so many doctors before that they were skeptical. Perhaps this time might be different.

. . .

Ricardo U. Sorensen, a Chilean, has a calm, quiet, giving manner with people. He is gentle with his young patients and urges parents to call him anytime. He has watched with joy as the tools of modern medicine healed many children. But so, too, he has shared the suffering of many families who watch helplessly as their children deteriorate from an incurable illness. Ashi had one of those rare, terrible illnesses he knew too well.

When Sorensen first came to Cleveland in March 1976, his boss, Dr. Stephen H. Polmar, then head of immunology at Rainbow Babies, already had been taking care of a boy with ADA deficiency. Polmar discovered that red cells, for unknown reasons, contain plenty of the ADA enzyme, and pioneered the use of red blood cell transfusions to treat the deficiency. Transfusions provided enough of the enzyme to detoxify the boy and restore his immune system somewhat, but transfusions didn't work for all ADA patients. It also caused complications, such as iron overload, and it carried the risk of blood-borne infections. It did, however, prove for the first time that enzyme replacement therapy was possible. Purified ADA protein alone, however, failed to help. Normally found inside the protective membranes of cells, the ADA protein proved too fragile when injected directly into the blood. It survived only a few minutes, not long enough to help.

Polmar's patient survived on the transfusion therapy from the time of his birth—doctors diagnosed the ADA disorder while he was still in the womb because an older sister had died of it—until he was eight years old. His immune system was never totally normal, but adequate. He fought off the regular colds and infections of childhood, at least well enough to go to public school. But just before going on a midyear school vacation, the boy contracted chicken pox. In a person without normal cellular immunity, the herpes virus that causes the pox spreads like wildfire. He was rushed to the hospital with a very rapidly progressing infection. Those were the days before modern antiviral drugs, such as acyclovir. Polmar and Sorensen had nothing effective to offer. The boy died in twelve hours as they watched helplessly.

That bitter experience sensitized Sorensen to the rare ADA deficiency, and to the catastrophic consequences of chicken pox in immune-suppressed individuals. He started to look for ADA deficiency
routinely.

. . .

Most children with ADA deficiency get sick soon after birth, usually by six months of age. Sorensen's first ADA patient was different.
Born on July 6, 1981, in Canton, Ohio, Cynthia Cutshall appeared to
be a normal baby. She occasionally contracted and fought off the usual runny nose—until she was about three years old. Then she started preschool, where she was around other kids with the usual range
of infections. She began to develop chronic sinus infections that
would frequently and rapidly deteriorate into pneumonia. "She was
never exposed to any sick children as an infant," said her mother, Susan Cutshall, a recovery room nurse at Aultman Hospital in Canton.
"I think that is what saved her from getting sick early."

With exposure to other children, Cindy began to suffer serious, potentially life-threatening infections that required powerful antibiotics
to quash. "She would have such constant, severe sinusitis that she
would get up to start her day coughing so hard that she would bring
up her breakfast," her mother said. "She would upchuck and then say,
'OK, I'm fine now. Let's go.' "

This went on for a year and a half until, suddenly, one of the sinus
infections with a strange form of *Streptococcus pneumoniae* infected her left hip. It caused a form of septic arthritis of the joint—an inflammation caused by the immune system trying to fight off the
bacteria attacking the hip. The unusual condition was serious, threatening to destroy the four-and-a-half-year-old's ability to walk. Surgeons cut away the severely infected tissue to remove the bacteria and
preserve the hip, and gave her four weeks of intravenous antibiotics.
Cindy remained bedridden while her hip healed.

Two weeks after she went off the IV antibiotics, Cindy contracted
another sinus infection. The Cutshalls, becoming frightened by the
unrelenting string of sickness, went to demand that their pediatrician
do something. Before they could even speak, Dr. Lawrence V. Hofmann, their Canton pediatrician, declared, "Enough is enough. We
gotta do something about this kid." In early 1986, Hofmann sent the
Cutshalls to Rainbow Babies in Cleveland to see Ricardo Sorensen.

"Cynthia was clearly unusual," Sorensen recalled. She lacked tonsils, adenoids, and a thymus gland, a critical part of the immune sys-

tem normally found in the center of the chest. The thymus teaches lymphocytes, the white blood cells, to recognize the body's normal tissue, which they should leave alone as the lymphocytes defend the body against infections. The relationship between the gland and white blood cells is synergistic: without a thymus, the lymphocytes cannot defend against invaders, and without chemical signals from active lymphocytes, the gland withers away. When the thymus fails to educate the white blood cells properly, they inadvertently attack normal tissue, causing so-called autoimmune disorders like multiple sclerosis and lupus. When the gland does not work at all, the body has little protection.

Sorensen, with help from Dr. Melvin Berger, also a pediatric immunologist at Rainbow Babies, studied Cindy's antibody production and found it was abnormal. In addition, her white blood cell count was low. As had become his standard practice, Sorensen tested Cynthia's cell-mediated immunity, the ability of the cytotoxic T cells to attack an invader. Hers was abnormally low. This surprised the doctors because Cutshall had not developed the clinical symptoms—fungal infections or unusual diarrheas—seen in individuals without immunity, such as people with AIDS.

Sorensen tested for the ADA enzyme. Cynthia Cutshall produced less than 1 percent of the normal amount. She had the rare ADA deficiency, and it was slowly destroying her immunity. Although she stayed healthier longer than most children with this disease, Cynthia Cutshall clearly had started on the downward spiral of constant, and ultimately lethal, infections.

Sorensen sent a sample of Cindy's blood to his old boss and mentor, Steve Polmar, who, by then, had moved to Washington University in St. Louis. Using additional tests, Polmar confirmed the diagnosis. Polmar suggested that Sorensen contact Hershfield at Duke. Hershfield, a biochemist, had the most complete capabilities for looking at metabolic abnormalities such as Cindy's. More important, Hershfield had just begun testing a new injectable drug in which cow ADA was encased in a chemical bubble that protected the medication from destruction in the blood. The drug was called PEG-ADA because its protective outer layer coats the enzyme with polyethylene glycol, or PEG, a chemical component of automotive antifreeze. The Duke doctor had just started treating ADA patients with the newly developed therapy, and it seemed to be working. At age five, Cynthia Cutshall would become the fifth person to receive the experimental weekly injections.

Her response to the drug was astonishing. "We were giving her less than one cubic centimeter of this stuff, once a week, and that changed everything," Sorensen said. "That was really awesome." Cindy responded to the treatment much faster than the first patients on ADA. The doctors didn't know why, but it might have been related to the same unidentified reason she got sick later than other SCID kids. With treatment, the biochemical abnormalities in her blood cells began to fade. Her cellular immunity improved significantly. And she began producing antibodies she previously could not make. Best of all, she stopped getting sick. The runny nose and the vomiting stopped. The bright, cheerful Cynthia Cutshall reemerged.

• • •

When Ashanthi DeSilva came to Sorensen two years later, he already had more experience with ADA deficiency than nearly any doctor in the world. Sorensen knew immediately by Ashi's history of chronic infections that she did not have allergies. She had been on antibiotics nearly all of her short life. Yet, remarkably, Ashi had never been hospitalized. When the DeSilvas came to see Sorensen in September 1988, Ashi was just turning two years old, an age by which most kids with the disease are dead. And although ADA deficiency is extremely rare, Sorensen had a hunch.

"Why don't we cut this short," Sorensen said to Velma Paschal, the pediatric allergist who brought Ashi to Children's Hospital. "Let's take a chest X ray and see if the child has a thymus." A few minutes later, the image came out of the darkroom Ashi, like Cindy before her, lacked a thymus. "I was sure this kid had something wrong with her cellular immunity," he said.

They took a blood test to count the number of lymphocytes and other white blood cells. The lymphocyte numbers were very low, almost nonexistent. With difficulty, Sorensen coaxed a few lymphocytes to grow in a laboratory culture so he could test their ability to function. They were not functioning at all. Ashi was immunologically defenseless. Sorensen moved rapidly. He called Hershfield at Duke and said he thought he had another ADA-deficient patient, and sent a blood sample to Duke by an overnight delivery service. The next day, Hershfield called back to confirm the diagnosis. Ashanthi DeSilva would become the ninth patient to go on the experimental PEG-ADA medication.

In November 1988, two months after her second birthday and a year and a half of sniffly, anxiety-filled hell for her parents, Ashanthi DeSilva began treatment with PEG-ADA. Sorensen succeeded in getting her into Duke University's study testing the effectiveness of the experimental drug, making her a guinea pig in the drug study that would later win Food and Drug Administration approval for Enzon, Inc., of South Plainfield, New Jersey. Ashi's contribution to science earned its own reward. Getting the drug through the study did save her life as Sorensen predicted.

PEG-ADA, however, was no miracle cure. In January 1989, two months after she started taking the drug, the whole DeSilva family got the flu. A few days later, on a Friday night, Raj noticed that Ashi was extensively bruised. "What are those blotches?" he said to Van. "What is going on here?" Every place he touched Ashi created another bruise. Something was terribly wrong.

Early the next morning, he put his child in the family van and made the familiar trip to the emergency room. Raj phoned Sorensen and told him to meet them there. A simple blood test showed the problem immediately. Ashi's platelet count had dropped dangerously low. Platelets are specialized blood elements that play a key role in clotting; without enough of them, a person bleeds spontaneously. Even the slightest pressure to Ashi's skin caused blood to leak out of her vessels and into the spaces between the cells (the essence of a bruise).

Sorensen didn't know why her platelet count had dropped. This could be a well-known reaction to the viral infection that would go away by itself. Or, more menacing, her immune system, revived by the PEG-ADA treatments, might be making antibodies against her own platelets. This would be a form of autoimmune disease, the worse possibility. If it was an autoimmune reaction, Ashi might begin making antibodies against PEG-ADA itself and the drug levels in her body would drop. Blood tests showed no sign of that. Sorensen began treating her with injections of antibodies—intravenous immunoglobulin shots that stimulate the body to make more platelets. Over the next six months, Ashi's platelet levels returned to normal, but for everyone, doctors and family included, it was a scary time.

To Raj DeSilva, this incident was a warning that showed PEG-ADA was no cure for ADA deficiency, but rather a high-tech bandage applied to his daughter's behind once a week. He would place her facedown on his lap, and Van would jam the syringe into Ashi's buttock. The solution of solvent and drug formed a painful knot in her gluteus maximus. It made Ashi cry, and Raj hated it.

When he asked if any other children had experienced platelet problems, Raj learned that an ADA-deficient child in California had developed an allergic reaction to the drug. It was a classic Catch-22. Without the drug, the child had no immune system and consequently lacked the ability to launch an allergic reaction. When given the drug, the Californian began to develop a competent immune system that promptly attacked the life-giving medicine.

Ashi ran into a different problem. Before she went on PEG-ADA, laboratory tests could detect little activity among her immune cells. After drug treatment began, her immune activity rose dramatically. The number of T cells, the white blood cells that mount the body's defenses, rose from fewer than 100 T cells per cubic milliliter of blood to 1,200 cells within six months of starting PEG-ADA. But the improvement did not last. Over the next six to twelve months, Ashi's T cell counts gradually declined, fading to between 300 and 400 per cubic milliliter of blood. Her Cleveland doctors, after consulting with company scientists, doubled Ashi's PEG-ADA dose. That helped. Her T cell count went back up, but only a bit, to between 500 and 600. Her immunity never got back to the initial high levels, and it seemed to begin falling again sooner than before. The meaning of these studies, however, remained unclear because clinically she still seemed healthy, at least for the moment.

Doctors who treat kids with various immune disorders have often watched a medical treatment improve enough to moderate the deadliness of the disease, only to have something go terribly wrong later in life. For example, medical treatments had lengthened the lives of kids with Wiskott-Aldrich syndrome, an immune deficiency less severe than ADA deficiency. Instead of dying in childhood, they could live into their late teens and early twenties with appropriate medical care, but then they would suddenly develop some form of lethal cancer that their defective immune defenses could not suppress. ADA children like Ashi had never lived long enough for doctors to know whether they would face similar cancers or other problems. PEG-ADA, their life-sustaining drug, was too new.

These were the kinds of complications, the kinds of unknowns, that worried Raj DeSilva. "I have seen so much good data and bad data that it is hard to tell the difference," he said. "The problem with PEG-ADA is that it is a young drug. There is no history to go by."

Although no one could prove it one way or the other, DeSilva became convinced that it was the PEG-ADA that caused Ashi's platelet count to drop. He would remain deeply suspicious of the drug, but stifled his concerns because he could see that his daughter had never-theless improved. Slowly, the severity of her infections had diminished. She began to keep down food and gain weight. The child, rapidly becoming a toddler, began to grow.

. . .

It was against this backdrop of guarded optimism that the DeSilvas first heard about gene therapy. Because the PEG-ADA treatment required frequent tests to monitor Ashi's blood levels, the DeSilva's regularly drove to Cleveland to see Sorensen for the tests. One day, after Ashi had been on PEG-ADA for about a year, Sorensen held up a tube of her blood before her mother's eyes.

"Look, Van," Sorensen said, holding up the extra tube. "For gene therapy."

Neither Van nor Raj really understood what Sorensen was talking about. He explained that he had been talking to doctors at the National Institutes of Health who had been studying ADA deficiency and decided that it was the best genetic illness on which to try a revolutionary treatment. Instead of injecting the enzyme every week, Sorensen explained, the NIH doctors wanted to put the gene into the patient's body. It would be a permanent solution. If it worked, it would eliminate the need for weekly ADA injections. Raj liked the sound of that, but dismissed it as some treatment for the dim and distant future.

Each time a package containing Ashi's chilled cells arrived at the NIH's clinical center, couriers hustled it up to the tiny sixth-floor laboratory of R. Michael Blaese and Kenneth Culver, both in the National Cancer Institute. They had shown it was possible to put the ADA gene into mouse and human T cells. They next wanted to show they could cure genetically defective T cells of their genetic defect. For that, Blaese and Culver needed blood cells from people with ADA deficiency, and Sorensen was a principal source since he had two of the world's few ADA patients. For reasons still not understood, Ashi DeSilva's white blood cells grew effortlessly compared to cells from other patients, including Cindy Cutshall. What's more, when Ashi's cells were genetically engineered with the ADA gene, they produced large quantities of the ADA enzyme. She would be the perfect candidate to go first if the NIH doctors ever received permission to test their gene treatment ideas in human patients.

. . .

In spring 1990, pediatric immunologist Mel Berger called the DeSilvas with some news. Berger had taken over Sorensen's patients after Sorensen moved to Louisiana State University in New Orleans in mid-1989. The NIH doctors working on gene therapy who had experimented with Ashi's blood cells wanted to come to Cleveland to talk about testing gene therapy in their daughter. The NIH doctors would be talking to the Cutshalls, too. Did the DeSilvas want to come in to find out about it? Yes, they said, though Raj continued to take for granted that this was something for the future, a technology that was ten or fifteen years away, not four months.

Mel Berger and Mike Blaese were buddies who had been colleagues at the NIH from 1978 to 1981, although they worked on different floors of the clinical center. Both are pediatric immunologists, a small clique among all the specialists of medicine. This ancient connection helped smooth the way for Blaese to reach the patients.

Blaese already knew the conditions of Berger's patients because he had reviewed their cases as a consultant for the company that manufactured PEG-ADA. The exercise meant Blaese knew all the patients identified with the disease in the United States and their physicians. They would be the potential patient pool for the gene therapy the NIH doctors had in mind. But most important, the review convinced Blaese that PEG-ADA was not working for everyone.

When *The New England Journal of Medicine* report described the first two patients on PEG-ADA, the drug looked like a miracle cure. Both children did well with no side effects. But then, nothing on the PEG-ADA patients was published for awhile. Blaese was astonished to learn that several children did not improve very much, and at least one was failing PEG-ADA treatment. (The patient, an Amish boy from Pennsylvania, eventually had a successful bone marrow transplant that cured him of his illness.) "I began to realize that the enzyme replacement itself had not resulted in the profound reconstitution of immune function as I had believed that it had, or would," Blaese recalled. "That left the door open for an alternative approach, like lymphocyte gene therapy. In retrospect, it is conceivable that if I had not seen that data, I might never have pursued the lymphocyte research. I would have presumed that PEG-ADA was the answer."

But it wasn't, and now in 1990 Blaese and Culver were hunting for a couple of patients to take the next step: gene therapy. They already had submitted their gene therapy plan—along with French Anderson—to several regulatory bodies that needed to approve their idea before it could be carried out. All they needed was permission and a couple of patients.

. . .

The DeSilvas met Blaese and Culver at 2 P.M., May 18, 1990, in a hospital room across the complex from Mel Berger's office at Rainbow Babies. The DeSilvas already were seated when Berger, Blaese, and Culver came into the room. Ashi, just three and a half, stood by the window, looking out. Stoic, she never turned to face or talk to anybody, but she was listening, taking it all in. She knew they were talking about her.

Blaese sat on the edge of the hospital bed and started reviewing the gene therapy proposal, immediately talking about how genes could be transferred into Ashi's cells. He tried to keep it low-key. They talked about what the treatment might do for Ashi and about its limitations. Blaese also talked about risks and, more important, what was not known about the risks. But the NIH doctors also wanted to give the DeSilvas hope, and they stressed the positive results they had had with Ashi's cells over the last eight months in the NIH laboratory, how her cells grew well—better than any of the other children's— and how the scientists were able to get the ADA gene into her cells and cure them in the laboratory dish. If the scientists could cure her cells in the lab, Blaese argued, they should be able to cure them in Ashi's body.

The DeSilvas had come expectantly. They already had seen one cutting-edge therapy, and they liked the results, though the long-range effectiveness remained unclear. Van immediately was optimistic: "We should try it."

Raj was more cautious, but not unwilling. "I would consider anything, especially knowing that [the girl in California] had developed antibodies against PEG-ADA. It could happen to Ashi." Still, Raj had scads of questions, particularly about the risks.

The meeting ended without any commitment on either side. The NIH doctors said they would stay in touch and keep the families apprised as the legally required review process went forward. Blaese suggested that the DeSilvas come to the NIH to look the place over, see the hospital, and to meet French Anderson, whom Culver had taken to calling the Father of Gene Therapy. The DeSilvas agreed to do that, perhaps after a planned vacation later in the summer put them on the East Coast.

• • •

Back in Bethesda, the review process ground on as French Anderson and company pushed for approval. They faced considerable opposition from some scientists who had their own reasons for complaining. After sometimes rancorous debate, a key committee, the Recombinant DNA Advisory Committee (RAC) of the NIH finally approved the Anderson-Blaese-Culver gene therapy protocol for ADA deficiency on July 31, 1990. At the very same meeting, the RAC gave permission for Steven A. Rosenberg, chief of surgery at the National Cancer Institute, to try using a form of gene therapy to fight cancer.

For the two teams of NIH doctors, an internal race among NIH collaborators had developed to see who would be the first to perform the historic experiment. It would be the revolutionary attempt to heal a sickness by changing genes in the patient. Anderson had been proclaiming for years that he wanted to be the first to conduct a successful gene therapy experiment. He had devoted the considerable resources of his large laboratory to solving the technical problems that blocked gene therapy. Now Rosenberg, who had been introduced to gene therapy by Anderson and Blaese, apparently had decided that he wanted to be the first. Although both protocols had been approved at the RAC, both teams still needed FDA approval before they could begin. Rosenberg proclaimed publicly that he was ready to treat the first patient when FDA gave its approval. Anderson, Blaese, and Culver decided they would be ready, too. The moment FDA blessed the procedure, they would treat a patient immediately. Ashanthi DeSilva was their first choice. With the RAC approval at the end of July, the NIH doctors decided to have the DeSilvas come to NIH for a final meeting.

There were many reasons to take Ashi ahead of all the other SCID kids. Perhaps most important, Ashi's cells grew better in the laboratory than cells from any of the other patients, and she had many of them in her circulation, probably due to PEG-ADA treatment. What's more, when the scientists corrected her cells genetically, her cells made more ADA than any of the other patients'. Unless the NIH doctors could get enough white blood cells out of the patient and grow them in the laboratory and get in a functional gene, gene therapy had no chance of success.

"Ashi is going to be the first," French Anderson told the DeSilvas when he initially met them. Van was thrilled until she started to read the protocol itself. It said Ashi would have to lie still while her blood was removed to gather the white blood cells that would be genetically corrected. Van panicked: Ashi would never be able to remain inactive, especially now that she felt so well on PEG-ADA, which she would continue to receive during the experimental treatment. Anderson reassured her that total immobility was not required.

Finally, in the excitement of a two-day visit to NIH in early September 1990, the DeSilvas decided that they would do it. They would let Ashanthi become the first person to undergo gene therapy. The decision brought both relief and anxiety. Anderson tried to be reassuring. "You have made the right choice," he told the DeSilvas. "If this were my daughter, I would have done the same thing."

The decision to go ahead brought everything into sharp focus, and established an agenda for the NIH team. Since the DeSilva family already was at the NIH, they decided to begin preparing for the eventual treatment by removing some of Ashi's blood and genetically correcting the white blood cells. It would take about two weeks to grow enough of her genetically modified cells in the lab for a treatment. Perhaps in that period, they would hear from the FDA and could give the cells back to Ashi and begin the treatment. If not, then this would be a dress rehearsal. They would walk through every step that would be taken in treatment except returning the genetically engineered cells to their new patient.

On September 5, 1990, three days after her fourth birthday, Ashanthi DeSilva entered the clinical center's blood bank to have her white blood cells removed. She did have to lie on her back for a long time. Van was right: Ashi didn't like it. But at least she could still snack, use her crayons to color pictures, and watch the tiny, movable TV hanging on a slender silver arm above her bed.

Her cells were carted to a laboratory in sterile plastic sacks, and stimulated with chemicals that make them grow. Then they were exposed to the genetically engineered virus that carried the correct ADA gene, and left in the warm darkness of a NIH incubator to multiply. The next morning, the five DeSilvas drove home to North Olmstead to await word that it was time to come back.

The next nine days were an anxious and chaotic time for everyone. Mike Blaese and Ken Culver hovered over Ashi's growing white blood cells, counting them, making sure they looked all right under the microscope. They carried out test after test on the cells, making sure some infectious agent had not crept into the culture and that the cells made the enzyme. They needed to make sure that the new ADA genes went into Ashi's white blood cells and that the genetically corrected cells were making enough ADA.

Actually, ADA production was not going well. "Transduction efficiency [the ability to put genes into the cell] was quite lousy," Blaese said. Apparently, the genes had not gone into as many white blood cells as they had hoped. Lousy or not, the NIH researchers had to show that the growing cells made a minimum amount of ADA, or the Food and Drug Administration would not let them proceed. "It was all arbitrary," Blaese said. "We were told [by FDA] to set limits, and so we made up numbers and set them. But once you set them, you have to meet them. I am awfully glad I set an arbitrarily low number." ADA production barely reached that threshold. It was low, not as much ADA as the NIH doctors would have liked, but it was enough for them to proceed.

While Blaese and Culver sweated over the cells in the blood bank, Anderson struggled with the game of regulatory chess. The NIH team still lacked Food and Drug Administration approval to treat Ashi. Legally, FDA has thirty days to rule on a request to conduct a clinical trial. Pressing the FDA for an answer, however, requires a delicate touch. The agency can just say no. Anderson played a gambit. He called Kurt Gunter, the FDA case officer reviewing the gene therapy protocol, and told him that they had been growing Ashi's gene-corrected cells in the laboratory, and that things were looking pretty good. Was there any reason why they couldn't have the family come back to Bethesda by week's end and give the cells back to the girl? Did Gunter think it was possible that FDA could give them official permission by then?

Gunter too was careful. As far as he was concerned, the NIH had provided FDA all the information it wanted. Gunter told Anderson that it looked like they could go ahead, but he still needed an official ruling from his bosses. The NIH doctors still needed final, written approval. It should be possible to get all the paperwork done by the end of the week, Gunter said. In Washington, paperwork can move with glacial speed. To Anderson and the others, however, the clock had started. They were on their way.

Ken Culver, who had primary responsibility for the family, called Raj DeSilva at work on Tuesday and told him that FDA permission was imminent. Could he get the family together and get to Bethesda by Thursday, September 13? An NIH social worker assigned to the gene therapy project called Van at home in the middle of the day:

"You guys have to come down for the treatment," Van remembers her saying. "The protocol was not really approved," Van recollected, "but they said we are very hopeful that the protocol will be approved. We think it will be approved. And that French was working behind the scenes."

• • •

Forty-eight hours later, exhausted from travel and anticipation, Raj and Van barely spoke to each other about what would happen in the morning. As they settled into the Children's Inn for the night, neither wanted the other to know how nervous he or she had become. Raj, the rock, had analyzed everything every which way. The decision to proceed was rational, he knew, but his heart ached. Although they had to be at the NIH hospital by 7 A.M.—which meant they had to rise an hour or more earlier to get all three children ready for the long trek up the hill—the parents remained awake well beyond 1 A.M.

Raj lay in bed that night running everything through his head. He still hated that Ashi would be the first to try this revolutionary experiment. Yet, the risks seemed small and the promise great. If gene therapy worked, one day she could stop taking the painful, weekly PEG-ADA shots that, at $4,000 an injection, threatened to consume her lifetime health insurance maximum, leaving her uninsured forever. And there was the hope that one day, because of gene therapy, she would live a normal life. "But at the back of your mind, you worry: 'Is it going to turn out right?' Nobody's done this before. It is not a proven treatment. There are risks involved. But you say, 'Well I have thought this thing out. I have talked to the doctors. I have a good feeling about them. It's going to be OK.' "

Eventually, he fell asleep.

• • •

French Anderson too spent a fretful night lying in his darkened Bethesda home. It was the eve of the most important experiment of his life, one he had been dreaming about for two decades. It was the experiment he had been pushing toward since 1983 when he first figured out what combination of laboratory techniques would be needed to try it.

Anderson had the habit of trying to think through a problem in advance. He stayed in his NIH office late that evening, until 9 P.M. or so, writing out all the possible problems and items that needed to be checked. What if the gram stain, a test for bacterial contamination of the cells, was suspicious? What if Ashi had a cold or a fever? What would they do if the FDA said no? Even after he got home that Thursday evening, he stayed up most of the night pondering the morning to come. "I do this so we wouldn't be caught in a situation where the problem had not been thought out, and we would be scrambling around playing catch-up," Anderson explained.

First-time experiments in human patients often had surprises, and they could be deadly. Anderson knew well the terrible experience suffered by Dr. William T. Shearer, chief of pediatric allergy and immunology at Baylor College of Medicine and the Texas Children's Hospital in Houston. For seven years, Shearer took care of David, the boy in the bubble. In 1982, researchers at the Dana Farber Cancer Center in Boston, Massachusetts, discovered a way to make mismatched bone marrow transplants work. Since David didn't have a matched bone marrow donor, this new approach offered the only hope of a cure, and the only hope that he would ever leave the plastic encasement that had protected him all his life. After much anguish and consultation, his parents decided David should attempt a marrow transplant, using his sister as the mismatched donor. In October 1983, she flew to Boston to have her marrow removed and treated to prevent the complications caused by unmatched marrow. David received a marrow infusion the next day.

At first, nothing happened. By Christmas, however, the treatment proved disastrous. Some of his sister's white blood cells, the B cells that produce antibodies, had been infected sometime in the past with the Epstein-Barr virus, EBV. Her normal immune system suppressed the virus, so it caused her no harm. When the infected cells entered David's body, however, the virus roared to life, spreading rapidly through the few B cells his devastated immune system managed to produce. The virus forced David's B cells to grow rapidly.

Just after Christmas, for the first time in his life, David began to make antibodies. This was a hopeful sign, except that simultaneously, he began to suffer a frightful fever and feel poorly. Then he developed bloody diarrhea. Shearer and the other experts could not figure out what was going wrong, and the plastic bubble impeded their ability to care for David. Finally, the doctors took him out his protective enclosure for the first time in his life—a prospect that frightened David immensely. After a dozen years beyond the embrace of his family, the gravely ill boy came out to face the world. Two weeks later, he died.

The uncontrolled activity of the contaminating Epstein-Barr virus had propelled the growth of David's B cells to the point where they became cancerous. The cancer spread to every organ system, all over his body. The bone marrow transplant never took. The doctors learned for the first time, but too late for David, that transplanted tissue could carry hidden virus.

"At the time, this was a unique observation," Shearer said. "It was a singularly important advance for medicine, but it was a terrible loss for his family and the people who loved him." David taught the world about the importance of an immune system, and the extremes to which doctors had to go to protect a defenseless body. To spend a dozen years of cost and hope in isolation only to have it end so badly was overwhelming. Doctors stopped putting SCID babies in germ-free incubators. David's bubble is now in the National Museum of American History. Instead of isolating their tiny charges, doctors now fight their illnesses with drugs and bone marrow transplants.

• • •

Anderson had no intention of allowing such losses for the DeSilvas or anyone else. David's situation had become desperate. He had been in the bubble all his life and could not go on that way. Ashanthi De-Silva was not in the same situation. PEG-ADA had stabilized her condition. Besides, she was living at home. There were problems, but she was doing relatively well. The NIH doctors had to make sure that nothing went wrong.

In addition, Anderson deeply feared the death of a gene therapy patient, especially the first one. Here was a medical procedure born amid political controversy and scientific uncertainty. If the first patient died, it would set back the entire field for years. Anderson had spent much of his professional career struggling to give gene therapy life. The last thing he wanted was to kill it aborning by some stupid mistake. Unable to think of anything else he had to anticipate, Anderson finally drifted off to sleep.

Normally, the Anderson household alarm goes off at 6 A.M. French Anderson goes to NIH early to do rounds with the clinical staff, a morning walkthrough to visit all the patients on the various services at the NIH where he consults. French's wife, Dr. Kathryn Anderson, a pediatric surgeon, heads downtown from their suburban home to the Children's National Medical Center. This morning, Friday, September 14, 1990, French Anderson got up before the alarm. After arriving at NIH, he walked into the clinical center's blood bank to check on Ashi's cells. Charlie Carter, the blood bank technician helping with the study, had been there since 5:30 A.M.

At 7 A.M., Anderson descended the seven floors from his office to the clinical center's lobby to meet the DeSilvas who were just coming through Building 10's doors. Ashi had to check in and get settled in 3B south, the pediatric wing.

An hour and a half later, FDA permission finally arrived. At 8:35 A.M., Jay J. Greenblatt, the National Cancer Institute's liaison with the FDA, called Anderson. He had just gotten off the phone with FDA's Kurt Gunter, who said OK in the usual, careful, bureaucratic way: Gunter really did not have any objections if the NIH team proceeded with the experiment. He still wanted the written safety test results on the solutions used to carry the genes into Ashi's cells, and at some point he wanted her medical history. But, Gunter said, the team could go ahead. Anderson was elated.

Greenblatt too was excited. Involved in the quest for approval since the beginning, Greenblatt wanted his own memorial of that historic day. He had been telling his family about the experiment for months. Now that it was about to begin, he wanted his six-year-old daughter Sarah to meet Ashi, the young genetic pioneer. Later that morning, although shy among all the towering strangers in white coats, Sarah Greenblatt mustered enough courage to give Ashi a purple silk butterfly, a token between children.

Ashi already had settled into 3B south, and was preparing to move up to the pediatric intensive care unit on the tenth floor. If anything went wrong, the NIH doctors reasoned, they wanted every piece of emergency equipment available to keep her alive. The ICU had the highest high tech of all. Its assemblage of aggressive young physicians and nurses could rapidly apply all the high-powered medicine NIH had to offer a fragile, failing child.

In the pediatric ICU, Ashi's gleaming dark eyes darted back and forth between "Sesame Street" on the television and the 20-gauge needle in Ken Culver's hand. She had been there before, sitting on the soft, pale white government sheets waiting for someone to put a needle into her body. No one could really tell how much the needles that had poked and prodded her since infancy bothered Ashi. She seldom made a sound, even when the cold steel pierced her skin, probing for the tiny veins beneath. She never fought back. She never cried. Her father had taught this round-faced girl with the brown skin and dark curly hair that she faced a lifetime of needles because of her illness. To call her stoic seemed insufficient.

Culver sat with furrowed concentration at Ashi's bedside and jabbed the hollow spike though the skin along the top of her right hand. She made no sound nor did she pull away. Her eyes stared impassively down at her seemingly disconnected hand, watching the doctor dig. Abruptly she shifted back to Cookie Monster singing the cookie song on the TV above her bed. Ashi ate a cracker with cheese.

Culver skewered the vessel, but movement dragged the needle out of the vein before he could secure the metal shaft with tape. He stiffened, acknowledged defeat, and switched to the left hand. Ashi frowned, but made no sound. Again Culver's needle pierced her skin, probing for the vein. Crimson drops filling the needle's clear plastic barrel heralded the connection. Culver puffed out his cheeks, exhaling loudly.

"OK, this one looks like it is real good," Culver said, never taking his eyes off the needle. "Looks like this is the magic hand." He had hit veins there before. Tape secured the portal into his patient's tiny body. A bag of clear fluid hung from the hook on a shiny pole next to the bed. Its tubing snaked down to a Medusa head of plastic coils attached to the needle in the girl's hand. Ashi seemed unaware of the historic event about to take place in her body. She was aware only of Big Bird.

French Anderson hung around in the background, near the white ICU wall, watching Culver install the intravenous line. The room was filled with tension and clatter as people talked, the TV blared, and staff came and went. Anderson and Blaese reviewed the last-minute safety tests. Everything looked fine. Now, there was only one thing left to do. The doctors had to sit down with the DeSilvas one more time and completely review every step they would take with Ashi. The team gathered with the DeSilvas in the small library across the hall from Blaese's laboratory.

The mood started somber as the DeSilvas reread the protocol and the informed consent form, a horrible listing of all the things that could go wrong. It described how this had never been tried in a child. Just returning the cells to Ashi might cause "chills, fever, nausea and/or body aches," it said. What's more, while the virus used to transfer genes is generally considered harmless, there is a small chance that it could again become infectious and cause an illness. In addition, merely putting the gene into a cell could cause it to become cancerous. It could kill her.

For an hour, French Anderson, Mike Blaese, Ken Culver, Mel Berger, and Raj and Van DeSilva went over everything one more time.

There would be no surprises for anyone. Everything would be done by the book. The atmosphere lightened, a mixture of serious and exuberant. The DeSilvas understood the risks; they understood the potential benefits. They knew these government scientists were about to make history in their daughter's body. They hoped it would work.

"I remember when I picked up the pen to sign the protocol," Raj DeSilva recalled. "They said, 'Sign here.' I did not hesitate. I just signed it."

· · ·

When the signing was all over, Ken Culver raced out of the library and down to the blood bank where Ashi's cells had been loaded into a small plastic transfusion bag. It would take half an hour or so, depending on the elevators, to make the round trip to the first floor of the cavernous clinical center, make the final preparations, pick up the cells, and get them back to the tenth floor ICU. Culver had been struggling with his own concerns about this experiment. He wanted to be considered part of the team, not just one of the technical staff. Although his boss, Blaese, was officially the principal investigator, Culver had done most of the hands-on work. He had grown Ashi's cells and put in the gene. He had prayed over them and made sure they multiplied. Now, everything was ready. When the cells reached their peak, they had to go back into Ashi's body. They couldn't freeze them for later if the FDA didn't give them permission, Culver had practically shouted at Blaese at one point. Too many cells would die in the process, and the treatment would be useless with thawed cells. They would have to start over.

Culver reached the first floor blood bank and scooped up the intravenous bag containing Ashi's cells. He wheeled around and headed for the tenth floor. He was not going to let anyone else carry the bag of cells he had spent the last two weeks on. And there was no way he was going to let anyone else give this historic injection. Culver had grown impatient with his colleagues and their concerns. They feared, for example, possible toxicity from the cell infusion. Culver, who had trained in bone marrow transplantation, had given children 2, 3, 4, 10 billion cells, without complication. They would give Ashi a mere 1 billion at most.

Everyone allowed now crowded into the room as Culver came steaming in with the intravenous bag of gene-modified cells. Ashi's mother and father stood to the left of her bed. Ashi's siblings sat in their wheelchairs just outside the room's door. Ashi quietly played tic-tac-toe with a nurse. The nurse let her win.

Mike Blaese paced relentlessly as Culver completed the connections. French Anderson sat down on the left side of the bed. He took Ashi's right hand and placed his finger on her pulse. In trying to anticipate anything that might happen, Anderson even worried about a power failure and all the monitors going off. "If everything went black, I would have hands-on control of Ashi," Anderson said. "Once I did it the first time, it became a superstition. Then it became a standing joke. Ken and Mike rib me about getting a callus on my finger from taking the pulse."

Mel Berger quietly leaned against the counter across from the foot of her bed and waited. A nurse stood on either side of Ashi, waiting. Ken Culver, his small wire-rim glasses shining in the TV light, sat on the stool to her right, and paused momentarily. Everything was ready to go. Ashi sat there impassively, barely blinking.

Then, impatient to begin, fearing that someone would take the needle from his hand or somehow preempt him, Culver lined up a separate syringe that contained a test dose of cells with the round rubber cap on the plastic Y-joint in the intravenous tubing. On one side of the joint, the intravenous fluid cascaded down toward the needle in Ashi's hand. Culver pushed the steel spike through the rubber nipple, immersing the needle into the cool saline solution. As his thumb pressed the plunger, the first of the gene-altered white blood cells squeezed out of the syringe and into the saline flow, heading for Ashi's blood stream. Culver pushed harder, emptying the 100 milliliters of white blood cells, about 5 percent of the total dose, into her body. It took less than five seconds.

Everyone winced.

Blaese stopped pacing and glared at Culver. He screamed inside his mind: "Geez, he gave all those at once. It wasn't even a minute. What are you doing?" He feared that the new cells would cause some unforeseen reaction. There was no need to rush the treatment.

Raj tensed up, too. "It was the most nervous time, waiting to see if she would have a reaction" to the test injection.

They all waited. Five minutes.

Nothing happened.

Ashi's heart monitor showed a steady pulse, 101 beats per minute. Anderson and the rest had just crossed the threshold into a new age of genetic medicine. They were now the first scientists to try relieving the symptoms of a human patient by adding new genes—a true molecular therapy. And there was no way to know whether it would work.

. . .

Culver told the nurse to hang the rest of the cells on the silver pole next to the girl's bed. Ashi watched as the bag was connected to its tubing, and then plugged into a port in the intravenous line.

"It's not one o'clock yet," Anderson interjected. The protocol said they would wait ten minutes between the test dose and the rest of the treatment.

"I was going to say I give him gray hair," Culver said about Anderson to the nurse, "but I can't give him any more."

After ten minutes, the nurse opened the stopcock, allowing the rest of the gene-changed cells to start flowing into their patient. Blaese leaned over and whispered to Anderson that he thought Culver was going too fast, that the rate at which the cells were given should be slowed down.

"How fast is that going in?" Anderson asked.

Culver looked up surprised, staring for a second, and said it shouldn't make any difference. They were only giving about 1 billion cells in total, less than $\frac{1}{10}$ of 1 percent of all Ashi's circulating white blood cells. Certainly that was not a large volume, and it was not going in faster than a normal blood transfusion. Besides, Culver was worried about the cells in a different way. They were sitting in saline solution, not their buffered growth media. They could begin to die, so he wanted to get them into her body as fast as possible. Anderson and Blaese worried about Ashi.

"Susan, why don't you slow it down just a little," Anderson said to the nurse controlling the flow of cells. Culver silently relented. It would take about twenty minutes to put in all the cells. It probably made no difference one way or the other.

Anderson was understandably cautious. He had fought through technical problems, political battles, and professional ridicule for more than a decade to reach this day. He did not want anything to go wrong. Anderson had been the human gene therapy visionary. At scientific meetings, in public and in the press, he had kept alive the dream of gene transplants through the 1980s when it seemed the technical problems would always make it tomorrow's treatment. Playing the role of advocate earned him more derision than acclaim. On that

September afternoon, however, all the controversy and complaints remained outside the ICU door. Anderson and the others had done it. They had successfully put new genes into Ashi's blood cells and then quietly put them into Ashi. It would take months, probably more than a year, and repeated treatments to determine whether gene therapy helped.

The last of the cells flowed down the plastic tube and into the patient. The day's treatment was over. Culver smiled at Ashi, hugged her, and walked away. Anderson finally stopped taking her pulse and looked down; she looked back up, unblinking. At first, no one said a word.

"That's it, Ashi," Anderson said to her. "What did you think of that? All this buildup, and that's it. All that will happen now is we will watch you for a while, and then tomorrow, you will go home. And that's it."

Ashi still didn't say a word. She was only five and didn't have any appreciation of history. Her parents, however, were visibly relieved. "It was an anticlimax," Raj said of the rest of the treatment after the test injection caused no problems. "That is the time I really began to get excited about the whole thing. We were in it now, whether we liked it or not. The step had been taken." Van took her cues from her child. "She was very happy," the mother said of Ashi. "This was the signal that she was going to be well. I was just happy that the treatment had started and that she was going to be well. I was thrilled that she was the first."

All the hoopla meant little to Ashanthi. She had only one request on that momentous occasion. She wanted to know if she could go back to the playroom in the Children's Inn. The treatment had gone so well that Anderson and the others decided that Ashi probably could, after a bit of observation on 3B south. By 4 P.M., Ashi was up to her neck in toys. The DeSilvas, elated that things had gone so well but exhausted and emotionally wrung out from worry about the experience, went back to their rooms with the other children. In the morning, the doctors would take a final round of blood samples, check her vital signs, and send her back to Ohio.

• • •

In January 1991, four months after the first gene therapy treatment, Cynthia Cutshall, the other ADA-deficient child from Ohio, would follow Ashi DeSilva into treatment. She would be the second child to have genes put into her body with the intent of curing an illness. There would be little fanfare for the second child. No reporters camped out at the NIH searching for a scoop. Most editors had wearied of the story.

. . .

Anderson walked down the seventh floor hallway toward his office when he remembered that he had not eaten anything for nearly twenty-four hours. He was exhausted by both the lack of sleep and the exhilaration of the adrenaline rush he had been riding for nearly two days. The excitement was over for now. Ahead lay the marathon of the first human gene therapy protocol. For the foreseeable future, Ashi's treatment would be repeated every month. She would fly in with her father to give up several bags of white blood cells. The cells would be genetically altered and grown in the laboratory. Two weeks later, she would fly in again to have the gene-repaired white blood cells squirted back into her body.

Anderson took off his white lab coat and leaned back into the broken chair he kept next to the table he used for a desk. It was an old government chair, worn comfortable by years of use. A bolt had fallen out so that whenever the chair was unoccupied, it tilted backward precariously. Still, it was his favorite.

Slipping off his shoes, Anderson tucked his feet into the soft moccasins he wore in the office whenever he had to do paperwork or answer the endless phone calls. Even now, a stack of pink message sheets hung on the hook by the door demanding his attention. First he phoned acting NIH director William Raub to tell him the gene treatment was over and that the patient was doing fine. Then he called his boss, heart institute director Claude Lenfant. Anderson had become too good a political insider to make the mistake of letting the boss hear about the day's events from anyone else.

After the political calls, Anderson started wading through the stack of media requests. *The Washington Post, The Chicago Tribune,* the *Los Angeles Times, Time* and *Newsweek,* the others. Leaning back in his chair, with the pile of message slips, he gave them descriptions of the treatment, Ashi's condition, and predictions of what would happen next. Yes, the parents were nervous. No, the child did not cry. She wore a turquoise top and green pants with a drawstring that her mother made for the occasion. Yes, she clutched a doll during the treatment. She only received about 1 billion blood cells, less than 1 percent of the total blood cell count. No, there were no side effects. The questions and answers went on for nearly two hours.

The network newscasts that night would announce to the world that the age of gene treatments had arrived. *The Washington Post* and *The New York Times,* as well as most newspapers around the world, put the story on their front pages. Ricardo Sorensen, Ashi's former physician who got her on PEG-ADA, was on his first visit to Japan that Friday in September. "I read [about the start of gene therapy] on the front page of an English-language newspaper in Tokyo."

When they started, no one would know for months, maybe years, whether the experiment launched that day would make a difference in Ashi's life. In many ways, the injection of a billion gene-altered cells into her body was more important as a symbol than it was as a real medical therapy. The victory was as much a social and cultural one as it was technological. Science had crossed a theoretical boundary. Researchers had, for the first time, changed the genetic makeup of a human.

People could be changed genetically, for good or ill. This offered the promise of solving many of medicine's most difficult problems— of treating, even curing the incurable. It also raised questions about the genetic improvement of people, the kinds of problems posed in 1932 by Aldous Huxley in *Brave New World*. The idea of changing the genes in people had power and simplicity, even elegance. Scientists and social reformers had speculated about the possibilities of genetically improving humans ever since Charles Darwin postulated natural selection in the mid-nineteenth century. Anderson, Blaese, and Culver had given that possibility a chance to become reality.

Yet, like many ideas that revolutionized the world and the way people see themselves, this bit of science occurred on a human scale. It was launched within a tiny, five-year-old girl, sitting in a bed with a simple intravenous tube connected to her hand. That simplicity belied the complex technological struggles that preceded the event. It obscured the mighty political battles Anderson waged to win all the necessary permissions to carry out the first therapeutic gene transfer.

In many ways, Anderson's most important skill was persistence. He did not develop the gene transfer technologies needed in the experiment, a fact for which he was criticized by many scientists. And he did not produce the key insight of putting the genes in short-lived white blood cells. Mike Blaese did that. But early on, Anderson perceived the importance of gene therapy. He displayed tenacity in ironing out the myriad technical difficulties that blocked the way, a kind of applied research. And there was his trailblazing drive through the wilderness of the political regulation of genetic research that took gene therapy from the lab bench to the bedside.

Anderson pushed back in the chair, closed his eyes, and tried to think back to the beginning, to the day he wrote his entrance essay to Harvard College and pompously predicted that he would be among the first doctors to cure illnesses at the molecular level. It was 1956 then. The double helical shape of deoxyribonucleic acid, DNA, had only been discovered three years before. Surgeons were still hacking diseased tissues out of patients with abandon and medical men heavily relied on toxic drugs to poison cancer cells. Doctors barely understood what molecules were then, but Anderson was going to cure diseases by changing them.

Yet Anderson, Blaese, and Culver had done it. They put a healthy gene into the sick cells of a living person for the first time. They had done the work. They had contributed many essential ideas. Like all scientists, however, they had the help of a hundred years of ideas, of scientists who had come before them and contributed bits and pieces to the mosaic that formed the increasingly clear picture of human genetic therapy. The quest to change people's genes long antedated available technology. Even with the technology, the NIH team had help. They stood on the shoulders of scientists from the early 1970s, who devised ways to manipulate DNA and put it into cells, the ones who ushered in the era of recombinant DNA or genetic engineering. And, more recently, the NIH doctors benefited from gene jockeys at the Massachusetts Institute of Technology, the Salk Institute, and Memorial Sloan-Kettering Cancer Center, who first reengineered mouse retroviruses—the unusual viruses that can infect a foreign cell's chromosome without killing it—to carry new bits of human DNA that could be used to fix sick cells.

Although fewer than a dozen people stood in the NIH ICU room when the cells began to flow into Ashi, a complex web of relationships, of scientists and supporters surrounded Anderson and the rest. A tangle of relationships that both made gene therapy possible and constantly threatened to bring it down. There was a decade-long struggle with confusion caused by technical problems. There were tremendous political battles over the quality of individual bits of research and a struggle over the right to go first, to exercise the hubris of changing cells genetically.

If Anderson, Blaese, and Culver had not done it, someone else would have conducted a successful human gene therapy experiment. Yes, they had opened the era of human gene therapy. If not them in 1990, then within the next few years someone else. Anderson himself once calculated that his work had sped up the gene therapy era by only three years. Yet, only by understanding the morass of connections between scientists, ideas, and the political struggle for social acceptance of a revolutionary technology can the importance be appreciated of that moment on September 14, 1990, when genetically altered cells first flowed in Ashi's body—the importance not only for Ashi and her family but for all patients who have suffered a sickness in the genes, and for those yet to come who will one day be cured within their genes.

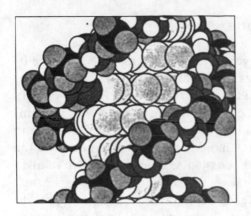

Chapter 2
The Idea of
Changing
Human Genes

"Improvement in social conditions will not compensate for a bad hereditary influence. . . . The only way to keep a nation strong mentally and physically is to see to it that each new generation is derived chiefly from the fitter members of the generation before."
— *Ethel M. Elderton,*
nineteenth-century eugenicist

In the summer of 1969, Neil Armstrong slowly dropped off the Apollo 11 lunar lander *Eagle* and took a large step for mankind. Back on earth, that same summer, a Harvard Medical School research team took a smaller step that had a much more lasting impact: They isolated the first single gene as a pure chemical entity. It was a major technical accomplishment, relying on biochemical manipulations for which three other Americans received the Nobel Prize that year. "This is a very significant achievement," a leading geneticist confirmed in *The New York Times* when the discovery was announced in November.

Exploiting fortuitous mutations in two viruses that infect bacteria, the Harvard team managed to extract a famous gene complex, called the *lac operon,* from the 3,000 or so genes residing in the single chromosome of the bacterium *Escherichia coli,* a normally harmless germ inhabiting the human gut. The *lac operon* comprises three genetic elements: two components that regulate gene expression and the gene itself, which produces an enzyme used to digest a particular sugar. French scientists shared a Nobel Prize in 1965 for describing how *lac*'s regulatory elements—a kind of molecular on-off switch for genes—controlled the enzyme's production. The French researchers made the discovery without ever actually working with the gene itself, inferring their conclusions from indirect evidence.

Now, thanks to the Harvard team, scientists could hold a gene in their hand, photograph it with electron microscopes, watch it chemically interact with other cellular components. For the first time in the 104 years since the Austrian monk Gregor Mendel proposed the concept of inheritable traits in pea plants, genes were more than an abstract concept. They were real.

Yet, when Harvard's team leader, Jonathan Beckwith, stepped to the microphone to announce their discovery, he did not dwell on the brilliance of their work. Instead he "made a rather awkward attempt to issue a political statement," as he later put it. He warned the public that a genetic revolution was at hand, and that society needed to protect itself from the misuse of this new science by the government, and even the industries, of the United States.

"The more we think about it [the new technique for isolating genes], the more we realize that it could be used to purify genes in higher organisms," Beckwith told *The New York Times*. "The steps do not exist now, but it is not inconceivable that within not too long, it could be used, and it becomes more and more frightening—especially when we see work in biology used by our government in Vietnam and in devising chemical and biological weapons."

This was the year that Vietnam—and the antiwar protests—reached their pinnacle. United States forces peaked at 543,000 in April 1969, three months after peace talks started. It was the year of the My Lai Massacre and major antiwar protests in Washington, D.C., where the police responded by filling the streets with swirling clouds of tear gas. The social turmoil, the general decline of trust in the country's institutions was well understood by the Harvard scientists.

"The work we have done may have bad consequences over which we have no control," warned fellow Harvard team member James Shapiro. "This use by government is the thing that frightens us."

Beckwith, Shapiro, and the others had become politically active at a time most researchers remained cloistered in their laboratories. They believed scientists could no longer feign ignorance of how their discoveries were put to use. Scientists, after all, had given the world's governments the terrible tools of mass destruction: the atomic bomb, chemical weapons, napalm, and Agent Orange.

Beckwith's group confidently predicted that one day researchers would pack isolated genes into viruses and then infect people with those viruses to heal inherited illnesses such as hemophilia. Given the government's adaptation of novel technologies to destructive ends, they worried that the methods used for gene transfer might be used to do "more evil than good."

. . .

The advent of genetic manipulation might lead to the next level of domination by the powerful few who controlled society's institutions. In revolutionary terms, Beckwith warned of the possible misuses of science unless society remained vigilant. What he feared most was the return of the biased and misguided turn-of-the-century social programs that were intended to improve the republic by ensuring the reproduction of superior humans while blocking the births of those considered inferior. Early twentieth century policies in the United States and England, based on the then latest scientific findings in the new fields of evolution and genetics, led to laws mandating involuntary sterilization for those judged unfit, and limited immigration of certain groups on the supposed basis of scientific evidence that linked them to crime, feeblemindedness, and other behavioral inadequacies. In Nazi Germany, these ideas led to the outright execution of undesirables, from mentally retarded Germans to healthy, normal Jews.

Could the isolation of the first gene, a bacterial gene, quickly lead to genetic engineering—perhaps even human genetic engineering? The speed of the on-rushing genetic revolution exceeded even the Harvard team's expectations: In a mere two years, new laboratory techniques would make it possible to begin performing the very experiments they feared most—the genetic engineering of living cells. Next would be entire organisms. And then people. It was the human experimentation that would come in the future—like the treatment of Ashanthi DeSilva by the NIH team in 1990—that worried the Harvard scientists the most. Researchers, for the first time, would give humanity the power to redefine itself, to change itself at its most fundamental level—the human gene pool.

The question was: Would society be ready? Only two years earlier, Marshall W. Nirenberg posed that very question in an August 1967 editorial in *Science* magazine. Nirenberg—the National Heart, Lung and Blood Institute scientist who cracked the genetic code in the early 1960s, and won a Nobel Prize in 1968, the year before the Harvard breakthrough—worried that rapidly expanding genetic knowledge would "greatly influence man's future, for man then [would] have the power to shape his own biologic destiny. Such power can be used wisely or unwisely, for the betterment or detriment of mankind."

Nirenberg predicted that cells would be "programmed with synthetic messages [gene fragments assembled in the laboratory and then inserted into the cell] within twenty-five years. If efforts along those lines were intensified, bacteria might be programmed within five years." What's more, he worried, "Man may be able to program his own cells with synthetic information long before he will be able to assess adequately the long-term consequences of such alterations, long before he will be able to formulate goals, and long before he can resolve the ethical and moral problems which will be raised."

If anything, Nirenberg wasn't brazen enough in his predictions, though they were within a first approximation: The first bacterial cell was programmed in just four years, and in mammalian cells only a decade later. The first genes were put into humans a decade after that, twenty-two years after Nirenberg's editorial. As scientific advances go, genetic engineering developed at a blistering speed. As the technologies moved toward reengineering people, the questions persisted about whether society possessed sufficient wisdom to rewrite its own evolution.

Yet, there seemed little doubt that society would if it could. In a 1992 national survey, 87 percent of adults in the United States said that they would be willing to use gene transfer technology to cure a fatal disease. More astonishing, nearly half said they would use genetic modification on their children if it would improve the child's physical or intellectual characteristics. Yet, in the same survey, only 13 percent said they were either very or somewhat familiar with how gene transfers worked.

The public was more willing to use the technology for human improvements than were the scientists, or the bioethicists who scrutinize their work. To the public, to parents, enhancing a child's genes apparently is just another privilege, like ballet and French lessons, already enjoyed by the children of the economic elite.

· · ·

How far from creating a genetically flawless child would society have to move before it used its new genetic tools to launch large-scale attempts at creating a perfect culture filled with Friedrich Nietzsche's *Übermenschen*? The dream of an ideal community has a hoary call. The ancient Greeks, of course, created the first fairy tales of a perfect world almost five centuries before Christ. Aristophanes, in *The Birds,* described a model city of the clouds whose exemplary citizenry contrast with the banal ways of the Athenians. Plato, in *The Republic,* concocted his complete community in which "natural rulers" (philosophers, of course), backed by the warrior class, organized the workers in cooperative arrangements and assigned social roles based on natural abilities.

The genre evolved through various religious and secular stages until the sixteenth century, when Sir Thomas More named the movement in his book *Utopia.* While utopian writers spun grand schemes of harmonious living, pioneers, and other utopian adherents coming to the New World in the Americas, attempted to turn literature into life. Dutch Mennonites set up the first communitarian society in what is now Lewes, Delaware, in 1663. From then until 1858, religious and secular visionaries established nearly 150 utopian communities in the United States, including Brook Farm in Massachusetts and the secular Oneida Community in New York State.

American utopianism faltered after the Civil War. Elsewhere, it was overtaken by social change and other idealistic movements that embodied some of its precepts in economic terms, such as the 1848 *Communist Manifesto* by Karl Marx and Friedrich Engels. But even as the communist rebellion raged against capitalism, Charles Darwin launched an even more profound revolution that shook humanity's view of itself in the world—and with its God.

After his five-year, round-the-world voyage on the H.M.S. *Beagle* that started two days after Christmas 1831, Darwin finally released his masterwork concerning evolution, *On the Origin of Species by Means of Natural Selection, or the Preservation of Favoured Races in the Struggle for Life,* on November 24, 1859.

Like any new ideology, Darwinism quickly came to be so broadly interpreted that it could be used to justify virtually any viewpoint or social program. Within a decade of *Origin*'s publication, the principles of natural selection were applied to people, as groups and even whole races. Theorists of Social Darwinism freely used extrapolations of Darwin's arguments to explain various phenomena in politics and economics. According to this distorted concept, weak members of society were unable to compete and successfully survive, so they were destroyed, and their culture with them. At the same time, the strong individuals grew in power and cultural influence. Although social competition could be brutal, those that survived would be the most fit, improving society overall. By this logic, millionaires were the fittest members of society, and deserved their privileges, wrote Social Darwinist William Graham Sumner of Princeton, New Jersey. Andrew Carnegie and John D. Rockefeller, not surprisingly, agreed with Sumner, arguing that this gave scientific justification for what were essentially the excesses of capitalism.

Social Darwinism provided scientific cover for all sorts of frank brutality. The British government used the concept to justify colonialism and its subjugation of indigenous, so-called primitive peoples. It was also used to justify child labor practices. The natural extension of the theory suggested that humanity itself had undergone evolutionary change, that societies of people evolved just like species. If this was true, then it was a law of nature that the fit survived, even thrived, while subjugating the weak. By this logic, some argued, the white-European civilization had evolved to be the most fit, justifying its dominance of other cultures. Darwinism gave racism respectability.

• • •

There was, however, a difference between using Darwin's ideas to justify past social practices and using them to "improve" society through some form of directed evolution. And there were precedents for just such a notion. When Darwin first hit on the idea of natural selection, he sought to verify its central principles by going to the only people who actually performed evolution: plant and animal breeders. Humans have been remolding animals since Mesolithic times, some 9,000 years B.C., and plants, at least through organized farming, since the Neolithic era, about 7,000 years B.C. Breeders only reproduce those plants and animals that have some desired characteristic—thicker fur, more meat or milk, more grain. Under the constant pressure of human selection—which mirrored the effects of natural selection—it was easy to see that domesticated plants and animals had changed over the centuries. The new traits made them considerably different from their wild ancestors. Darwin concluded that the artificial pressures brought by breeders could indeed transmute a species. It provided confirmation for natural selection.

That observation had a profound effect on Darwin's cousin, Francis Galton. If Darwin's ideas suggested that mankind did not fall from grace, as religion taught, but rather that humanity was rising up from a lowly beginning through evolution, then perhaps evolution could be aided. In a two-part article in *Macmillan's Magazine* in 1865, Galton raised a question that would dominate the rest of his days, and change the lives of countless millions: If breeding worked for plants and animals, "Could not the race of men be similarly improved? Could not the undesirables be got rid of and the desirables multiplied?" Instead of letting the laws of nature decide which people were the fittest to survive, Galton urged society to take an active role in deciding who would be allowed to marry and have children, and who would not. By approving marriages among those deemed most fit, while preventing people with less desirable traits from having children, the characteristics of each succeeding generation would be enriched with desirable traits, improving the race. "What Nature does blindly, slowly and ruthlessly, man may do providently, quickly and kindly," Galton wrote.

In 1883, Galton coined the word *eugenics*—from the Greek root meaning "well born," "of noble race"—to describe his vision. It was based on a simple premise: There was positive eugenics, in which people judged to have desirable characteristics were urged to reproduce, and negative eugenics, in which those judged inferior were prevented from having children. As a science, however, eugenics was built on sand. Galton and his followers treated the inheritance of complex human behaviors—including vaguely defined failings like feeblemindedness, pauperism, criminal tendencies, alcoholism, degeneracy, seafaringness, and wanderlust—as though they were single inherited characteristics, like hair and eye color. They were quite wrong.

If Galton's eugenic notions had remained merely a fascinating concept, it would have passed harmlessly into history. Galton, however, had written books and pamphlets that gathered a small following over the last decades of the nineteenth century. None of this ever really amounted to much until the turn of the century when Galton, ailing at seventy-six, took to the stump again as a spirited enthusiast of eugenics. Although his eugenic message had been around for years, the world had changed. Social views were evolving. Competing concepts—from capitalism to communism and socialism—were all struggling to win the hearts and minds of Britain's masses. Large audiences turned out in England to hear Galton. In 1904, doctors, scientists, and writers stuffed the Sociological Society in London to listen to Galton talk about eugenics. H. G. Wells attended the London talk, and then published *A Modern Utopia,* filled with eugenic notions, a year later. The Irish writer George Bernard Shaw heavily laced *Man and Superman* with concepts also gleaned from eugenics. Galton and his ideas were suddenly in season.

A social movement began to form. The English Eugenics Society, originally founded with Galton's support as the Eugenics Education Society, was established in 1907 to spread the eugenic message. The organization was, said one founding member, "propagandist." The eugenics movement also crossed the Atlantic Ocean to the United States where, in 1904, Charles Davenport, a Harvard-trained biologist who initially studied mathematics, convinced the newly founded Carnegie Institution of Washington to fund an evolution research center at Cold Spring Harbor, Long Island, New York. In 1910, he opened the Eugenics Records Office with support from John D. Rockefeller, Jr., and the family of railroad magnate Edward H. Harriman.

The money paid for a small army of eugenic field workers who went door-to-door filling out a standardized form Davenport had developed to gather statistical information about inherited characteristics among humans. The family pedigrees created by these reports did lead to the discovery that some diseases are inherited, such as hemophilia and Huntington's chorea, a fatal brain disorder that does not strike until the person reaches the fourth decade of life, and that some abnormalities, like albinism and the propensity to have too many fingers, were passed in the genes. On the other hand, the surveys discovered ridiculous "genetic disorders," including nomadism, shiftlessness, and "thalassophilia," which Davenport defined as a love of the sea that occurred only in men, and "innate eroticism" in "wayward" girls.

· · ·

Even though eugenics was not based on sound science, it had the ring of scientific authenticity, and that gave it authority and respectability. And that allowed the emergence of eugenics' dark side as foretold in a bleak counterutopian fiction penned by Aldous Leonard Huxley. Steeped in a tradition of great science, Aldous Huxley was the grandson of Thomas Henry Huxley, the Victorian scientist, essayist, agnostic, and vocal public defender of Charles Darwin. Aldous's brother Julian was a major figure in British biology, and a leading critic of eugenics, along with his friend, the British biologist J. B. S. Haldane. Haldane, a brilliant, nasty freethinker, "one of the great rascals of science," published a bit of fiction in 1923 called *Daedalus, or Sci-*

ence and the Future in which he predicted, and described, test tube babies.

Between Julian's antagonism toward eugenics and Haldane's macabre view of mass-produced humans, the young Aldous found a bleak picture of how science could be used to control people and society. He discovered a "brave new world."

Huxley's imaginings, no matter how dark, never compared to the awful reality that occurred in the 1920s and 1930s when eugenic ideas were turned into social policy. The United States enacted the first eugenic laws that required mandatory sterilization for the unfit. Indiana passed the first in 1907, with fifteen states following suit in the next decade. By the end of the 1920s, twenty-four states had sterilization laws. By the mid-1930s, more than 20,000 sterilizations had been legally performed in the United States, and by 1941, that number had climbed to 36,000. Most of those sterilized were inmates in state institutions for the insane, were convicted of drug addiction or sexual offenses, or suffered epilepsy.

Eugenics also was used to justify the Immigration Act of 1924, a law that severely restricted immigration of each nationality to a small percentage of those already in America as of the 1890 census. Americans who once saw their country as having endless space and open borders now perceived those seeking freedom from Eastern and Southern Europe to be hordes of feebleminded thieves that would prey on the rest of society, require state handouts, and generally degrade the nation's protoplasm. With the "scientific" eugenic evidence of the dangers presented to Congress, the borders snapped shut.

But nothing in America compared to the horrors ordained in the eugenic edicts of Nazi Germany as Adolf Hitler and his henchmen tried to create their own utopian world order. Beginning with the Eugenic Sterilization Law of 1933, the country required compulsory sterilization for anyone "who suffered from allegedly hereditary disabilities, including feeblemindedness, schizophrenia, epilepsy, blindness, severe drug or alcohol addiction, and physical deformities that seriously interfered with locomotion or were grossly offensive." Being deemed ugly could result in sterilization. In 1934, new laws required German doctors to report the unfit to Hereditary Health Courts. In the next three years, the Nazis had sterilized 225,000 people, half reported to be feebleminded.

But German eugenics went far beyond sterilization. In 1939, euthanasia became the new eugenic tool. Inmates in German asylums for the mentally ill were the first targets, with more than 70,000 eventually killed by bullets or gas. From there, it was a smaller step to begin eradicating anyone who threatened the gene pool. The Jews came next, along with the other ethnic and social undesirables who were simply exterminated. The early eugenic laws that the Nazis instituted to protect the purity of the Aryan race led directly from sterilization of the mentally ill to the holocaust at Auschwitz and the other death camps during World War II. Mainline eugenics would never recover from the blow dealt by the Nazis' logical extension of their notions to barbaric state policies.

•   •   •

The memories of the Nazi pogroms would haunt the development of genetic engineering in the 1970s, and go hand-in-hand with any discussions of changing genes in people, whether to heal lethal disease, or to enhance some personal characteristic like height or intelligence.

Even as the pseudoscience of eugenics began to influence social policy in the early twentieth century, biologists were abandoning it in droves. At the turn of the century, they rediscovered the work of an obscure Moravian monk named Gregor Johann Mendel and his report in 1866 that the characteristics of an organism were inherited. In 1906, the British zoologist William Bateson coined the term that described the physical material that carried Mendel's inherited traits: he called them *genes*.

The unearthing of inheritance set off a quest among scientists to find which chemical actually carried the genes. There really are only four classes of chemicals in a living cell to choose from: fats, sugars, proteins, and nucleic acids. Scientists quickly ruled out fats and sugars as the carriers of genetic information. That left proteins and nucleic acids. Proteins are the body's most intricate and complex compounds. They are made from twenty different subunits called amino acids that can be assembled in an infinite number of combinations by linking them together, like railroad boxcars. The boxcar amino acids can come in any order within the protein, and can make a protein train of any length, long or short. And they are as diverse as snowflakes. Most early twentieth century scientists believed proteins carried genes.

Nucleic acids, on the other hand, were long, repetitive polymers, especially deoxyribonucleic acid, DNA. This molecule was made up of only four subunits, called nucleic acids, usually arranged in a chemical form called a nucleotide. The nucleotides seemed to repeat randomly over and over again in the long DNA fibers that were found nestled in the chromosomes in the center of the cell's nucleus. It was hard to see how that might encode genes, or even participate in biological activity. Few early twentieth century scientists thought DNA was the stuff of life.

Still, it was a candidate ever since Friedrich Miescher purified DNA in 1869. He was only twenty-five at the time, and an unknown Swiss scientist working at the University of Tübingen. Consequently, the discovery went unreported for more than two years. Miescher initially isolated a material, which he called nuclein, from white blood cells in pus. It later proved to be DNA. By August 1869, he had isolated the material from many cells, including yeast, kidney, liver, testes, and nucleated red blood cells. This, however, did not at all prove that nuclein, or DNA, carried genetic information, though it was considered during the early 1900s.

The first real clue that DNA might hold the genes came in 1928 when Frederick Griffith, an English Ministry of Health microbiologist, discovered that one form of the bacterium *pneumococcus,* the nonvirulent R type, could be turned into a pathogenic S type when extracts from heat-killed S pneumococcus were mixed with the benign, living R form. Something from the dead S, some chemical or some factor, transformed the R into its deadly cousin. Griffith had no idea what the transforming factor might be. He apparently did not suspect that DNA from the S pneumococcus was getting into the R pneumococcus and genetically changing it. That, of course, is what was happening. Unfortunately, after describing his discovery for the first time in 1936, he never really advanced it, dying five years later when his London lab took a hit during the German bombings of World War II.

Across the Atlantic, in the relative safety of New York City, Oswald T. Avery, a respected bacteriologist at Rockefeller University, and his group had been conducting a number of experiments with several species of related pneumococcus bacteria. After initially disbelieving Griffith's results, Avery, along with Colin MacLeod and Maclyn McCarty, set out to identify the substance that transformed one type of pneumococcus into another. Through laborious biochemical purifications, they systematically separated protein from sugar, fats from nucleic acids. Each class of compounds was carefully tested for its ability to convert one form of pneumococcus into another. Only the nucleic acid fraction transformed the cells. They published their results in the *Journal of Experimental Medicine* in 1944. Astonishingly, their conclusion received a cool reception. Biologists, long wed to the notion that proteins carried the genes, were reluctant to believe the Rockefeller data. Later, the importance of their work was recognized by the Nobel committee that gave them the prize.

Over the years, however, experimental evidence accumulated that Avery and his colleagues were correct. In 1952, Alfred Hershey and Martha Chase presented convincing evidence at a Cold Spring Harbor Laboratory meeting that the viruses that attack bacteria only inject DNA into the host cell, and that no protein from the virus enters the bacteria. Viral DNA alone was sufficient to direct the production of new viruses.

A year later, the American biologist James D. Watson and the British physicist Francis H. C. Crick, both then working in the Cavendish Laboratory at Cambridge University, figured out that DNA had a helical shape, with the nucleotides pointed inward to form the rungs of its twisted molecular ladder. Their model led to several important deductions: how genetic material reproduced itself in a semi-conserved fashion so its genetic information could be passed on to future generations in the patterns Mendel described; how genes were encoded on DNA; and how genetic information flowed through the cell so it could be converted into proteins. Crick would use these deductions to develop what would later be called the Central Dogma of biology.

• • •

In the end, genes do only one thing: They store the blueprints for making proteins. From one viewpoint, the early twentieth century biologists were right when they said proteins were the most interesting molecules in the cell. They do everything that gets done in a cell and in the body. Proteins make the girders and trestles that support structures within cells, and they form the conveyor belts that move material around inside the cell. Every enzyme is a protein, and enzymes regulate every chemical reaction in a cell; there is one unique enzyme for each unique reaction. It is the protein's specific shape that gives the enzyme its specificity. A protein's shape is determined by the order of amino acids in the protein. The order of the amino acids is controlled by the genetic code, by the order of nucleotides in the DNA fiber.

Crick's Central Dogma described how this genetic information flowed from the DNA to the proteins, and how that controlled what went on in every cell in the body. Boiled down to its simplest principles, the dogma says that biological information flows from DNA to RNA to protein.

Genes are encoded on the long fiber of DNA that is shaped like a tightly twisted ladder. The vertical risers on either side of the ladder form the backbone of the DNA molecule. Inside the twisted ladder, the rungs are made from two nucleic acids, each firmly attached to one of the vertical risers. For the ladder to hold together, each of the nucleic acids on either side of the rung must be attracted to the other, like the north end of a magnet being attracted to the south end of another magnet.

To get that magnetic attraction, the two nucleic acids must be complementary. There are four nucleic acids in DNA, usually designated by the first letter of their chemical names: A, T, G, and C. The A nucleotide is complementary to, and magnetically attracted toward, the T nucleotide. They pair up to form a rung in the DNA ladder. And the G nucleotide pairs up with C. This complementarity allows the DNA ladder to reproduce itself by pulling apart into two separate halves. Each half acts like a template for assembling the missing half. Enzymes match up loose As, Ts, Cs, and Gs on the intact half of the ladder, then they chemically weld the new pieces, assembling them into an entire new molecule of DNA.

What's more, the genetic code is stored in this sequence of As, Ts, Cs, and Gs along the DNA backbone. Every three nucleotides on the coding side of the DNA ladder specifies the position of one amino acid in the protein manufactured from that set of genetic instructions. But the relatively gigantic threads of DNA, all wrapped up in chromosomes safely ensconced in the cell's nucleus, are unavailable for directing the production of protein. Crick predicted that there had to be an intermediate molecule that carried genetic information to the cell's protein-making machinery, structures in the gooey cytoplasm called ribosomes. Later work proved that the intermediate between genetic storage and protein production was ribonucleic acid, RNA— a close chemical cousin of DNA itself. Actually, three forms of RNA are involved in protein synthesis: messenger RNA, transfer RNA, and ribosomal RNA.

The process of protein synthesis works like this: Enzymes copy the genetic information stored on the DNA molecule onto a strand of messenger RNA. By analogy, the DNA contains a picture of the protein, much as a negative on a piece of 35-millimeter film holds an image. The DNA image is transferred to messenger RNA, in a process similar to developing a photographic print. The negative image in the DNA becomes a positive image in the RNA. Biologists call it transcribing the gene, since the information on one molecule is copied onto the another molecule—like copying words from one page onto another—and the language of DNA is the same as the language of RNA. But where DNA is a gigantic molecule that cannot move out of the nucleus, messenger RNA is tiny by comparison, and can easily slip out of the nucleus to the cell's cytoplasm where proteins are made.

The messenger RNA, now stamped with the picture of the gene, is ready to create a protein in that image. To do that, the messenger RNA travels to specialized structures called ribosomes. Ribosomes are like machine tools capable of grinding out some complicated shape from chunks of metal when the right instructions are fed in. For these biological machine tools to work, they need the right raw materials : the twenty amino acids from which all proteins are made, and the right set of instructions.

Each one of the twenty different types of amino acids gets attached to a specialized form of ribonucleic acid called transfer RNA. Transfer RNAs are like forklifts on the factory floor whose job it is to deliver parts to the assembly line. Transfer RNAs carry amino acids to the ribosomes, and then insert the correct amino acid into the growing protein chain at the proper place. The proper place is determined by the genetic information carried by the messenger RNA.

As the messenger RNA is fed into the ribosome, a long string of amino acids in the proper order for the protein being made comes out the other end. Once the entire amino acid string is assembled, the linear protein snaps into its functional, three-dimensional shape. Ultimately, the messenger RNA is destroyed, and synthesis of that protein stops.

· · ·

This basic understanding of how DNA and the genes it carried work revolutionized biology. It gave the science a mechanistic model, much like the physicists' understanding of atoms and their subatomic parts. Finally, biologists had a framework for understanding what went on inside the cell. And with that understanding came ways of intervening. First, the interventions were aimed at learning more, but soon scientists began to think about ways to repair the system if it broke down.

In the back of the minds of many was the notion Avery first expressed in the opening line of the 1944 paper announcing genetic transformation: "Biologists have long attempted by chemical means to induce in higher organisms predictable and specific changes which thereafter could be transmitted as hereditary characters." Hermann Muller, while working at the University of Texas, had become the first to change genes in a living cell deliberately, by causing mutations with radiation. Although he received the 1946 Nobel Prize for the effort, the technique caused only random changes in the DNA. It could not be used to direct genetic alterations for a specific purpose, such as gene therapy.

Avery, who was at retirement age when he and the Rockefeller group proved that the transforming factor was DNA, retired soon after and died in 1955. But the work at Rockefeller continued, principally by Rollin D. Hotchkiss, who had come back to the university in 1938 after a year in Copenhagen, to work in Avery's lab. Athough he was not working on transformation directly at the time the initial discovery was being made, Hotchkiss carried on the research, which stimulated him and others to begin thinking about trying to transfer genes to mammalian cells with this approach. Changing bacteria genetically was interesting, but if the technique was going to make a difference for human ills, scientists would have put genes in eukaryotic cells.

Even in the hands of a master like Hotchkiss, who had been doing transformation work for nearly twenty years, things went poorly. First of all, detailed quantitative studies showed that the efficiency of bacterial transformation was rather low—usually around 15 percent, though occasionally as high as 50 percent, he reported in a 1965 talk in which he coined the term "genetic engineering." Starting in the mid-1950s, workers in his lab launched "a series of attempts to introduce into mouse embryonic cells a simple pigment marker" by transferring DNA from pigmented mouse strains to nonpigmented embryos. "The treated cells were 'plated' under the skin of young white mice," making it easy to detect any cells that picked up genes for pigmentation.

They never found a single transformed mouse cell. "After we had exhausted all the arts we had acquired . . . we had been able in one whole year to test and follow up only some 20 million mouse cells, but not one seemed to acquire the pigmented marker," Hotchkiss concluded. "This was enough to tell us that DNA transformation of mammalian cells was not going to be easy, and we went back to the more malleable bacteria."

For Hotchkiss and the field of genetic transfer, this was an understatement. For the next decade, mammalian cells would resist easy alteration of their genes. What would follow in the short term was a series of failures as scientists across the country, now convinced that it should be possible to engineer cells genetically, searched for ways to do it.

A group at the National Institutes of Health took a transforming approach similar to Hotchkiss's, but with a significant twist: They had developed a way to identify the mammalian cells that were genetically transformed. In 1962, Waclaw Szybalski, then at NIH, later at the McArdle Memorial Laboratory at the University of Wisconsin, developed a kind of nutrient soup in which mammalian cells, including human cells, could not grow if the cells lacked a certain enzyme, hypoxanthine guanine phosphoribosyl transferase, or HGPRT. The soup, or growth medium, contained hypoxanthine, aminopterine, and thymidine, and was called HAT medium.

HGPRT is a housekeeping enzyme, one that the cell needs all the time to prevent the production of uric acid from building up in the body. Babies born without HGPRT are essentially born with a type of gout; there is so much uric acid in their blood that the crystals of it form in their urine. But their symptoms are much worse than mere gout; males missing the HGPRT gene suffer a miserable mental illness called Lesch-Nyhan disease. Although they are born apparently normal, they gradually decay into a state of mental retardation where they have profound, irresistible urges to mutilate themselves. Unless they are physically restrained, they will eat their own lips and fingers.

J. Edwin Seegmiller, an NIH scientist, discovered the link between HGPRT deficiency and the Lesch-Nyhan syndrome in 1967. At the time, Theodore Friedmann, a young pediatrician rapidly becoming interested in the prospects of gene therapy, was working in Seegmiller's lab, and decided to see if he could transfer the HGPRT gene into Lesch-Nyhan cells. The gene itself had not been isolated; this was before the Harvard discovery in 1969. Instead, Friedmann, working with Seegmiller, tried to mash up bulk DNA extracted from normal cells and to transform the cells as Avery and the others had done with bacteria. Since other scientists had tried it and failed, they presumed that it was a rare event in mammalian cells, and they needed some way of spotting the occasional cell that took up the proper gene and was genetically repaired.

Seegmiller and Friedmann decided to use Szybalski's HAT medium. The principle is simple and would be used repeatedly during the genetic engineering revolution that was to come. They exposed the Lesch-Nyhan cells to purified DNA, and then grew the cells in HAT medium. Only those cells that had taken up the new gene would survive. The results were very disappointing. "We found some cells," Friedmann said, "but we could not grow them. The cells were not permanently changed by the manipulation."

To take it a step further, the NIH scientists attempted to detect the growth of genetically repaired Lesch-Nyhan cells by using the relatively new technique of autoradiography in which a radioactive atom is physically linked to some chemical needed in a reaction. The reaction can be followed by measuring the rate at which the radioactive atom is incorporated into the new chemical that results from the reaction.

The NIH team decided to label one of the nucleotides that would have to be added to the genes of newly dividing cells. "We could see an occasional cell that was now clearly HGPRT positive," Friedmann said. "This was in 1967. It was the first demonstration of a [mammalian] cell adding foreign DNA."

But even among themselves, the group could not agree whether the "occasional cell" that seemed to be incorporating radioactive material really represented genetic transformation in mammalian cells. They presented it as an abstract at the American Society of Human Genetics meeting in 1968, but never published the results beyond that.

Transformation of mammalian cells by bulk DNA just wasn't going to work. Scientists would have to find another route. Hotchkiss already had decided that "a more effective mode of introduction of DNA would be through viruses. . . . " It made sense. In 1956, Joshua Lederberg and his colleagues at Rockefeller discovered that the viruses that infect bacteria had an unusual property: when they infected a bacterial cell, they sometimes picked up bits of DNA from the host chromosome that became incorporated into the new virus progeny. The viruses would then go on to infect new bacteria, and deposit the bacterial genes it picked up elsewhere in the host cells. Some of those host cells would survive the infection and become permanently changed by the new genetic material they received. Lederberg called the process transduction. He won a Nobel Prize for the discovery in 1958.

Lederberg's discovery suggested that viruses could be used to transfer genes into target cells intentionally, but its usefulness was limited by a simple problem: No one could determine which bits of DNA the virus picked up. What's more, the viruses Lederberg studied were only able to infect bacteria. If viruses were going to be used for therapeutic purposes, researchers would have to find a way to get DNA into eukaryotic cells, like those that make up mammals. Carl R. Merril, a researcher at the National Institute of Mental Health, part of the National Institutes of Health, believed he had discovered evidence that the viruses of bacteria—called bacteriophage—were able to infect human cells growing in tissue culture. He also became convinced that the bacteriophages could transfer genes into the human cells just as Lederberg's viruses could in bacteria.

In a 1971 paper in *Nature,* Merril described how he had infected human fibroblasts growing in culture with a lambda virus that was carrying an *E. coli* gene galactose operon—the same one that Jon Beckwith and the Harvard team isolated in 1969. The fibroblasts came from patients that suffered galactosemia, an inherited disorder that destroyed their ability to metabolize milk sugar. Merril claimed that the bacterial virus not only got into the human fibroblasts but cured them of the inherited defect by providing the *lac* gene that made the enzyme needed to digest the milk sugar. This was major news—if true.

Merril's claims were controversial because most researchers did not believe that he could infect human cells with a bacterial virus. The targets for those viruses, prokaryotic bacteria, are just too different from eukaryotic mammalian cells. There should not be any way for the bacterial virus to get into the human cell. What's more, many other research teams tried similar approaches in an effort to get bacteriophages to transduce eukaryotic cells. None of them ever produced consistently positive results.

It made sense that researchers had begun testing viruses for their ability to transfer genes into target cells. After all, nature already had turned viruses into a reliable gene delivery mechanism for their own genes. What's more, it was now known that viruses could pick up and carry extraneous bits of genetic material, insert it into a cell without killing it, and force the cell to manufacture the protein coded for by that gene product. But could a virus really be used to heal a genetically defective cell?

Early work suggested that it could, perhaps even as naked DNA. Several research teams disassembled virus particles and isolated the viral DNA. They then succeeded in inserting viral DNA alone into target cells. They knew it worked because they could get out intact virus or proteins unique to that virus. But the effectiveness was low, and the approach did not seem like a likely solution.

The first solid clue about the usefulness of viruses for gene transfer came from the Oak Ridge National Laboratory in Tennessee, a government laboratory more concerned with blowing things up than with biology. Carved out of the Tennessee hills during World War II, Oak Ridge began as a factory for making fissionable materials needed in atomic bombs.

Oak Ridge also had a biology department, which mostly studied the effects of radiation on living things. And the biology department had a character named Stanfield Rogers, M.D. Although Rogers was a doctor, he also was a biochemist who liked to do research. In the 1950s, he was studying an unusual virus that caused hornlike warts to appear on wild cottontail rabbits roaming the Tennessee countryside. The virus, the Shope rabbit papilloma virus, causes usually benign skin tumors. In 1959, Rogers reported that the strange virus carried the gene needed to make arginase, an enzyme that breaks down the amino acid arginine, keeping its levels in the blood under control.

He was growing rabbit skin cells, epithelium, in the laboratory, and when he infected them with the virus, the cells produced the arginase enzyme. Apparently, the virus—which can infect cells without killing them—was transferring its arginase gene to the skin cells, giving them the ability to make an enzyme that they normally lacked.

At first, Rogers did not appreciate the significance of what he had found. With time, he began to realize that the rabbits, mice, and rats infected with the Shope virus all had significantly lowered levels of arginine in their blood. "Then, much to our surprise, we discovered that of the people who had worked in the laboratory with this virus, about half could be detected from studies of their blood arginine alone," Rogers recounted years later for a 1975 New York Academy of Sciences conference on the prospects of human gene therapy. "Clearly, the virus had infected the laboratory workers, increased the ability of their cells to make arginase, and had significantly lowered the level of arginine in their blood. No other harmful effects could be found," Rogers reported. "It was clear that we had uncovered a therapeutic agent in search of a disease."

In 1969, Joshua Lederberg helped Rogers find a disease for his rabbit virus. Lederberg had spotted an odd paper in *Lancet,* the British medical journal, that described two German sisters who had extremely high levels of the amino acid arginine in their blood. No other cases like this had been reported in the scientific or medical literature before. Although having high concentrations of arginine was not itself harmful, the excess arginine impaired the body's ability to process urea. Normally, urea is degraded and used to make arginine. Because their bodies already had too much arginine, the enzymes responsible for these conversions did not work properly, and the children could not eliminate the urea. This caused ammonia to build up in their bodies. The ammonia, apparently, was toxic, causing the children to be "epileptic, spastic, grossly retarded, and progressively becoming worse," Rogers said.

Apparently, the girls had inherited a genetic defect that blocked their ability to make the enzyme arginase. Both their parents had only half the normal level of the enzyme, as did their other two siblings. Rogers reasoned that if they lacked the ability to make arginase, and the Shope virus conferred the gene needed to make the enzyme, then perhaps the Shope virus could be used to treat these children. Theoretically, it should lower their arginine levels as it had done to his experimental animals and his lab workers. With the arginine levels down, urea could then be processed more normally and the ammonia levels would decline, preventing further brain damage. On paper, it all made sense.

Rogers contacted the German authors of the *Lancet* article, and suggested that he had a possible treatment for the children. The German doctors agreed it was worth a try. As a preliminary test, the German doctors sent Rogers cells from the girls to see if he could infect them with the Shope virus in the laboratory while the cells grew in tissue culture. He did that successfully. The next step would be the girls themselves.

In preparation for treating the children, Rogers purified the virus and checked it with the electron microscope to make sure the samples contained only Shope virus. In 1970, Rogers packed the virus in ice and boarded a plane for Cologne, Germany, where he would treat the girls. The oldest was five years old, and by then was suffering mental retardation and convulsions. Her younger sister, then only two, was in much better shape, though starting to show signs of chronic ammonia poisoning.

Because none of the researchers knew how much virus they should use, Rogers and the German physicians decided to be conservative: They gave one twentieth the dose Rogers had previously shown to be harmless in mice. They decided to err on the side of safety, but they were probably too conservative. "As might be expected, no effect whatever was found either in the condition of the children, or in their arginine or blood ammonia levels," Rogers reported. Since the Shope virus does not reproduce itself in humans, there was no chance that it could make new copies of the arginase gene. That increased the safety of the experiment, but it lowered the chance that enough enzyme would be made to help the girls.

About a year later, the older child worsened. She was given a second treatment with about as much virus as Rogers had used in rabbits to demonstrate a reduction in blood arginine levels. The researchers determined that "the blood arginine level dropped transiently, but it subsequently returned to its original high levels. Thereafter, however, the children gradually improved for a time. It is not certain, however, whether this was related to the virus or was an effect of the low-protein diet, which was known to lower ammonia levels."

Eventually, even as the oldest child grew worse, the family had a fifth baby. It too had the disease. Rogers and his German colleagues decided that treating the child early in life might prevent the disease. The virus was prepared in Tennessee and sent to Germany, but family complications delayed the use of the virus for months. During that time, the relatively unstable viral preparation deteriorated and became inactive. By the time the child was treated, there were no infectious viruses remaining, so the therapy failed to help.

The overall results, Rogers concluded, "were at best disappointing."

But they were not without unexpected effects. The experiment put Rogers at the center of a controversy. As word of the genetic treatments trickled out, a growing number of scientists objected to the treatment as potentially unsafe and certainly premature. Ted Friedmann, who by now had moved from the NIH to the University of California at San Diego, and Richard Roblin, then at the Massachusetts General Hospital, complained the experiment was hopelessly premature, possibly dangerous, and certainly of questionable ethics.

First, the basic idea may be wrong, they wrote in a letter to *Science,* and later in a full article. The virus may not carry the arginase gene after all, but may have lowered the arginine levels in lab workers by overactivating their natural enzyme. If that were true, then giving the virus to the children could not stimulate arginase since the girls had no functioning copy of that gene. What's more, they wrote, the Shope virus may not be as harmless as Rogers believed, arguing that "portions of virus-induced papillomas in both domestic rabbits and cotton tail rabbits developed into invasive malignant skin cancers." Conceivably, the virus could turn a cell it infected into a tumor, giving the patient a lethal disease. Until these kinds of issues were sorted out, Friedmann and Roblin believed, no one should attempt gene therapy with viruses.

Rogers, however, disagreed. He defended the decision to treat the children, saying, "The chance to prevent progressive deterioration in these children's condition was the only ethical route to take." What's more, he added, "should other children be found to have this disease, they should be so treated."

After the attacks, however, Rogers abandoned any more attempts to treat arginase-deficient individuals with the Shope virus, though he remained interested in genetic transfer using viruses. Recognizing that there would be few future lucky accidents in which a virus carried a gene useful in treating a disease, Rogers began experimenting with ways to insert genes into viruses intentionally. For his model, he chose the tobacco mosaic virus, an RNA virus. He spent the rest of his career working with plants and their viruses, never again venturing into human studies.

Nevertheless, Stanfield Rogers of the Oak Ridge National Laboratory became, in 1971, the first physician-scientist to attempt human gene therapy using a virus that just happened to carry a gene that might be useful to the patient. It was a historic, if unappreciated, experiment.

It also put gene therapy on the map, if not in the public mind, at least in the minds of researchers. Through the 1960s, more and more scientists began thinking about the possibility of correcting genes in sick individuals by somehow transferring the right code into the patient's cells. Although it was clear existing techniques were inadequate, the influx of people willing to think about it and do the work could make that change, and maybe change rapidly.

At the same time, others besides Beckwith and Nirenberg were beginning to talk about the consequences of genetic transfers in people. Often, the debates came back to ethics and eugenics. Would gene therapy lead to governmental programs that used the new technology for harm, as Beckwith feared? Would the eugenicists, armed with the new technology of genetic transfer, reemerge from the dustbin of history to attempt once again to improve humanity with these powerful new tools?

Ernst Freese, a respected scientist in the Laboratory of Molecular Biology in the National Institute of Neurological Diseases and Stroke, a part of NIH, decided that there was enough activity in the field to bring everyone together to see where things stood. A meeting might identify what needed to be done to ensure the studies were conducted properly and ethically—and successfully.

In early 1972, Freese gathered the invitees at the Stone House, the NIH headquarters of the Fogarty International Center. It went as conferences go, with experts describing the experiments they had been doing and the outcomes. Because there was so little work actually attempting to put genes in human cells, the discussion quickly turned to theories about the possibility of using genetically engineered viruses to insert genes into human cells and ethical concerns about the eugenic implications of being able to detect genetic defects *in utero* and then aborting the fetus.

In 1972, at the dawn of the genetic engineering revolution, the prospects of human gene therapy looked bleak to those gathered in Stone House. No new ideas about how to conduct human gene therapies came out of the conference. The techniques were simply too primitive. Instead, some of those in attendance abandoned the field, others drifted back to their labs to begin basic research to find ways around the gene transfer problems they had all encountered.

Although the intent of the researchers was to find a mechanism to transfer useful genes and relieve human suffering, ethical concerns always floated in the background. They would focus on a medical model of disease and its treatment, but none failed to understand that the same techniques that could heal might also be used to harm. Society would have to decide how gene therapy would be used, and given its track record, that would bear watching.

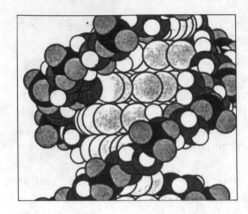

Chapter 3
Lineage of
a Lab Rat

*"Among the scientific elite . . . chains of 'hereditary' influence are
no aberration. They are the norm."*
—*Robert Kanigel,* Apprentice to Genius:
The Making of a Scientific Dynasty

French Anderson's first clue to his future quest for genetic cures
came in the mail. It was 1952 when the catalog of undergraduate
courses offered by Harvard College hit the mailbox of his parent's
home in Tulsa, Oklahoma. The brochure described Harvard's world
of intellectual possibilities, but one in particular caught the eye of the
high school senior: Biophysical Chemistry 184, a physical chemistry
course taught by John T. Edsall, an expert in protein chemistry who
studied biologically important molecules like those involved in muscle contraction, or the ones that carried oxygen in the blood. To Anderson, a boy searching science for unifying ideas, it made sense to
study the body's function from the viewpoint of molecules. High
school science taught him that all matter, living and nonliving, was
made from atoms. Like Russian dolls, atoms made molecules; molecules—proteins, carbohydrates, and fats—made cells; and cells made
the body. Visualizing the innermost doll might say something about
the structure and function of the largest. Understanding the body's
molecules might lead to insights about its diseases and thence to the
discovery of new therapies.

Anderson already had decided to make his career in medicine. Before he was ten years old, he announced to his mother that he was going to be a doctor. The Harvard catalog now showed him what kind
of doctor: a doctor of molecules for people with molecular diseases.

In his application essay to Harvard, he declared his intention to develop a molecular understanding of human organs and molecular cures for human sickness. It was a hazy idea at best then, but it seemed to hold great potential. Decades later, gene therapy would fulfill those vague notions.

In characteristic style, once the idea settled in, Anderson clung to it. Though Anderson applied to and was accepted by several other universities, including Princeton, Dartmouth, and Yale, Harvard was really his only choice. He wanted to study biophysics with John Edsall; he wanted to become a scientist. In the thirty five years between college admission and the first genetic therapy in September 1990, French Anderson went from adolescent to adult, both personally and scientifically. Along the way, a personality emerged with a dogged determination to do something unusual and important, and to do it first. Anderson would endure disappointment, failure, and ridicule by scientific peers who believed his talk of genetic treatments—which started in the late 1960s—was premature by decades, even goofy. He never wavered.

The direction may have been set by the Harvard catalog, but the single-mindedness with which he pursued it came from his Oklahoma roots. Born in Tulsa on New Year's Eve 1936, William French Anderson was a precocious, hyperactive kid. By five months, he shook the crib continuously. By one year, he walked and talked. By two, he could recognize written words, according to family stories.

By the time he was five, his mother thought he was ready for school. Tulsa schools accepted first graders at age six. But his mother, LaVere Anderson, a book editor for *The Tulsa World,* persisted, and took her son to see Mrs. Knappenberger, the first-grade teacher. William French Anderson—his mother called him Bill—was small for his age. Mrs. Knappenberger was skeptical. She handed the child a reading primer. He read it out loud. She accepted him into her classroom, first grade, a year early. For years, he was ashamed to admit he had skipped kindergarten.

For Anderson, first grade was a shock. He had thought that he was a normal kid, though he tended to be reclusive, read all the time, and not have any friends. In class he discovered the other kids in first grade, those who were a year older than he, could not read or do much arithmetic. Bill looked down on his peers; clearly, they were not as smart as he was. He quickly began to think he was better than other people and told them so. In the social milieu of elementary school, where children are learning to work in groups and follow the rules, Anderson's arrogance earned him only scorn among the students. He didn't care. Instead of altering his approach to classmates, he rejected them, withdrawing into his books and later to his room, where he had a chemistry set.

By age eight, science had become an abiding passion for Anderson. He never knew why. Maybe it was the influence of his father, Daniel French Anderson, who was a civil engineer with the Southwestern Power Administration, a Department of Interior agency set up in 1943 to control flooding and to distribute hydroelectric power from its dams to three states: Oklahoma, Arkansas, and Texas. Maybe it was just one of those innate phenomena, like a four-year-old sitting down at the piano and playing Mozart. All he knew was that science consumed him to the exclusion of most else.

By fifth grade, however, Anderson's arrogant ways began to catch up with him. A classmate, Mary Dunn, stunned him one day as they were walking home from school. She turned to him and said, "You are the most unpopular boy in school." The news was crushing. Bill Anderson began to realize that if he was going to do something special, he was going to have to get along with people. To show he was going to try to change, he hit on something symbolic. He decided to change his name.

For six generations, the Anderson family had a tradition of altering first names between father and son, while keeping the middle name the same. If the father's name was William French Anderson, the son would be named Daniel French Anderson. As a child, the boy would be called by his first name, William or Daniel. At age twenty-one, he had the option of changing his name to W. French Anderson or D. French Anderson. Bill Anderson's father long had been called D. French Anderson. Bill, at age ten, broke with tradition. He changed his name to French eleven years early. He became almost irrational about it. He refused to answer anyone who called him Bill. Except for his mother, of course. To her, he would always be Bill.

As a child, growing up in a book-oriented family, Anderson learned to pursue knowledge single-mindedly, as well. When he was ten, his mother received a review copy of the new *World Book Encyclopedia*, and decided to donate the set after her story was published. Distressed that he was about to lose the knowledge contained in his new reference set, young Anderson carried all twenty or so volumes up to his room and began to copy them on his mother's Underwood. When his mother came upon him a short time later, he already had transcribed several pages of volume A. She got him his own set.

Anderson also acquired his father's skills of logic and mathematics. French considered mathematicians the true intellectuals, working in an abstract world where controlled experiments did not exist. Results relied only on logical proofs. In 1953–54, Anderson encountered a bit of mathematical illogic as a high school senior when he took Mary J. Barnett's Latin class where they read Virgil's *Aeneid*. During class, Ms. Barnett described the Roman counting system, and how the Romans had never devised a way to use their numerals in arithmetic. Historians, she said, believed that Romans added and subtracted on their fingers, or used an abacus, and then wrote down the answers. Unlike the sleek, Hindu-Arabic numbers used today, the bulky Is and Vs and Xs and Cs and Ms of the Caesars must have been too clunky to run through formulas. The Romans never used their numerals to add or subtract, multiply or divide.

To Anderson, who had played with numbers since childhood, this historical belief made no sense. "A system of notation merely symbolizes abstract quantities called numbers," he would later write. It should be possible to carry out a mathematical operation no matter what symbol is used. Consider algebra, where letters substitute for numbers. Even a high school student has little trouble multiplying X times Y to come up with the correct answer. Anderson posed the problem to his family at the dinner table that evening and declared that he intended to solve it. His father, the engineer, made suggestions. His mother offered support and agreed to help him write up his results. But it was Anderson himself who hunched over the desk in his bedroom and began the laborious task of creating a mathematical system out of the ancient numbers.

Adding and subtracting was a snap, easier, actually, than the same operations with the more familiar Hindu-Arabic system. The rest— multiplying and dividing, exponents and square roots—was tough. For one thing, Roman numerals do not have a symbol for zero. For another, the historians were right when they said Roman numerals were difficult to tract in complex math operations. With some help on numerical systems from G. W. Hall, his high school chemistry teacher, and University of Tulsa math professor Ralph W. Veatch, Anderson developed a system for carrying out all the basic arithmetic operations with Roman numerals. He did not claim that this is how the ancient Romans calculated the percent return on a conquest investment, but he did prove that Roman numerals could have been used in such calculations.

Before he made too many claims for his new mathematical system, Anderson wrote to some experts in antiquity, including Prof. Sterling Dow at Harvard University and A. W. Richeson of the University of Maryland. They agreed that Anderson had hit upon something novel, something that had not been shown before in any of the known literature of the world. Anderson was seventeen.

After Anderson got to Harvard, he wrote up the Roman numerals work and submitted it, with Dow's help, to a journal called *Classical Philology,* which was devoted to classical scholarship, especially in linguistics. His work ran as the lead article in the July 1956 edition. Even before the *Philology* article appeared, *Time* magazine ran a small piece in March 1956 about how Anderson had "produced the first theory of how an ancient Roman did his multiplying, dividing and square-rooting with Roman numerals." *Time* called the eighteen-year-old sophomore one of Harvard's prodigies.

• • •

Anderson arrived at Harvard in cowboy boots. It was the first time he had ever been out of the Southwest. Harvard would be a massive, life-changing adventure. He jumped in, trying to do everything—often all at once. In his freshman year at Lowell House, Anderson recorded how he spent his time, "a full record of his studying, sleeping, and athletic habits, complete with graphs, totals, and averages," wrote the *Harvard Alumni Bulletin* when it discovered his practice and insisted on publishing an analysis of his daily life as a freshman. The diary began at the beginning: "(Sept. 16) Boarded train [in Tulsa] at 9:45 A.M. First trip on train. Spent night in day coach. Got up at 4:30 A.M., toured train. Went back to bed. (Sept. 17) Came into St. Louis at 7:45 . . . lunch and dinner on train. Another night. (Sept. 18) Met five good guys going to Harvard. First names—Steve, Bill, Mike, John, and Dick. Arrived at 11:35. Caught cab with guys. First room."

The record, more a clipped list of events than a diary, described average days like this: "(Sun. Nov. 14) Rose 10:15. No breakfast or shower. Math (one and a half hours). Lunch at 12:30. Math. Studied German grammar and words (three hours). Dinner. German (one and a half hours). Read Bio up in Perry's rooms, horsed till 11:45 (one and a half hours). German with frequent interruption. Bed 12.00." The daily descriptions were averaged at year's end: "sleep accounted for 34.7

percent of Anderson's time; classes and study, 32.1 percent; and miscellany (including meals and athletics) 33.2 percent. Anderson was high on the Dean's List last year." He graduated magna cum laude.

In the years that followed college, Anderson kept a similar list on a calendar—two calendars, actually. One was a government-issue appointment calendar, going back to his first day in government service at the National Institutes of Health, on which he recorded every appointment he had. The second he used for tracking his scientific productivity. Anderson recognized the truth about his days at Harvard as observed by James Watson, the double helix's codiscoverer: "He was the kind of person I avoided," Watson said. "He was always excited about everything. He didn't discriminate about the kinds of ideas and questions that were important."

Knowing he could get sidetracked, Anderson set up a system of coloring half days on the calendar red, yellow, or green. A half day was colored green if he spent more than four hours working on science, such as working at the lab bench, attending seminars, reading papers, or reviewing data. The block got colored yellow if he spent less than an hour on science, and red if he did not spend any time on science. The color calendar provided a graphic picture of his productivity. If the overall picture remained mostly green, then he maintained his present course. As the colors turned yellow and red, he reevaluated his activities. "If I get too distracted, and there are too many yellows or reds in there, then it makes me settle in again and concentrate on what is important," Anderson said.

There was one exception from continuous academic work: athletics. At summer camp as a teenager, Anderson learned he could run fast, faster than almost anyone else. In high school, he became a track star. As a college freshman, Anderson captained the first-year squad. As a sophomore, he pushed Harvard to unexpected victories. "Y'know, one of the major factors in our success this Winter was the way French came out of nowhere to help us," Harvard coach Bill McCurdy was quoted in a newspaper report in 1956. "He caught two men in the final 10 yards for a key second place in the 600 at Dartmouth; he won the 600 against Yale, took second in Heptagonal 600, and proved the surprise addition we needed on our relay team. I wish I had a whole team of Andersons." As a senior, he won the Wilcox medal for his performance in the quarter mile.

Anderson considered trying out for the 1960 Summer Olympics, but his body could not keep up with his unbounded determination to work hard. The more he pushed his legs, running flat out during competitions, the more he injured them. Finally, his hamstrings simply ruptured. Although he continued to run throughout college and graduate school, he would never again have the speed of a contender.

• • •

While Harvard opened a whole new world to Anderson, intellectually and culturally, it also marked the beginning of his career as a research scientist. Science is a social, collective enterprise. Few work alone. The ideas themselves are so complicated, and the experimental technology so complex, that no one person can master every nuance of every needed skill. To be successful, a scientist must associate with other scientists. They become supporters, collaborators, competitors, and even enemies. The web of social connections begins in college, where professors and mentors play a central role in shaping the student's scientific career.

The status of the mentors can be as important as the science that is studied. A scientist's pedigree shapes the trajectory of his or her career. Genealogy—in whose lab in what university the fledgling scientist is formed—often determines a researcher's very connection to the social fabric of science, just as the status and privileges of a child's family influences a person's place in the world.

" 'Genealogy' plays as central a role in the careers of scientists as it once did in the alliance-by-marriage diplomacy of the royal houses of Europe," Robert Kanigel wrote in *Apprentice to Genius: The Making of a Scientific Dynasty.*

A scientist's early reputation rests as much on whose lab he has worked in—on whose scientific progeny he is—as on what he has discovered. There are "schools" of science, in particular disciplines, just as there are in art and music. There are scientific "families," each of whose members can be traced to one or a few original Adamlike figures. By one reckoning, more than half of all American Nobel Prize winners have worked as graduate students, postdocs, or junior colleagues of other Nobel laureates.

Harriet Zuckerman, a sociologist at Columbia University, did the counting in her 1977 book, *Scientific Elite: Nobel Laureates in the United States.* Of ninety-two American Nobel winners who had claimed their prize by 1972, forty-eight, more than half, had worked for, or with, a previous Nobel winner. Anderson would follow that pattern.

His first chance to join the scientific elite came in Harvard's Gibbs Chemical Laboratory, in the second-floor space of George B. Kistiakowsky. The Ukrainian-born chemist grew famous during the bomb days of the Manhattan Project. Tall and outspoken, Kistiakowsky became an explosives expert early in World War II while experimenting with bombs for the National Defense Research Committee. At Los Alamos, New Mexico, the center of the Manhattan Project, Kistiakowsky perfected a type of explosive lens, a way to focus or shape conventional explosive charges so that the force wave generated by the explosion would smash together radioactive material to achieve a critical chain reaction and atomic detonation. After the war, Kistiakowsky returned to Harvard before moving to Washington, D.C., as Pres. Dwight D. Eisenhower's science adviser.

During this time, between the war and Washington, Anderson slipped into Kistiakowsky's lab with three other adventurous undergraduates. That they did so at all was unusual. Undergraduates seldom did independent research projects. In 1936, Kistiakowsky had experimented with the rate of thermal isomerization of cis 2-butene—the ability of heat energy to make the molecule take on different shapes. Butene is a simple molecule, four carbon atoms in a line with a double bond in the middle. The chemical can be found in coal gas or isolated out of petroleum oils, and can be quite inflammable. Kistiakowsky had shown that the 2-butene could take on two different forms called isomers. An isomer shuffles its atoms of hydrogen, oxygen, and carbon in different arrangements as it undergoes changes in temperature, but keeps the same number of atoms.

In the 1930s, Kistiakowsky had tried to sort out what happened during all that atomic shuffling, and how fast it proceeded. His measurements, however, put 2-butene in a class by itself, conflicting with the measurements of similar chemicals undergoing isomerization. This caused problems with the theories that explained the isomerization phenomenon. One possibility was that Kistiakowsky's measurements were wrong. Anderson and the others decided to find out.

From the 1930s when Kistiakowsky made his first measurements to the 1950s when the band of undergraduates invaded his lab, a new measurement technique using infrared light became available. Called spectroscopy, the method bounces a beam of infrared light off the vapor of 2-butene, and measures how much light of a specific wavelength is absorbed by the chemical. With this approach, the undergraduates could accurately measure the rate at which one form of 2-butene turned into the other form.

Kistiakowsky agreed to allow the students to work on the problem, gave them some lab space, but didn't expect much. Nor did he give them much support, allowing the students to learn their first useful scientific skill: scrounging and improvising. They borrowed a 1,000-watt constant voltage transformer from the physics department, which they connected to an electric furnace they had swiped from somewhere, and came up with glass stopcocks from all over campus to build the glass tubing system that carried the different liquids and gases into the 500-milliliter carbonized Pyrex flask that served as the reaction chamber.

It was a Rube Goldberg contraption at best. Anderson built it. He learned to blow glass, weld glass conduits to glass valves, and poke hot tubing through the wall of the Pyrex flask. The central flask was key to the entire experiment. Nestled in the coils of the electric furnace, it held the bubbling butene as it was heated up to 443 degrees centigrade, more than four times the boiling point of water, and pressures ten times greater than atmospheric. As the student team added more and more components to the experimental line, segments kept breaking. Though Anderson thought the central Pyrex flask was a thing of grace and artistry, it also was fragile. As he added five, then six glass tubes to the central bulb, the bulb's glass wall got thinner and thinner. Finally it broke, and he couldn't fix it. There were simply too many glass tubes punched through the Pyrex wall.

"I had no idea how to rebuild it," Anderson recalled of his defeat. "I went to Kisti, and he came down and looked at it." What he found surprised Kistiakowsky. Anderson and the three others had designed a more refined experimental device than he initially told them to build. It was better, sort of. Certainly it was more sophisticated, or did it just have more parts? More important, it showed the undergraduates were serious about doing a serious experiment. He had never stopped in to see what they were doing before.

Kistiakowsky examined the fractured Pyrex globe, and quickly hit on a solution. Then he said to Anderson, "I'll show you how."

But Anderson leaped in saying, "No, just tell me what to do, and I'll do it *because it's mine*."

Taken aback, the man who helped make the bomb that ended the war, and changed the course of history, cut Anderson off and said, "No, I'll do it *because I am the professor.*" And he did.

Anderson sulked.

After that, Kistiakowsky started dropping by at the end of the day to see how his fledgling scientists were doing. They did just fine. By the end of the project, they had carried out refined isomerization measurements, contradicting their professor's past measurements. "The 1936 value for the frequency factor [of isomerization] is completely outside the range of error," they politely wrote in a 1958 paper in the *Journal of the American Chemical Society.* "The theoretical difficulties which the butene-2 isomerization formerly raised may thus be considered as eliminated."

It didn't turn the world of chemistry on its head, but the four undergraduates demonstrated their technical skills, and it gave them a certain level of credibility. What's more, it opened for them the doors to virtually any other laboratory at Harvard. Anderson, who was first author on the paper, decided to work with Paul Doty, a biophysicist with a passion for peace. Doty would work for disarmament, becoming an organizer of the Pugwash, Nova Scotia, meetings, formally called the Conference on Science and World Affairs, which began in July 1957 and at which scientists from around the world searched for ways to reduce the risk of nuclear war. That same year, Anderson entered Doty's lab as an undergraduate. The times were exciting, scientifically as well as politically.

Doty was a scientist with international standing. In 1946 and 1947, he had been at the Cambridge University laboratory, working with Max Perutz, the x-ray crystallographer who would decipher the structure of the hemoglobin molecule and win the Nobel Prize. Through Perutz, Doty would later come to know Francis Crick and Jim Watson, of double helix fame, and become convinced of DNA's importance. Doty was there when the foundations were laid for molecular biology, and added his own bricks: He described the atomic glue that binds together the two railings of the double helix to form the chemical's twisted-ladder shape. In scientific parlance, he worked out how the four chemical subunits of DNA "hybridize" with each other, like pairs of magnets with a north pole attached to one half of a rung of the ladder attracting a south pole on the half rung on the opposite side. The magnets attach the half rungs together to form a complete rung. His work expanded to RNA, and to the three-dimensional structures of proteins like collagen.

Anderson joined Doty's lab to do a senior thesis, an honors course for a premed honors student, at a time when DNA meant everything. Anderson learned nucleic acid chemistry at the feet of the master. In his own project, Anderson studied the effects of ultraviolet radiation on DNA. Later, when he was a senior in medical school and had come back to Doty's lab to work as a technician, Anderson helped with the work on polynucleotide phosphorylase, an enzyme critical in the processing of genetic material.

Anderson stood out in Doty's lab. "He was inner driven, and extremely hard working, and bright, deferential," Doty recalled of Anderson. "He had a real gleam in his eye. There was no doubt he knew where he was going. What was extraordinary was his continued alertness to everything around him."

One of the more interesting items to pique Anderson's alertness arrived with Julius Marmur, a postdoctoral fellow from Rockefeller University who had just arrived in Doty's lab with stories about genetic transformation. Scientists had known since 1944 that genetic material in the form of the chemical deoxyribonucleic acid, DNA, could be extracted from bacteria of one type, and used to transform bacteria of a second type into the first type. Work by Oswald T. Avery and others at Rockefeller University proved that DNA carried genetic information. Rollin Hotchkiss, who joined Avery's lab in the late 1930s, eventually took over the experiments on transforming factor at Rockefeller. In the mid-1950s, Hotchkiss had taught Marmur all about genetic transformation in bacteria. Marmur brought those techniques to Doty and his lab. Anderson was in the lab; he learned transformation from Marmur.

One of the other things that made Anderson stand out in Doty's mind was his aggressiveness when pursuing ideas. "In our group seminars, which were weekly," Doty recalled, "French questioned what was being said more than most graduate students would. And he was an undergraduate. It was a precociousness that was useful, but at times it did approach combativeness. He would not easily give up a view he held. His views were mostly on target. In his not understanding something, he would pursue it until he did, sometimes taking more time than he should out of the seminar."

One of those seminar sessions would have a serious influence on the rest of his life. In the winter of 1958, Anderson attended a lab meeting put together by John B. Edsall, the biophysical chemist whose course description persuaded Anderson to attend Harvard. Anderson took Edsall's courses, joined his tutoring program, even helped proofread one of his textbooks, and generally fell into the orbit of the older scientist. Edsall's lab meeting, which included people from Doty's lab, had been called to review the latest results from England on the structure of the protein hemoglobin and its chemical cousin, myoglobin. As the scientists around the table reviewed the physical data of biology's most studied proteins, an idea began to emerge for Anderson. By now, Marmur had taught Anderson about bacterial transformation, and how it could cause permanent, inheritable changes. At the same time, Anderson already had decided to go to medical school, so he was starting to think in medical terms.

"If it is possible to work out the structure of normal hemoglobin," he wondered out loud during the meeting, "then maybe you can work out the structure of sickle cell hemoglobin, and then you could determine what the defect is. And, if you could use transforming factors, then maybe you could put in the gene for normal hemoglobin, and correct sickle cell hemoglobin."

The scientist leading the discussion turned to Anderson and snarled, "This is a serious scientific discussion. If you want to daydream, keep it to yourself."

Anderson retreated into silent humiliation. He just wanted the session to end so he could get out of there as gracefully, and as quickly, as possible. When the meeting finally broke up, John Edsall encountered Anderson on the way out the room. In his low, grumbly voice, Edsall said to Anderson, "Interesting idea," and walked out.

Anderson nearly fell over. He quietly swore to himself, "If Dr. Edsall thinks it is an interesting idea, then I think it is an interesting idea." Years later, Anderson pinpointed that seminar as the moment he decided to find a way to put genes into cells. There would be endless distractions to keep him from that goal, and it would take nearly twenty years before he would focus his energies exclusively on gene transfer techniques.

First he would have to finish Harvard College. After struggling to decide whether he wanted to be a physician or a research biochemist, a third, unexpected opportunity arose: study abroad. Anderson had been so intensely focused on biochemistry and research that some of his professors worried he was too narrow-minded. A trip abroad might be helpful, so Anderson was nominated for several fellowships. He won the Lt. Charles Henry Fiske 3rd Scholarship at Trinity College at Cambridge University in England. The one-year program would give him an opportunity to study in England, and travel throughout France and the Continent.

Anderson ended up as a graduate student in the same Cavendish lab where James D. Watson and Francis Crick discovered the double helix. Watson was only recently gone, beginning his own career at Harvard. But Crick, John C. Kendrew, and Max Perutz, all of whom would win a Nobel Prize one day, were still there, as well as Sydney Brenner, another founder of modern molecular biology, and the only one in that group not to win a Nobel, though many thought he should have. It was a research powerhouse, and Anderson just walked through the doors as if he belonged—very much thanks to the interventions of Paul Doty, who had his own connections to the Cavendish from a previous sabbatical.

Anderson came as a graduate student, and studied under Leonard Lerman, an American scientist from the University of Colorado who was trying to understand how certain dyes caused alterations, mutations, in the chemical structure of DNA. Though Anderson was able to get in, resources, especially space, were scarce. He got two feet of bench space next to the water bath used to keep culture flasks at a constant temperature during viscosity measurements. Whenever someone was making a measurement, he had to move. It wasn't much, and he wasn't there much, but the year-long fellowship exposed Anderson to the highest levels of biological thinking. In the end, he earned a master's degree from Cambridge University, ran on the track team, traveled in France, and decided to stay on for an additional twelve months in England to start his first year of medical school.

That's when his life took another dramatic turn. During the first-year anatomy course in medical school in England, over a head and neck dissection, French Anderson met Kathryn Dorothy Duncan, a pretty, slight, determined Brit with clear eyes, nimble hands, and a penetrating mind. Within months, they were engaged, though their marriage would have to wait for a year while she completed required course work in England. Anderson, meanwhile, returned to Harvard to take up his second year of medical school in Boston. They were married in England during the summer of 1961.

Marriage and medical school was a difficult combination. Neither had a job, so they lived on loans and scholarships. French graduated a year ahead of Kathy, and entered his internship at Boston's Children's Hospital Medical Center. He earned $83 a month. "We would budget for a month to go to a movie, and then we would have to walk to it." After they would see a film in midwinter, the fierce cold got to Kathy's ears, making them ache with tinnitus. "It got so bad that we had to spend money on the subway," Anderson recalled. "It was all the money we had."

They lived in a tiny first-floor flat in a run-down Roxbury-Brookline neighborhood, between the medical school and Fenway Park. Their double bed barely fit in the bedroom; to open the chest of drawers, they had to sit on the bed. The cinder parking lot out back raised so much dust that everything was dirty all the time, no matter how much they cleaned.

The stress and poverty of two burgeoning medical careers was too much. They argued constantly, to the point that their marriage was foundering. Anderson decided rather than risk divorce, he would drop out of his residency after the internship and get a job. He could always do his residency after Kathy finished.

When Anderson decided to drop out, Children's physician-in-chief offered financial help. An attending physician offered to take Anderson into his practice. Anderson declined. "We needed something more than a handout," he said. "We needed a fundamental change, and that meant, basically, fending for ourselves and pulling ourselves up by our own bootstraps."

Anderson landed an American Cancer Society research fellowship that gave him the princely sum of $7,000 to work for a year in the Department of Bacteriology and Immunology at the Harvard Medical School while Kathy finished her internship at Children's Hospital. It also allowed them to move to better accommodations, a third-floor apartment in Brookline, where the landlords, a husband and wife, lived on the first floor, and their sons, both police officers, lived on the second. The stability probably saved their marriage.

The crisis also made them look at another aspect of married life: parenthood. Both recognized that each was ambitious, that their devotion to their work could be all-consuming. They also recognized that children might not thrive well in an environment where both parents were so preoccupied with their own careers that they seldom came home. French and Kathy Anderson decided children were a luxury they could not afford. Instead, they would adopt. Not legally, but spiritually. They formed protective relationships with a series of young people whom the Andersons drew into their lives, helped financially, and supported emotionally.

In some ways, this decision grew out of the frontier generosity Anderson learned in Oklahoma, a nurturing attitude he displayed as an undergraduate at Harvard when he acted as tutor and protector of a twelve-year-old freshman from Berkeley, California, named Fred Safier, Jr. Anderson helped Safier, considered a prodigy by many, survive the rigors of college science courses, and negotiate the social life of Harvard.

There would be others. Some were patients or former patients. Others were young people they encountered along the way. Lauren Chang was the last. Born in Hong Kong, Lauren came to the United States with her mother and sisters. In high school, Chang worked part time at NIH, and met French Anderson. He took an interest in the young woman struggling to learn science and English and preparing for the college entrance exams. Anderson began helping her study, quizzing her on concepts and on English. "At one point," Chang recalls, "we were talking, and we sort of decided to adopt each other. He doesn't have any children, and I don't have a father." Now, more than a decade later, Chang is a doctor, a pediatrician, overseeing research projects in the molecular hematology branch Anderson founded at NIH.

· · ·

The National Institutes of Health was the natural place for French Anderson to spend the bulk of his career. He arrived on July 1, 1965, when both he and it were younger and full of vim. The Bethesda campus was *the* hotbed of biomedical studies in 1960s America. There was no place like it on Earth. NIH began in 1887 as a tiny, one-room laboratory in the attic of the Marine Hospital on Staten Island in the harbor of New York City. Its scientists were supposed to find a cure for cholera and yellow fever. Years later, the lab moved to Washington, D.C., but remained a small, sleepy federal facility focused on microbiology and public health. In the late 1930s, however, it began to grow and, after World War II, it exploded.

A powerful collaboration formed between Dr. James A. Shannon, U.S. Sen. Lister Hill of Alabama, and Rep. John Fogarty of Rhode Island. Shannon was the driven, hard-nosed, easily excited Irishman who directed NIH from August 1, 1955, to August 31, 1968. Hill and Fogarty were each chairman of key appropriations subcommittees who worried that American biomedical research suffered from insufficient funding. Their congressional largess poured cash into Shannon's NIH, supporting all types of research aimed at curing disease. It also laid the foundations of the molecular biology revolution. During Shannon's reign, NIH's budget soared twenty-fold, from $54.3 million to $1.1 billion. It's staff, too, nearly tripled, from 5,412 to 13,105. Buildings to accommodate all these scientists and support staff sprang up like mushrooms on the institutes' parklike, 306-acre campus near downtown Bethesda, Maryland. Eventually, some sixty buildings would be devoted to research, including the massive, red-brick Warren Grant Magnuson Clinical Center, Building 10. Opened in July 1953, the 500-bed clinical center became NIH's hospital, and brought together the research bench and the bedside. Ideas developed in the morning could be tested on patients across the hall by afternoon. It revolutionized research.

NIH, however, owes its biggest boon to war. In World War II, the development of antibiotics and blood transfusion techniques, including lifesaving blood expanders such as plasma, convinced Washington's political leadership that medical science was worth supporting. That gave Hill and Fogarty the political clout they needed to push through large funding increases for NIH.

War also brought raw talent. After the Berlin Wall went up in 1961, and all during Vietnam, the nation had a doctor draft. Every newly minted M.D. owed the government some form of military service, either immediately after medical school or after additional residency training. The doctors had few choices: the Army, Navy, Air Force, Marines, Coast Guard, *or* the Public Health Service, the country's sixth uniformed corps. For those who didn't want to spend two years in an Indochina MASH unit, or giving physical exams at some induction center, the Public Health Service offered a decent alternative. Although the young doctors risked being sent to a public health clinic in an inner city or on an Indian reservation, those who landed at NIH enjoyed all the benefits of being in the nation's capital: good restaurants, entertainment, and a decent social life. They also enjoyed first-class research facilities where they could learn the latest medicine, much of it invented in Bethesda, and develop their own research skills while serving their country.

Fledgling doctors—they called themselves Yellow Berets during Vietnam—flocked to NIH. The institutes' research managers had the pick of the litter. This critical mass of smart, young, challenging minds made NIH vibrant. It was a time of great scientific advancement. They celebrated its vitality. Work carried out on the NIH campus throughout the 1960s led to government scientists' winning four Nobel Prizes.

The first Nobel went to Marshall Warren Nirenberg, a biochemist who trained at the University of Michigan. He came east in the summer of 1957, just a few months before Sputnik I went beeping over America. Nirenberg joined the lab of DeWitt "Hans" Stetten, Jr., a venerated NIH figure, then scientific director of the arthritis institute. Tall, lanky, and shy, Nirenberg came as a postdoctoral fellow with definite ideas about what he wanted to study.

Two years after his arrival, Nirenberg was allowed to set up his own small lab to study protein synthesis, the process in which cells convert genetic information stored in DNA into the proteins that provide the cell's scaffolding and metabolic machinery. This process of protein production was only dimly understood, and messenger RNA was completely unknown when Nirenberg started. The NIH scientist just wanted to see if he could dissect out the components needed for protein synthesis, and understand how they were controlled.

In August 1960, Nirenberg, thirty-three, and his partner, Johann Heinrich Matthaei, a thirty-one-year-old plant physiologist from the University of Bonn, in Germany, began by making a so-called cell-free system for synthesizing proteins. The idea of cell-free systems to manufacture proteins was only a few years old, the initial work done in the mid-1950s. It was more art than technology. The NIH scientists made their cell-free system by grinding up bacterial cells and extracting just the components—such as ribosomes and amino acids—needed to make protein. For the method to work, however, the scientists had to add messenger RNA—the genetic material that carried the genetic code from the DNA to the protein-making machinery—to direct the production of the new protein. So they could understand the results, the scientists selected a unique type of messenger RNA.

Typically, messenger RNA is a long, threadlike molecule that the cell makes by connecting four different chemical subunits together. The linear messenger RNA kind of looks like a train made from boxcars that come in four colors: red, blue, yellow, and black. In the cell, the boxcar colors in the RNA train appear random: red, blue, blue, red, red, yellow, black, black, yellow . . . The order of the colored RNA boxcars is controlled by the information in DNA.

While testing their new protein-making technique, Nirenberg and Matthaei decided to use a synthetic RNA composed of a single type of subunit, uracil. It was like making a train of only red-colored boxcars. When they put the synthetic polyuracil RNA into the test tube with their cell-free protein-making system, they got out a synthetic

protein made exclusively from a single amino acid, phenylalanine—
a protein made of only one color amino acid. Normally, proteins contain twenty types of amino acids found in the cell, and each one of
the twenty is a different color. A hemoglobin protein has one pattern
of colored amino acids. A protein enzyme would have a completely
different pattern of colored amino acids. Somehow, the synthetic
RNA made from only one subunit specified that the synthetic protein
would be made from only one amino acid—phenylalanine. This experiment was the key to deciphering the genetic code, the holy grail
of modern biology. Not only did the experiment give them the first
bit of code, but it pointed to a method for discovering the rest of the
code. Nirenberg and Matthaei would be able to write out nature's
Rosetta Stone.

Other biologists had already suggested that genetic information
was coded by the sequence of DNA and RNA nucleotides (the rungs
of the double helix ladder), and that every three nucleotides, or half
rungs, determined which of the twenty amino acids would next be
added to a protein, and that the order of the amino acids determines
how a protein functions. But no one was able to test the idea or work
out the code.

Nirenberg announced the first clues to the triplet nature of the genetic code in a fifteen-minute talk delivered to a nearly empty room
at the International Symposium on Biochemistry in Moscow during
the summer of 1961. No one came because no one knew Nirenberg, at
least no one in the inner circle of biology—not the Jim Watsons or
the Francis Cricks or Sydney Brenners. Nirenberg was so junior and
so unknown that when he applied to attend the June 1961 meeting at
Cold Spring Harbor Symposium on (genetic) Regulation, he was refused admission, even though he had, by that time, deciphered the
first bit of the genetic code.

"There is a terrible snobbery that either a person who's speaking
is someone who's in the club and you know him, or else his results are
unlikely to be correct," Matthew S. Meselson, the Cal Tech biochemist who demonstrated that DNA reproduced itself semiconser-
vatively, told author Horace Freeland Judson. "And here was some
guy named Marshall Nirenberg; his results were unlikely to be correct, because he wasn't in the club. And nobody bothered to be there
to hear him. . . . I heard the talk. And I was bowled over by it."

Meselson went to Francis Crick and got Nirenberg added to a larger session at the end of the meeting. Hundreds of scientists heard his

presentation, and none of them missed its significance—and after that, all the biologists knew who Nirenberg was. The announcement, however, had an undesired effect. It pitched him into a feverish race to be the first to finish deciphering the complete code. Other scientists, recognizing the importance of the question, jumped into the field. Nirenberg and Matthaei had only discovered the first two combinations in the code at that point. There were sixty-four possible arrangements of triplets, and they were a long way from having them all.

Severo Ochoa, who ran a large nucleotide chemistry lab at New York University, immediately dived in, inserting synthetic RNAs into a cell-free system his groups had developed. Ochoa threatened to overtake, and overwhelm, Nirenberg's tiny lab. Less than a year after the competition began, Matthaei left Nirenberg, in April 1962, to return to Germany to establish himself as an independent scientist. NIH colleagues from several different laboratories rushed to Nirenberg's aid, providing him with manpower and material to advance the code work, saving Nirenberg from defeat at the hands of Ochoa. Completing the code would not take one year, as many had predicted, but five.

• • •

In the middle of the fury, in 1963, French Anderson came to Nirenberg's lab looking for a job. Anderson's work on nucleic acids in Paul Doty's lab at Harvard made him just the kind of person Nirenberg needed. Indeed, many scientists wondered why Doty had not jumped in, or even done the cell-free experiment with synthetic RNA himself. Doty had polyuracil, but never put it in. In the middle of a race with other scientists, Nirenberg desperately needed Anderson's skill in working with nucleic acids. Nirenberg asked the young Harvard doctor to join his group.

Anderson said yes, but not right away. He had to wait for his wife Kathryn to finish her internship in pediatrics at Children's Hospital in Boston. Gutsy and aggressive, Kathryn wanted to venture into the field of pediatric surgery, a field dominated by men who opposed her very presence. She eventually did a surgical residency at Georgetown University, and would go on to become a nationally respected pediatric surgeon. The Andersons wouldn't move to Maryland for another two years.

When French Anderson did finally arrive in Nirenberg's lab in

1965, the deciphering work was nearly complete. Nirenberg and the NIHers had beaten Ochoa and the others. Clean-up operations were all that remained. Nirenberg and a postdoctoral fellow named Philip Leder, later to become chairman of genetics at Harvard University, had developed a new technique that made the decoding work far simpler. All that remained was some final analysis and confirmation of previous conclusions.

Nirenberg's team, flush with government-provided resources, had grown large. Nestled in the seventh-floor D wing of NIH's clinical center, the Laboratory of Biochemical Genetics took up an entire corridor and still didn't have enough room. Nearly a half dozen lab modules ran down each side of the hall. A jumble of refrigerators and freezers line the fluorescent-lighted, institutional-tile hallway, along with filing cabinets, boxes, and unused scientific equipment. An instrument room crowded with centrifuges and incubators stood opposite a cold room—a walk-in refrigerator where coat-clad researchers painstakingly separated proteins and other molecules in tall glass columns stuffed with beads and other filter material.

Nirenberg's space was so crowded that he had none for Anderson, not a desk top or a lab bench. Anderson would have to wait for someone to leave and clear out some space. Impatient to begin, Anderson made himself a portable desk out of a slab of wood connected to a big tin can. He carried it around with him. Whenever he had to write something down, he would set up his luggable office wherever there was room. After two and a half months of this, Phil Leder left the lab for the Weizmann Institute of Science in Rehovot, Israel, and Anderson commandeered Leder's bench space on the right side of room 7D09. For the next quarter century, Anderson would not leave the seventh-floor hallway, D wing of the NIH clinical center. Eventually, he would take over the entire corridor.

Competitive and hardworking, Anderson would come in early and stay late—even when there was no reason to. After a brief training period to bring Anderson and other new arrivals up to speed on the lab's main activities, he immediately launched his own project, with another newcomer, to study the mechanism of protein synthesis. Before their work progressed very far, however, in late fall 1965, Nirenberg ran into a problem with the genetic code work. One of the triplet codes being tested seemed ambiguous. Either the preparations were contaminated or the code specified more than one amino acid, which would violate the emerging rules about how the genetic information worked. The question became urgent because Nirenberg was to give a plenary session on the genetic code to the January 1966 meeting of the American Association for the Advancement of Science, the country's largest general science organization. Its annual convention was the biggest science gathering in the country. Everyone wanted to know about the genetic code. Nirenberg wanted the talk to be right. The ambiguity had to be resolved before then, lest he make a mistake and later be proved wrong.

Because of Anderson's nucleic acid work with Doty, Nirenberg assigned the young researcher the task of making a purified form of the RNA triplet so the ambiguity could be resolved. Nirenberg liked Anderson, whom he considered bright, quick, and able to get things done. "He is a very intense person. He puts his heart and soul into it. He did some beautiful things related to work on the code," Nirenberg recalled. "He rapidly became my right-hand man."

Making the purified triplet proved difficult. Most of the RNA chemicals Nirenberg's group tested had never been synthesized before. It was technically painstaking, laborious work. It should have been a three-to-four-month job. Anderson had a month. He dropped everything and virtually moved into the lab, spending eighteen-hour days at the bench—relishing every moment of the intensity and sense of importance.

"For the next two and a half weeks, I hardly slept, working literally around the clock to synthesize [the material], verify that it was pure, and providing it to Marshall for testing. It coded for only one amino acid [alanine]! No ambiguity," Anderson reported in a written account of the code work.

Good work gets rewarded, though the beneficiary doesn't always see it that way. Nirenberg put Anderson in charge of the repetitive and toilsome task of synthesizing all the RNA needed in the lab. Anderson quickly became frustrated, even agitated, because it really wasn't science in the sense that he was discovering something new. He was simply cranking out new chemicals. And this kept him from his own project. To pacify Anderson, Nirenberg conferred a major perk: He asked Anderson to give the lab's paper at the Federation of American Societies for Experimental Biology meeting in April 1966. It would be the first time that the complete genetic code would be described publicly. Some scientists in Nirenberg's group grumbled about the choice. Most were senior to Anderson, and many had been working on the code much longer.

Two months later, Nirenberg would present the finished code to the leaders of genetics assembled at the Cold Spring Harbor Laboratory. For Nirenberg, the talk was a delicious moment, one of those times in any competition when the winner stands before the vanquished to receive their adoration. Nirenberg's presentation would become the official version, the final form of the genetic code for all living things. It had taken five years to complete, thousands of person-hours. It would be written up in a Cold Spring Harbor report on the genetic code symposium. It would be a classic.

But, for Anderson, his federation paper was *the* major event. Anderson, with Nirenberg's help, began to groom for the meeting. In his written recollection, Anderson prepared his talk and rehearsed it—relentlessly. He had Nirenberg question him, anticipating any possible audience query. "But a week or so before the meeting, I answered one of his questions incorrectly. I was very concerned and spent the next several days and nights rereading everything written on the genetic code, and rethinking all the data obtained in Marshall's lab over the previous three years," Anderson wrote. The talk went without a hitch. No one asked a question.

That exaggeration, that he would be "rereading everything written on the genetic code, and rethinking all the data obtained in Marshall's lab over the previous three years," is classic Anderson. With "the data" scattered over the lab books of dozens of scientists who had worked for, or with, Nirenberg since 1961—many of them gone by the time Anderson arrived—he would not be able to review and rethink all of it. In the research community, where every claim must be supported with experimental results or raw logic, Anderson frequently made eyebrow-raising statements. And it cost him personally. Peers took him less seriously, considered him an undisciplined and unclear thinker. Anderson ignored such criticisms.

In 1968, Nirenberg won the Nobel Prize for the code work. Press calls deluged the lab. The laureate took some, but ignored others. Anderson, who was still in Nirenberg's lab some of the time, manned the phones to help out. A story circulated on the NIH campus that as Anderson answered journalists' questions, he sang Nirenberg's praises. Anderson described how he had worked with five other Nobel winners and that Marshall Nirenberg was the most brilliant. "Marshall is a sweet, lovely guy," said one longtime NIH hand who had helped Nirenberg during the race with Ochoa. "I am fond of him. Nonetheless, he is not the most brilliant guy you ever met. He works hard. He has good ideas, but he is not a Sydney Brenner, or anything of that ilk."

In the 1970s, Judson, the writer, asked Meselson, the Cal Tech scientist, whether Marshall Nirenberg had made it into the club comprising biology's leadership. "He paused for a long time," Judson reported, "and then he said, 'I don't know.' Then he said, 'Anyway—sure; in a sense, he has a club of his own.' " Clearly, Meselson didn't believe Nirenberg had made it.

In some ways, Nirenberg's failure to enter the inner circle of the scientific elite tainted Anderson. When French Anderson committed his own mistakes through arrogance or social missteps, he had no protector, no network of senior, interlinked scientists to forgive him his transgressions. People just talked about him behind his back.

"It amused people that French would say all this stuff [about working with five Nobel winners] when in fact he had very little connection with them," an NIH scientist said. At least, that was the perception among many of Anderson's peers. It seemed he sought reflected glory. If Anderson worked with the most brilliant, then he too must be pretty brilliant. But other scientists didn't see him that way.

Compounding that problem, Anderson didn't really get along with his NIH peers. Just as he looked down on his Oklahoma schoolmates and had openly and aggressively challenged upperclassmen during Harvard seminars, Anderson conflicted with NIH scientists who were at his level. He thought he was more intelligent, and he didn't care if they knew that's what he thought. "I always had the feeling that the vast majority of people in the world were not very smart," Anderson once said. As a result, he frequently challenged what people said, questioning their science and arguing about their logic without concern about how it made them feel toward him.

"Being liked I didn't care about," Anderson said. "Respect, I was obsessed with." He often acted as though he could extract respect by dint of his intellect, as though people would be awed by how much he knew, or the elegance of an experiment he conducted. That might have worked in Tulsa, where many of his old friends remember him as brilliant, but at Harvard and NIH, the strategy failed.

"There is much resentment of me in the scientific community because I have always gone my own way," Anderson acknowledged. He recognizes that his own behavior has often gotten in the way. "I don't go to meetings, or do any of the things you are supposed to do. I don't go to Gordon Conferences, or belong to organizations. They bore me.

They are social meetings. I am not a social person. I am an antisocial person. If it weren't for my wife, who taught me to be social, I would never talk to anybody. I am very shy. I used to be a chronic stammerer. I am a maverick." And through it all, he knew that people talked about him behind his back, but he didn't care. He only cared about what he was working on.

"He had the fire in the belly, the competitive hunger that all good scientists have at that stage in their career," said Donald S. Fredrickson, then Nirenberg and Anderson's boss as the heart institute's director, and later NIH director. "And French had the ability to move on his own. He was capable of picking his own projects, and developing techniques."

What's more, Anderson wanted to be independent. The federation paper Nirenberg allowed him to give was nice, but he wanted the freedom to pursue his own projects. Besides, he never quite forgot a stinging comment from the Nobel-winning French geneticist Jacques Monod. While introducing Anderson to Monod at Cold Spring Harbor in front of the cafeteria, Blackburn Hall, during a meeting in 1967, a mutual friend waxed eloquent about all the good scientists Anderson had worked with or knew, including Doty, Watson and Crick, and Nirenberg. Monod sniffed, as they shook hands, "It's about time he got a job of his own."

Anderson wanted that—a job of his own—more than anything else. He wanted to prove himself, to show that he could work independently and even make a major discovery, like Nirenberg. He wanted to solve some basic biological problem, and prove to the critics and doubters that he really was brilliant. Certainly, he convinced Fredrickson. "French came and said he wanted to set up his own small section. Marshall [Nirenberg] did not have enough resources for French," Fredrickson said. "There was some space that came open, and we gave French a module or two [a module is a lab with two workbenches and a small office that two scientists must share], that being a lot at NIH at the time, and he became an operator."

. . .

Even as Anderson was about to begin his life as an independent researcher, he and many others on campus were beginning to think about the implications of the new genetics for treating human diseases. "It was clear that the field was going to go in that direction," Nirenberg said, "even though it was not possible to do [genetic transfers] then. I thought it would become possible. I was very much aware, and so was he [Anderson], of the possibilities of doing gene therapy." Nirenberg was sufficiently convinced of its likelihood that he worried out loud about it at a symposium, and again in August 1967, in an editorial in *Science* magazine, asking whether society would be prepared for the day when it would be possible to change genes in people.

A year later, in June 1968, Anderson gave a talk in Chicago at an international scientific symposium on mental retardation sponsored by the Joseph P. Kennedy Jr. Foundation, and told the audience that "the first attempts to correct genetic defects *will* take place within the next few years." Anderson described several ways in which normal genes might be inserted into a patient's cells, including the use of nonpathogenic viruses that could transfer genes into target cells. He acknowledged that technical problems seemed insurmountable with available technology, but that studies with human material might point the way. What's more, the treatment would not only help the individual but would change the person's genes so that his or her offspring also would be spared the genetic harm. He was predicting the advent of germ-line therapy, changing genes in sperm and eggs. What the eugenicists could only dream about in general ways through breeding people, Anderson envisioned more directly, with doctors intentionally modifying the future makeup of humans.

Around the same time, Anderson laid out his ideas for genetic therapy in an article entitled "Current Potential for Modification of Genetic Defects," which he sent off to *The New England Journal of Medicine*. Franz J. Ingelfinger, the journal's legendary editor, wrote back on August 19, 1968, to say that the editorial board had decided to reject the article, but not without heated debate. One reviewer called it "medical prediction," rather than "medical progress," the category of the journal for which it was intended. "In fact, one of our Board argued we should undertake to publish what he rated 'a worthwhile adventure in pure speculation.' "

This early talk of genetic treatments made Anderson stand out. Whereas many scientists were thinking about it, even trying tentative gene transfer experiments, Anderson was beginning to beat the drum for gene therapy. He was becoming a proponent. Most scientists who bothered to listen considered him a loony. Hardly any genes had even been identified, let alone useful genes like the ones that made hemoglobin. And certainly, no one had developed the techniques needed to put genes into people. The gene engineering revolution was only just underway, and the best that researchers could do was get a gene into one out of every million or 10 million bacterial cells. No one had gotten a gene into a mammalian cell, and certainly 1 in a million efficiencies would not be good enough to think about treating patients. As a result, Anderson was dismissed by most scientists.

. . .

In 1968, Anderson had to focus on science that was possible, not gene therapy, if he was going to make it professionally. In academic research, survival depends on being able to attract grants, and grants usually come to people with good ideas. At NIH, staff doesn't have to worry about attracting grant money because they receive taxpayer support directly through their institute. But to stay in the institute, the standard is the same as any university: produce.

Anderson started with a small group: a technician and a handful of postdocs. He couldn't decide exactly what he wanted to focus on: hematology or endocrinology and metabolic diseases. He only knew that he wanted to somehow combine the basic biochemistry research with something aimed at treating a human illness. In the end, he went with what he knew: biochemistry. Anderson's first projects were extensions of the work he had done with Paul Doty and with Marshall Nirenberg: a basic research project to understand the initiation and control of protein synthesis. It was a hot topic; labs all over the world already were hard at work on the problem.

Anderson began by extrapolating from Nirenberg's cell-free systems that were made from bacteria to a cell-free system made from eukaryotic cells, the kinds of cells that make up mammals and other complex plants and animals. Eukaryotic cells are profoundly different from bacteria, the prokaryotics, and their cellular mechanism might be different. For example, eukaryotes pack their DNA into many chromosomes that are tucked away within a separate compartment called the nucleus. Prokaryotes typically have a single chromosome and no nucleus. A bacteria's protein-making machinery can lie right next to the DNA, making it easy for messenger RNA to migrate to the ribosomes; in eukaryotes, a membrane separates the genes from the protein factories.

Anderson set out to see if these two types of cells differed in any fundamental way. What's more, if he was going to turn basic biochemical discoveries into medical treatments, Anderson would have to make sure that his findings applied to the same kinds of cells from which humans were made.

• • •

Biology always has been a messy science. Living cells and organisms are complicated. Unlike a machine with its clearly discernible components, a cell is not easily reducible. So when biologists divine the activities of some class of chemicals they cannot visualize or consistently identify, they frequently give them a vague designation, a word like "factor" that indicates there is something there, even if it cannot be precisely named. Anderson was hot on the trail of one of these, the so-called initiation factor. It was some chemical—probably a protein, maybe more than one—that played a critical role in attaching messenger RNA to ribosomes to start the process of making a protein. By working out this bit of basic biology, Anderson would help explain how the normal process of protein production works, and might even find a defect that led to human disease.

Several years earlier, Severo Ochoa, Nirenberg's New York nemesis, had used a cell-free system made from bacteria to show that initiation factors did, indeed, exist, and that they were essential for starting protein synthesis. Ochoa showed that by grinding up the cells, and then washing away most of the gooey mess that contained the chromosomes and other cellular components, he could use a salt wash to pull out the ribosomes and the other proteins associated with it. These included initiation factors and the messenger RNA that the ribosome was using to make protein at the time of the salt wash. It turned out the protein initiation factors in bacteria were attached to the outside of the ribosomes during protein synthesis. Ochoa's technique allowed them to be discovered in bacteria, but it had not worked for eukaryotic cells.

In his new quest, Anderson first assumed that the initiation factors were there, that they were essentially the same as in bacteria. He also assumed that it was not going to be easy to find them. It would take persistence and hard work, something he was good at. In fact, it took a year and a half. He also assumed that there were probably many factors required to start protein synthesis, not just one.

What's true for *E. coli* is true for elephants. At least, that is what molecular biologists generally believe about the relationship between the bacteria *Escherichia coli,* the bacteria most widely used for studies in biology labs, and organisms higher on the evolutionary scale.

If initiation factors existed for bacteria, then they must exist for higher organisms. No one, however, had been able to find the initiation factors in eukaryotics. It could be that something was blocking their discovery. Or it could be that eukaryotics were different. No one knew. Anderson and his tiny band decided to take it on, using a modified version of Ochoa's procedures.

Purifying molecules out of a cell's innards is much like breaking open a plastic bag filled with coins and running them through a series of sieves with progressively smaller holes. The problem, however, is that many interesting molecules of the same size have significant biological differences. It's like trying to separate dimes minted in 1960 from dimes minted in 1970. The sizes are the same so the sieve doesn't work. Something else is needed.

To start, Anderson and his team ground up rabbit reticulocytes, the cells that become red blood cells, and extracted their ribosomes. As they ran the ground-up cells through their sieves, they could collect pools of molecules that came through at different times. These pools of proteins would then be tested for their ability to stimulate protein synthesis when mixed with ribosomes. At the same time, the scientists constantly varied the conditions under which the ribosomes and the proteins were isolated and mixed together again, hoping to separate out the proper molecules. After many months, in frustration, they left out the element magnesium, which was thought to be needed for the ribosomes to function. Suddenly, they got a signal, a weak one but a sign nonetheless, that they might have found a bit of initiation factor that apparently had been inactivated by excess magnesium. Anderson was ecstatic. It proved he could do it, that he was a serious, competent scientist. He wanted to tell everyone.

The 1969 Cold Spring Harbor Symposium on Quantitative Biology was coming up, and Anderson was going. By then, his group was able to produce small amounts of protein in a cell-free system in the test tube. Despite the seeming importance of the advancement—since many other labs were working on the same problem, but had had no success—Anderson could not get on the Cold Spring Harbor program to present his results. Finally, at the meeting itself, Gordon Tomkins, a well-liked and respected NIH scientist, convinced the chairman of a session that he wanted to hear Anderson's data, and persuaded him to give French a chance. The chairman was a German who didn't like being pressured. He introduced Anderson before the session really started, while people were still filing in and taking their seats, saying: "OK, we are going to have a two minute presentation by French Anderson, who is going to tell you about an experiment he did." Anderson got to show two slides on protein initiation, and then the chairman cut him off. No questions. Anderson was devastated, convinced that his career was over.

That afternoon, the field leaders like Ochoa were talking about how cell-free protein synthesis still could not be done in eukaryotes. Anderson was about to jump up and start shouting that he had done it, but at the last minute he realized that if they hadn't, and he had, then he had the lead in the race to be the first to synthesize proteins in a eukaryotic cell-free system. So he sat down and shut up. After the meeting was over, Anderson rushed back to NIH to keep working on his system.

Over the next year, Anderson and his group isolated three separate eukaryotic initiation factors. By spring 1970, they felt their data was solid enough to present it publicly. Anderson sent the lab's abstract to the Federation of American Societies for Experimental Biology for presentation at their upcoming meeting, one of the more important scientific gatherings for biologists doing basic research. Anderson's abstract was flatly rejected.

Floored, Anderson rushed to Don Fredrickson, who had given up his administrative position as heart institute director to go back to the lab. He was Anderson's branch chief. Fredrickson consoled the distraught scientist and suggested that he send the abstract to the American Federation of Clinical Research meeting a month later. The federation had no problem with Anderson's work, and asked him to present it to a plenary session. That attracted the attention of *The New York Times*.

The subsequent publicity precipitated lots of problems with his peers: They didn't like it. Not only had this basic biological advance been delivered to the clinical people and not his peers—basic scientists are about as separate from clinicians as the church is from the state in America—but it was simultaneously released to the public without being peer-reviewed first. Anderson rushed to get published in *Nature* and in the *Journal of Biological Chemistry* even as the word spread. Others working in the field either dismissed Anderson's results, or else claimed that it already had been done in their labs and that it was no big deal. Nonetheless, a spate of papers appeared in the quick publication journals all trying to establish precedence. Anderson, however, had beaten them all.

• • •

At the same time that his work in the basic research lab was going well, Anderson began to acquire patients. After all, he stood between two camps: the basic science community, dominated by Ph.D.s, who considered themselves the superior scientists and indifferent to clinical matters, and the clinical community dominated by M.D.s, who daily faced dying patients with inadequate treatments and technologies. Anderson wanted to bridge that gap, to get clinicians the tools that only basic research could provide.

If the basic research aimed at understanding the control of protein synthesis was going to make any contribution to medicine, it would have to be applied to diseases. Two Columbia University physician-scientists, Paul Marks and Arthur Bank, had shown that there was a discrepancy in the ratio of molecules that make up hemoglobin in patients with thalassemia, a kind of anemia in which the person cannot make enough normal red blood cells to sustain life. Normally, hemoglobin is made from four subunits: two protein subunits called alpha globin and two protein subunits called beta globin. To make normal hemoglobin, there has to be one alpha globin for every beta globin. The Columbia pair found the ratio was off in thalassemia patients. Instead of a 1:1 ratio, there was only one molecule of alpha globin for every two molecules of beta globin, or vice versa, depending on which form of the disease the patient suffered.

Thalassemia is a miserable disease. First described in 1925 by a Detroit physician named Thomas B. Cooley and later called Cooley's anemia in his honor, thalassemia comes from the Greek word for "sea" because it is common among peoples living around the Mediterranean. Without normal forms of the oxygen-carrying molecule, the patient's weakened red blood cells disintegrate far short of their normal 120-day life span. Victims suffer a sort of chronic, slow-motion suffocation from the inside out. No matter how much they breathe, it's never sufficient; the air cannot get from the lungs to the body's brain and heart and kidneys. There simply are not enough functioning red blood cells in the body—a condition called anemia. The person feels constantly exhausted. Organ systems from the heart to the liver to the brain to the kidneys fail to work properly. To compensate for the constant destruction of blood, the bone marrow produces red cells at a furious rate. This causes the bone marrow to swell, expanding painfully within the rigid bones, fracturing and deforming them, leaving the person hobbled and disfigured. Typically, thalassemics seldom live beyond their early teens.

Thalassemia's cause would eventually be deciphered. To understand it, think of hemoglobin as an automobile carrying oxygen as its passenger. The car has four tires, and each must be inflated for the car to move. Hemoglobin is a complex molecule made from four protein subunits, like the sum of the car's tires. Two of the hemoglobin subunits are called alpha globin, and are made by the alpha globin gene. The other two subunits are called beta globin, and are made by the beta globin gene. The alpha globin subunits are like the two front tires; beta globin, the back tires.

A person inherits one alpha and one beta globin gene from the mother, and one alpha and one beta globin gene from the father. It's like the tires on the left side of the car—one alpha and one beta, or one front and one back tire—come from the mother, and the tires on the other side come from the father. If a person inherits four normal tires, then his red blood cells carry oxygen normally.

Thalassemia occurs when a person inherits defective genes. Instead of bequeathing round, bouncy tires, the parent passes on a flat. In this biological automobile, however, it takes two flats to stop the car—a flat in both front tires or a flat in both back tires. This happens when the child inherits a defective beta globin gene from the mother and a defective globin gene from the father. They have to be the same defect, such as a broken alpha globin gene.

For a parent to pass on a single genetic defect, either alpha or beta, the parent too must have slightly abnormal hemoglobin—one of the parent's four tires is flat. The parent is said to be a carrier of the genetic defect. A carrier is not sick, but is not normal, either. If one of the two alpha globin genes is defective, for example, half the person's red blood cells are normal and half are defective.

It turns out that people with only three tires on their hemoglobin automobile have a certain advantage when it comes to another disease: malaria. This mosquito-carried microorganism causes this illness by burrowing into normal red blood cells, causing cycles of chills, fever, and sweating—and ultimately, in many cases, death. But when a person carries a hemoglobin defect, the malaria organism can't establish an infection in the person's blood. The hobbled hemoglobin molecules just won't support malarial life. The gene defect, in a sense, makes the person resistant to malaria.

This protection against malaria explains why thalassemia is so common around the Mediterranean Sea and parts of Asia and Africa:

They are, or were, regions with lots of mosquitoes and lots of malaria. A large percentage of the population carries a defective thalassemic hemoglobin, giving them some protection against malaria. The price of protection, however, is a trade-off with thalassemia: A quarter of the children of parents who both carry one defective hemoglobin gene will develop full-blown thalassemia, and half will become carriers themselves. More than 100 genetic defects have been found in the alpha and beta globin genes. The person's disease is labeled alpha thalassemia or beta thalassemia, depending on which gene mutated.

Before the advent of blood transfusions in the first third of the twentieth century, thalassemic children died young. Now, the patient receives frequent blood transfusions, sometimes as often as every two weeks. The injected red cells minimize the chronic exhaustion and organ damage. The transfusions, however, do not cure the disease. At best, they are a holding action. And they have a serious biological price: iron poisoning.

At the center of each hemoglobin molecule is an atom of iron. With each blood transfusion, a biological ton of iron is injected into the body and accumulated. In the normal person, red blood cells break down, and the iron is stored in the liver until it is needed to make new blood cells in the bone marrow. Regular transfusions, however, shut down the bone marrow's production of red cells, the ones with the defective hemoglobin. As a result, iron begins to accumulate in the liver until it cannot hold any more. Then iron begins to spill out into the rest of the body where it builds up in other tissues.

The heart suffers the worst effects. The iron interferes with the normal transmission of electrical signals throughout the heart muscle, signals needed for the heart to contract normally. An iron overload can cause the heart to beat erratically, making it unable to pump enough blood through the body. In the end, it may cause the heart to stop suddenly.

In the 1970s, the thalassemia patient faced a difficult choice: take the regular blood transfusions and suffer the side effects of iron overload, or avoid the transfusions and suffer the lethal effects of the thalassemia. For those with severe disease, the choice is Hobson's.

Although—by the early 1970s—the disease was shown to be inherited, no one understood the nature of the defect. Either the genes that carried the blueprint for alpha or beta globin were defective, or there was something wrong with the protein-synthesizing machinery that made hemoglobin. Marks and Bank's experiment could not determine which was the case.

Now that Anderson could look at the initiation of protein synthesis, he wanted to figure out whether thalassemia was due to a defect in the gene or the messenger RNA or protein production itself. To do the study, he needed cells from a thalassemia patient. He got two, a brother and sister named Nicholas and Julia (Judy) Lambis. Nick and Judy were born in Washington, D.C., and lived with their large Greek family. Since 1965, they had been coming to NIH regularly to receive the frequent blood transfusions that kept them alive, and to participate in a study of liver disease caused by the constant transfusions. The National Cancer Institute researchers who were treating the Lambises had given up the liver project and left NIH, so NCI was about to discharge the siblings.

In a panic, Anderson said he would take them, even though he lacked authorization or legal standing to do so. Over the next several weeks, Anderson and his heart institute bosses worked out the details, and the Lambises were admitted to 7 west, heart institute beds in the clinical center, in September 1968. Judy was his first patient because J came first in the alphabet. Nick was his second. Eventually, Anderson's group would have between fifteen and twenty thalassemia patients under their care.

Nick, sixteen when admitted to Anderson's studies, was diagnosed with beta thalassemia at age six months. Not much was known about his early medical treatment, other than that surgeons took his spleen when he was three years old. "The precise indications for this procedure are not known at this time," his medical record observed in 1970. He needed continual transfusions to live, receiving them every six weeks to three months. As he neared the time for the next transfusion, mild exertion would make him breathless, and he suffered "rather marked fatigue." He would perk up after the transfusion treatments, which allowed him to grow normally. But the long routine of transfusions already was turning his skin bronze as iron from all the injected red blood cells accumulated in his body. His heart also showed signs of enlargement, as did his liver, which had to process the excess and the abnormal red blood cells. His bones too showed changes related to the disease. Despite his illness, Nick had progressed through school at a normal rate. He worked part-time during school and full-time during the summers. Later in his life, he would work at NIH as a clerk.

Julia was in somewhat worse shape when she entered Anderson's service. She was diagnosed when she was three months old, and received continual transfusions every six weeks to three months, just like her brother, and had her spleen removed when she was four. Judy, as everyone called her, had signs of congestive heart failure and an enlarged liver as early as 1969, according to her medical record, which was written out by Dr. Joseph L. Goldstein, then a clinical fellow at NIH who would later win a Nobel Prize for working out the molecular biology of high cholesterol and its consequences. Judy also had bronze skin from the ongoing accumulation of iron from the transfusions.

Judy was "a very pleasant, cooperative girl appearing three or four years younger than her stated age of fourteen, mainly because of her short stature and lack of secondary sex characteristics [a common phenomenon in thalassemia patients]," wrote Arthur W. Nienhuis, a young physician fresh from training at the Massachusetts General Hospital, who joined Anderson's lab in 1971 and became head of its clinical service.

For these children, who came from a family without a father or much money, the NIH was a godsend: It offered free medical care. For Anderson's studies of protein production in the cell-free system, Nick and Judy provided thalassemic blood cells from which to extract ribosomes and messenger RNA. Anderson took their cells apart, and turned them into cell-free systems to probe for defective protein synthesis. He found none. The Lambises' ribosomes were just as able to make proteins as Anderson's (he used his own cells for comparison). But when he isolated their messenger RNA and ran that through his cell-free system made with rabbit ribosomes, he found defective hemoglobin. It was the messenger RNA, not the ribosomes, that was defective. Next, he turned his studies back one step, to the DNA itself: Were the hemoglobin genes themselves defective, or was there something wrong with the way the messenger RNA was made?

. . .

Even as those studies started, politics intruded into science. In January 1971, Michael Iovene, a thirty-year-old chemist from New Haven, Connecticut, called his congressman, Rep. Robert N. Giaimo, a Democrat, to complain. Iovene suffered from thalassemia, and as executive director of the Cooley's Anemia Blood and Research Foundation's Connecticut chapter, he was unhappy that the government was doing so little for people with this disease. That year, Pres. Richard Nixon was gearing up to launch the War on Cancer. A similar effort was being organized for sickle cell anemia, a blood disease similar to thalassemia that disproportionately afflicted African-Americans. It too had been singled out by Nixon for special attention; a bill to boost sickle cell research was pending in Congress.

Giaimo, who had never heard of Cooley's anemia, was moved by Iovene's account of the problem. Most of his constituents were of Mediterranean ancestry, as was he. Giaimo decided to do something for thalassemics, and tried to amend the sickle cell bill to include Cooley's, but was rebuffed by congressional leaders. So he introduced a parallel bill, substituting Cooley's anemia for sickle cell anemia. The Cooley's foundation mobilized its members in a lobbying effort, picking up support from congressmen and senators along the way. By summer, the National Cooley's Anemia Control Act of 1972 had passed both houses, and in August, Nixon signed it, making Cooley's anemia "the second ethnic disease to rate special national attention," according to one account.

French Anderson got caught up in the political machinations. He was singled out during congressional hearings as one of the few people at NIH whose studies had direct bearing on thalassemia. But Anderson's studies could hardly be considered a major effort: Congress was told he spent about $125,000 a year on thalassemia research (he said it was more like $500,000). Backers of the Cooley's bill authorized spending $11.1 million over three years for research in the diagnosis and treatment of Cooley's.

Once named publicly before Congress, Anderson could not avoid getting sucked into a political quagmire. The NIH leadership put Anderson in charge of NIH's intramural effort that was part of the Cooley's anemia program Congress had just authorized. There was one hitch: He wasn't going to get the $11 million. In fact, neither he nor anyone else was going to get additional financial support. As so often happens in Congress, an authorizing committee writes a bill that says something should be done, but a completely separate committee allocates the cash to do that something. The appropriations committees of the House and Senate declined to fund the Cooley's law. Anderson would have to carry on his part of the program with no money.

Ever inventive and anxious to please his superiors, Anderson asked for the spare change left over at the end of the year from the heart and arthritis institutes' on-campus programs. Both came up with some money. Anderson used it to organize workshops for setting research priorities. He didn't have to look hard to figure out the most urgent priority: Nick and Judy Lambis were getting sicker. As they entered their teens, they faced the problem faced by all thalassemics: iron overload.

From birth to age fifteen, the children received more than 100 transfusions with two or three or more bags of red blood cells. Each bag contains billions of red cells packed with thousands of hemoglobin molecules. Each hemoglobin contains a single atom of iron at its center. The cells flood into the body, live something less than the normal 120-day life span for a red cell, and then die. The liver captures the dying red cells and their debris as they flow through the organ, absorbing and storing trillions of iron atoms for later use by the bone marrow to make new red blood cells. But when the person is receiving endless transfusions, the body cannot get rid of the iron fast enough. Iron begins to build up within the liver, swelling the organ until it can hold no more. Soon, the iron starts overflowing to the other organs. The problems get serious when the iron levels in the heart start interfering with its normal function, causing galloping heartbeats and ultimately heart failure. Judy Lambis was starting to have heart trouble. If nothing was done, it would kill her eventually.

Anderson needed a way to pull the iron back out of their bodies. That would have to be the research priority. "These kids were dying from iron overload, not from their hemoglobin disease," Anderson said. "They were dying from iron, so take care of the iron."

It was easier to say than do. Drugs had been around for years to detoxify someone acutely poisoned by iron, but the drugs themselves had nasty side effects and were unsuitable for long-term use. At the organizing and priority-setting workshops, Anderson started pushing for new work on iron chelation therapy, as the iron-removing process is called. He also started raising money to do the work. That made all the difference.

The attention brought people into the field; the money supported their research. They started with some of the older drugs first. The breakthrough began when a new way was found to use an old drug, Desferal. It appeared to be effective in removing iron from tissue without debilitating side effects when used at lower doses. Anderson had pushed its development, including getting the Food and Drug Administration to relabel the drug for chronic use; Ciba-Geigy, the drug company that manufactured it, gave Anderson the first production vial.

As important, David Nathan at Boston Children's Hospital showed the drug could be used in a continuous intravenous treatment with a pump. With time, the iron levels in the bodies of his patients began to decline. It started to look like a useful treatment might actually come out of Congress's aborted program to beat thalassemia. For Anderson, it was a time of success. He would win the Cooley's Foundation's award for overseeing the development of iron chelation. Patients with a disease would actually benefit from the advances Anderson fostered.

• • •

Within NIH, Anderson was riding the tide of his successes. By 1974, he had taken over the entire D corridor on the clinical center's seventh floor, and his laboratory had become a branch, the biggest unit an individual researcher could control. In NIH terms, he had a thriving empire with space, postdoctoral scientists and technicians cranking out data, patients, and administrative support. Anderson hated it. To him, the lab had grown too large, too much oriented toward patient care, and too little directed toward basic research.

"We got to be fifty people, and all I was was an administrator," Anderson complained. "It went from me getting data myself, to me not working in the lab, but at least evaluating the original data as it came off. Then, it went to me evaluating data at lab meetings, to getting summaries of the data to the point where there was so much happening that I was getting summaries of summaries."

What's more, the genetic engineering revolution was taking place without him. Beginning in the 1970s, basic scientists had created a number of new techniques that allowed them to manipulate genes directly, isolate them, cut them up, and move them from one cell to the other. It was all work being pioneered by old cronies and competitors. Yet Anderson was a bystander, an observer. Scientists were starting to transfer genes into bacteria, not through random transformation as he had learned to do in Doty's lab after Julie Marmur arrived, but intentional, willful transfer of specific genes. These were going to be the key technologies needed to put genes into patients for therapeutic purposes. If he ever wanted to develop human gene therapy, he needed to learn to transfer genes into cells. By comparison, the good work going on in his lab with the clinical treatment of thalassemia and the basic work in initiation factors and protein synthesis was irrelevant to his long-term goal: gene therapy.

Anderson made an astonishing decision. He decided to give up the major portion of his laboratory space. Art Nienhuis got the lion's share, enough for him to eventually create the heart institute's clinical hematology branch. Others working on basic protein synthesis biochemistry kept a portion, and Anderson retained the rest. The move was so unusual, the other NIH scientists suspected Anderson was being reprimanded, that he must have done something wrong because no one gives away the kind of kingdom Anderson had built.

Anderson saw it differently. "If I wanted to do gene therapy, I had to give up all this stuff," Anderson said. "No one could conceive of me giving up an empire. It had to be taken away, and no one could find out why. When it came clear that I had given it up, then everyone thought I was just a weirdo. And to do that, to start a concentrated effort to do gene therapy in 1974 [he shakes his head]. . . . In retrospect, I was weird."

Chapter 4
Genetic Fear
Goes Public

"We're in a pre-Hiroshima condition. It would be a real disaster if one of the agents now being handled in research should in fact be a real human cancer agent."
—*Robert Pollack, Cold Spring Harbor Laboratory*

North of San Diego, the hulking, square concrete structures of the Salk Institute for Biological Studies perch on a bluff overlooking the ocean along Torrey Pines Road in La Jolla, California. Founded in 1962 by Jonas Salk, the developer of the Salk vaccine for polio, the place looks more like a bunker than one of those rare research boutiques where scientists can concentrate on their studies and avoid the distractions of academic life. Cramped laboratories, stuffed with equipment, line the outer buildings of its fortresslike edifice. Scientists have their offices in concrete towers in a courtyard crisscrossed with canals that carry water to picturesque fountains pouring toward the Pacific.

It is an ivory tower, one of the most productive research centers in the world. Besides Jonas Salk, who in the 1980s turned his prodigious mind to finding a vaccine to stop the human immunodeficiency virus (HIV) that causes AIDS, the institute has had the best and the brightest among its alumni, including Nobel winners Francis Crick, who helped discover the shape of genes, geneticists Salvador Luria and Jacques Monod, and other notables including Jacob Bronowski, who plotted the ascent of man, Sydney Brenner, who traced the pathway of genetic information in the cell, Maxwell Cowan, a neurologist who once showed Congress what a real human brain looked like, and physicist Leo Szilard, who helped make the atom bomb and then turned his attention to the study of living things. And the institute's leader, lab president Renato Dulbecco, a Nobel laureate, has a scientific pedigree that begins at the very roots of modern molecular biology.

While French Anderson was working on protein initiation and thalassemia back at NIH, it was here at the Salk, in many ways, where the genetic engineering revolution of the 1970s began and the first small steps were made toward the tools that would be needed to transfer genes into the cells of mammals—and ultimately humans. It started at the Salk because Renato Dulbecco, an elegant Italian who emigrated to America in 1947 to study biology, became—for a few brief, critical years—the world's expert in growing animal viruses and maintaining the eukaryotic cell cultures needed to sustain them. It was to the Salk that a steady stream of the country's best researchers came to learn the techniques that Dulbecco had pioneered. Paul Berg was among those that came to make the shift from bacterial viruses to the more complex and potentially more useful viruses that invaded eukaryotic cells, including human cells. Out of Berg's lab and the labs of other researchers working on related questions came even newer methods that allowed scientists to cut up DNA into gene-sized pieces, package the pieces into viruses, and use these cellular parasites like molecular delivery trucks to infect mammalian cells and deliver the recombinant DNA. The cells would then make the DNA part of its own chromosomes; the genes could function and the cell would be changed. A cell missing the gene that made a critical enzyme, for example, could be restored to normal by inserting that gene.

The new laboratory systems gave scientists powerful new tools to manipulate genes, to put them in places nature never intended, and to change the characteristics of organisms—even humanity itself. As the new science emerged from the laboratory, it spilled over into the public consciousness, and it was frightening. The clarion call in 1969 that genetic engineering could be used for evil, sounded by Jonathan Beckwith and the other Harvard scientists, fell on deaf ears—partly because many scientists did not believe the techniques to carry out Beckwith's Orwellian vision existed. But as the new field of recombinant DNA emerged, the technical abilities became reality, scaring a significant number of thoughtful scientists. When their concerns were magnified by the media, and Congress, then the public too became fearful.

Genetics became a front-page issue.

Like any new field, however, genetic engineering did not spring full-born. Rather, it had a long antecedence in the days of the Phage Group, a small band of biologists and physicists who, in the dark days of World War II, decided to dissect the process by which one virus can enter a cell, and multiply itself into hundreds or thousands of progeny a short time later. Phage is short for bacteriophage, the viruses that live only in bacteria and were the main tools for the group's studies.

The midwives were Max Delbruck, a German physicist who switched to biology after coming to America to escape the Nazis, Alfred D. Hershey, a Michigan-born microbiologist, and Salvador E. Luria, an Italian physician who studied physics and radiology, and who fled the Fascists to the United States in 1940. Delbruck once described them as "two enemy aliens and one social misfit." Together, they organized what was probably the most important tribe in molecular biology, and became legendary figures in the history of science. They shared a Nobel Prize in 1969 for their discoveries.

The Phage Group was born in 1941 when Delbruck invited Luria to spent the summer at the Biological Laboratory of Cold Spring Harbor on Long Island. It was then a sleepy, pastoral place with ancient, barely adequate buildings set on the hillside over Oyster Bay. What the facilities lacked in equipment, the scientists made up for in energy and ideas. These were the years of World War II. The group remained small; travel was difficult. Hershey didn't join until the three held the first "phage meeting" in St. Louis in April 1943. Although work on bacteriophage had been underway for twenty-five years, Delbruck decided that they would "get to the bottom of what goes on when more virus particles are produced upon the introduction of one virus particle into a bacterial cell." They would exploit this dangerous minuet between bacteria and the bacteriophages to understand the genetics of both the virus and its primitive cellular host.

Their work revolutionized the approach to the study of molecular genetics and suggested, for the first time, how genes might be moved from one cell to another. At that time, in the 1940s, scientists were still arguing what genes were and which molecules carried genetic information. Oswald Avery and his colleagues at the Rockefeller University answered that question in 1944. Hershey helped end the debate with additional work conducted at Cold Spring Harbor. In 1952, he and Martha Chase proved that only the bacteriophage's DNA, but none of its protein, entered the bacteria during infection, confirming that DNA, and only DNA, carries genetic information. Other studies showed that phage could pick up DNA from infected cells and carry it along to the next cell it infected. This discovery led scientists to begin thinking about using viruses for intentional genetic transfers.

The Phage Group's apprentices proved equally powerful. A young James D. Watson—a somewhat eccentric, Chicago-born and -educated ornithologist—joined Luria's lab at the University of Indiana. His labmate was Renato Dulbecco. Dulbecco became a charter member of molecular biology's scientific establishment. While working with Luria on bacteriophage, Dulbecco independently discovered photoreactivation of bacteriophage, a process in which virus that had been killed by ultraviolet light could be "brought back to life." It was a strange discovery. At the time, scientists used measured doses of ultraviolet radiation to kill a portion of a bacteriophage population, and then measure what percent of the virus population survived. During the test, the batch of ultraviolet-treated viruses, now presumed to be inactivated, were poured into a petri dish with a jellylike layer of nutrients on the bottom and a "lawn" of bacteria living on the jelly's surface. The bacteria lived on the jelly, and the viruses lived in the bacteria. A round glass cap fit on top of the dish to keep out other germs. Any living viruses would attack the bacteria and create a hole, or plaque, in the lawn where the reproducing virus killed the host bacteria. If there were no plaques, then all the viruses had been killed by the ultraviolet light.

But experiments often go astray, giving unpredictable results. Sometimes researchers would get virus survival where they expected none because the doses of ultraviolet radiation were strong enough to kill all the viruses. The explanation of viral survival came from a trivial observation. The glass petri dishes full of bacteria and viruses were often stored in stacks, sometimes near unshaded windows, sometimes not. Dishes on top received more light than those below. Dulbecco noticed that virus survival rates were consistently greater for the plates on top. As he investigated further, he discovered that exposing the "dead" viruses to visible light could reactivate them. When an amazed Max Delbruck learned about photoreactivation in 1949, he attributed Dulbecco's discovery to "the principle of limited sloppiness." If a scientist is too sloppy, Delbruck said, his results are unreliable. Yet if he is too precise—if the petri dishes always had been kept in a dark incubator, for example—photoreactivation would never have been discovered.

The discovery gave Dulbecco independent standing in the phage group and an invitation in 1949 to come to the California Institute of Technology to work on phage with Delbruck. In 1950, after holding a seminar on animal viruses, Delbruck decided the field of animal virology, about which he knew little, was ripe for rapid advances. He asked Dulbecco to go on a several-month tour of the country's best animal virus labs to learn techniques and assess the hottest areas in this branch of virology. The trip was disappointing: The cell-culturing techniques had not progressed much since he had worked on the problem in Italy, and animal virus research was not quantitative. Where the Phage Group had applied the statistical techniques of physics to studying bacteria and their viruses, animal virus researchers came out of a more traditional, biological background where little was measured statistically.

The problem, however, was not just the viruses but also the cells in which they grow. The Phage Group had it easy: Bacteria are free-living cells that easily survive laboratory manipulation and growth conditions. Animal cells, accustomed to cooperative living with many other cell types in a whole animal, were much more difficult to grow in laboratory tissue cultures. Dulbecco returned to Cal Tech to create, from scratch, the techiques needed to grow animal cells in culture. That work, using chick embryo cells on which he grew equine encephalitis virus (a fairly dangerous germ that reproduces by breaking open the cells it infects, including human cells) along with other work, led to his 1975 Nobel Prize.

With the tools for culturing mammalian cells in hand, Dulbecco set out to characterize a number of mammalian viruses, including the polio virus. In the late 1950s, two Cal Tech collaborators who had studied under Dulbecco, Harry Rubin, then a postdoctoral fellow, and Howard Temin, a doctoral student, had become interested in the Rous sarcoma virus, which causes cancer in chickens. Where the equine encephalitis virus caused infected cells to rupture, the Rous sarcoma virus reproduced more quietly: It appeared to enter a cell, integrate its genetic material into the cell, and then slowly release new virus particles over the normal life span of the cell. Although bacteriophage had shown that it could integrate virus genes into a bacteria's chromosomes, the phenomenon had not been seen in mammalian cells. What's more, genes in the Rous sarcoma virus were in the form of RNA, not DNA. For Rous to integrate its genes, it would somehow have to convert them from RNA into DNA.

Dulbecco decided to find a DNA virus that might integrate its genes into the chromosomes of an infected cell. He chose a small tumor virus called polyoma. It carries enough genetic material to encode five to eight genes on a small, circular molecule of DNA. Polyoma had many properties comparable to bacteriophage. Its circular DNA suggested that it could integrate its genes just like bacteriophage. Polyoma also could pick up bits of genetic material from a host cell, and carry them along, giving new properties to the virus and the next cell it infected.

Because of Dulbecco's successes, scientists started flocking to his laboratory to learn the new techniques of animal virology. "Originally, phage were the big virus. Then animal cells became big," Dulbecco recalled. "The shift happened in my lab. That is why people came. We were not the only lab, but we were the first to do it, and for a number of years, we were deep in the virus work, and ahead of everyone else."

Dulbecco would show that polyoma did integrate its genes into the target cell. With that question answered, however, he lost interest in the virus, and drifted off into other lines of research, including searching for growth factors and studying breast cancer. But for those brief, few years in the early to mid-1960s, Dulbecco's lab was the hottest in the world when it came to growing animal viruses. It attracted some of the best and the brightest.

And while Dulbecco's lab made intriguing observations about the integration of viral DNA into host cells, suggesting that animal viruses might be usable for selectively transfering genes, researchers would still have to figure out how to manipulate genes in the laboratory. Then, they would have to somehow paste the gene into the virus. Every geneticist recognized that if it could be done, it would revolutionize research and the treatment of disease. From the vantage point of the mid-1960s, however, this was science fiction. The genetic engineering revolution had yet to begin; no one yet could manipulate individual genes. Isolating the first genes had yet to occur. But the idea was compelling, alluring, promising. The way would be difficult, but the first steps were about to be taken.

• • •

In 1967, Paul Berg, a biochemistry professor at Stanford University, decided it was time to redirect his research efforts. He had been working on the basic biochemistry of DNA, using bacteriophage and bacteria. Now it was time to move on to something new. He turned his eyes southward to the Salk, to Dulbecco's lab where something new was going on. Berg was about to join the flow through Dulbecco's lab. He would learn about animal viruses, and that would change everything.

Born in Brooklyn in June 1926, Berg was a scientist's scientist. Tall, quietly intense, and exceptionally smart, he had graduated from high school early. He studied biochemistry at the Pennsylvania State University; there was an interrupting stint in the navy during the end of World War II (1944–46). After earning a Ph.D. in biochemistry from Western Reserve University in Cincinnati in 1952, Berg moved overseas to the Copenhagen lab of Herman Kalckar, a biochemist connected to the Phage Group through Delbruck's summer courses at Cold Spring Harbor Laboratory, and tantalized by the central importance of DNA in biology. Just before Berg arrived in Denmark, James Watson had fled Kalckar's lab for Cambridge, England, and a historic collaboration with Francis Crick. From Copenhagen, Berg moved to Washington University in St. Louis to work on DNA replication with biochemist Arthur Kornberg, not yet a Nobel winner but clearly on his way.

When Stanford University decided it needed a first-rate biochemistry department to go with its new medical school, it elected to buy a good one: Kornberg's. Like a baseball team bought by new owners and moved to another town, Kornberg relocated his research franchise in 1959 from the middle of the country to the foothills of the Santa Cruz Mountains, just dozens of miles south of San Francisco, taking with him many of his lab mates, including Paul Berg. Stanford had chosen well: Kornberg won a Nobel Prize for sorting out the mechanisms of DNA replication in the same year he and the others arrived.

During his time with Kornberg in St. Louis, Paul Berg had made significant discoveries of his own in the way genetic information stored in DNA was converted into new proteins by the cell's synthesizing machinery. In 1956, he was first to show that amino acids had to be activated before they could be added into a growing protein chain. The complete activation step hooked the amino acid to a small bit of RNA, later dubbed transfer RNA because it carries individual amino acids to the protein being manufactured.

After eight years at Stanford, working with the same bacterial systems he started with in St. Louis, Berg, at forty-one, was ready for a change. With the cracking of the genetic code in the early 1960s, most of the major questions about how genes direct protein production were answered. Berg was looking for a new, more fertile field.

Dulbecco's work with polyoma virus seemed to offer that opportunity. Maybe the successes of the Phage Group could be repeated by using animal viruses to study the genetics of eukaryotic cells. But even back then, Berg had bigger game in mind: He wanted to find a way to transfer genes between cells at will. He, like many others in the field, recognized that such abilities might turn the new genetics into a treatment for human diseases. The scientific community's article of faith said that discoveries by basic scientists about the fundamental processes of life often lead to something practical, such as a useful medical treatment. It's an article of faith supported by countless examples.

One classic case is the basic breakthrough that led to the polio vaccine. Although the polio virus had been isolated, there was no practical way to grow large volumes of the virus to make a vaccine. Early studies in the 1930s found that the polio virus could only be grown in whole monkeys, a costly procedure that ended the animal's life and produced few viruses. There were not enough monkeys on the planet to make all the doses of vaccines that would be needed to end the American polio epidemics of the 1940s and 1950s.

In 1936, Albert Sabin and Peter Olitsky at the Rockefeller Institute in New York City succeeded in growing polio virus in human embryonic nervous tissue cultures growing in the laboratory, but not in other types of cells. While the discovery suggested polio could be grown in tissue culture, the finding was useless because injecting nerve cells into a human causes an allergic reaction that damages the brain, often fatally. Tissue culture with nerve cells did not seem to be a feasible way to go.

By accident, the team of John F. Enders, Thomas H. Weller, and Frederick C. Robbins at Boston's Children's Hospital discovered that Sabin and Olitsky were wrong: The polio virus *could* grow in other cells. In 1948, Weller, a young research associate working in Enders's lab, had made up a tissue culture of monkey kidney cells to grow mumps and chickenpox virus. But he made too many tubes. Enders suggested Weller try seeding the extra tubes with polio viruses from the National Foundation for Infantile Paralysis—organizers of the March of Dimes campaigns. To everyone's astonishment, the polio virus grew in the monkey kidney cells. Jonas Salk, then at the University of Pittsburgh, immediately retooled his laboratory to exploit the newly developed tissue culture technique for growing polio virus. Salk, and later Sabin, used the approach to produce a viable polio vaccine that ended the annual summer epidemics of the infectious paralyzing disease.

Could the same be done for genetic illnesses? Could biochemical techniques designed to manipulate genes in viruses and tissue culture be used to fight maladies from within? Where turn-of-the-century eugenicists had spun fantastical schemes of human improvement from a pitiful understanding of human genetics, and where the bacterial workers in the 1950s and 1960s who discovered transformation failed to transfer genes into mammalian cells, Berg and others believed biochemistry could provide solutions. He decided to plunge into the hard work of designing a system for manipulating genes directly. Maybe it would turn into a medical treatment, maybe not. But Berg recognized these would be the first tentative steps toward a genetic treatment that he believed would take decades to develop.

But first, Berg, the biochemist, had to learn the biology of working with mammalian cells. Unlike bacteria, animal cells require gentle handling. The liquid solutions in which they grow must contain just the right blends of minerals and nutrients, acids, salts, and gases to sustain life. In the mid-1960s, it was an evolving art, and Dulbecco was one of the chief artists. So Berg went to the Salk.

Once there, he quickly learned the fundamentals of growing animal cells and working with animal viruses. The year-long Salk sabbatical was important for another reason, too: It would solidify the connection between two powerful research centers: Renato Dulbecco as director of the Salk and Paul Berg, an ambitious and talented scientist destined to run the well-endowed Beckman Center for Molecular Biology at Stanford University. Berg became a nonresident fellow of the Salk Institute in 1973; this gave him the power to influence which scientists would be invited to come and stay at the Salk. What's more, Salk linked Berg to a *Who's Who* of molecular biology, including David Baltimore, a former research associate of Dulbecco's.

Berg and Baltimore forged an instant friendship. They were much alike. Bright. Competent. Both came from New York: Berg from Brooklyn, Baltimore from Manhattan. Both were intrigued by viruses and the lessons viruses had taught the Phage Group. Baltimore, like Berg, had been a precocious youth, stimulated by science. His mother, an experimental psychologist, primed his fascination with physiology—in the body, not the mind. During the summer between his

junior and senior year in high school, Baltimore took the biology
course at Jackson Laboratory in Bar Harbor, Maine, where he got his
first taste of research biology, and saw biologists at work. He decided to study biology when he entered Swarthmore College, but later
switched to chemistry so he could do a research thesis. In the late
1950s, between his junior and senior years, he took a National Science Foundation–sponsored course at Cold Spring Harbor Laboratory. It was the height of the Phage Group's activity, and he was exposed
to its intellectual intensity and social values. The Cold Spring experience led him to molecular biology.

Baltimore moved to the Massachusetts Institute of Technology to
do graduate work in biophysics, and considered experiments with
bacteriophage. Knowing he could not compete with the Phage Group's
power, and looking ahead, Baltimore reasoned that perhaps animal
viruses and animal cells might teach the same kinds of lessons as bacteriophage. That "funny idea" carried Baltimore back to Cold Spring
Harbor for a summer course on animal viruses. There, he met Richard
M. Franklin, a Rockefeller University virologist who was trying to
figure out how viruses commandeered an infected cell's protein-making machinery to make new viruses. Baltimore joined Franklin at
Rockefeller to study mouse viruses, and showed how Mengo virus, a
picornavirus closely related to the polio virus, inhibits an enzyme
called RNA polymerase, which the cell uses to copy genetic information from DNA to messenger RNA to be used for synthesizing cellular proteins. After getting his Ph.D. and working as a postdoc at
Rockefeller and Albert Einstein College of Medicine, where he
learned enzymology, Baltimore got his first job as an independent investigator in 1965 at the Salk Institute, working on polio and other
animal viruses in association with Renato Dulbecco. Baltimore stayed
in La Jolla for two and a half years before returning to the Massachusetts Institute of Technology. Six years later, in 1974, he joined the
MIT Center for Cancer Research, working directly for its boss, Dr.
Salvador Luria, a Phage Group founding father.

• • •

In 1969 at MIT, Baltimore began to turn his attention from the double-stranded RNA polio virus to other types of RNA virus. First it
was the vesicular stomatitis virus, a complex RNA virus, which Baltimore and his team showed, for the first time, carried a special enzyme called RNA-directed RNA polymerase, an enzyme that

reproduced the virus's RNA genes. The discovery of a virus-built RNA polymerase led Baltimore to search for other examples among the RNA viruses.

An obvious place to search for the enzyme was among the RNA tumor viruses. One of these viruses, the Rous sarcoma virus, had been discovered decades earlier and shown to cause cancer in chickens. Howard Temin, by then an established researcher at the University of Wisconsin at Madison, had been working with the Rous virus and made a remarkable discovery: The virus could be inhibited by a drug that interfered with the actions of DNA. In 1964, Temin proposed that the Rous virus somehow converted its genes into a DNA intermediate that could be integrated into the chromosomes of the infected cell. Almost no one believed Temin because his idea violated the Central Dogma of biology. Genetic information just did not flow from RNA to DNA. Period.

But they were wrong. Temin had discovered that the Rous sarcoma virus was a retrovirus, a unique class of viruses that did violate the Central Dogma of biology. These viruses store their genetic information as a single strand of RNA inside the virus capsule. But once in the cell, a special virus enzyme converts the viral genome from RNA into a double-stranded DNA that can insert itself into the host cell's chromosomes. Even as Temin struggled to prove his notions and isolate the enzyme that carried out these reactions, Baltimore began searching for RNA-directed, RNA polymerases among the same retroviruses.

Baltimore knew Temin for years, ever since high school when Temin was the star at the Jackson Laboratory summer program Baltimore attended. The spiderweb of interpersonal connections reaches throughout science—often influencing the choices of individual researchers about what they study, and what gets learned.

When Baltimore dissected the Rauscher mouse leukemia virus, also a retrovirus, he failed to find his RNA-directed RNA polymerase. Instead he discovered that the Rauscher virus contained an enzyme that turned viral RNA genes into viral DNA, and that those genes could be plugged into the infected cell's genome. Simultaneously, Howard Temin and his colleague Satoshi Mizutani discovered the same enzyme in the Rous sarcoma virus. Their papers announcing the discovery were published side by side in *Nature* magazine in 1970. An anonymous correspondent from *Nature* magazine dubbed the enzyme "reverse transcriptase." The name stuck.

Like a translator rewriting a Greek text as a Latin book, reverse transcriptase rewrites genetic instructions stored in a single strand of the virus's RNA into a stretch of double-stranded DNA. The discovery explained how certain viruses could carry their genetic information in the RNA form, yet create stable, nonlethal infections of cells in which the virus seemed to tuck its own genes into the infected cell's chromosomes. This also provided clues to how certain RNA viruses caused cancer, ultimately leading to the discovery of oncogenes, normal cellular genes hijacked by viruses that caused uncontrolled growth.

Reverse transcriptase revolutionized biological thought. It also provided a key tool that helped launch the era of genetic engineering and biotechnology, the foundation for gene therapy. It also made Baltimore and Temin instant celebraties among scientists. Five years later, they would share the 1975 Nobel Prize in medicine with their old friend and mentor, Renato Dulbecco. David Baltimore was then thirty-five, the youngest person ever to win the Nobel.

A few years later, Inder Verma, a Baltimore associate at the Massachusetts Institute of Technology who later became a full professor at the Salk Institute, used reverse transcriptase enzyme to make the first bit of complementary DNA, or cDNA, from messenger RNA. This cDNA technique would become a powerful tool used by many scientists to rapidly isolate functioning genes from many different tissues. If a gene was turned on, it made messenger RNA that could be isolated and used as a primer for reverse transcriptase which then made a DNA version of the gene. Every cell contains all of the 50,000 to 100,000 genes estimated to be in the human body, but only a fraction of them are turned on in each cell. This selective expression of genes in different tissues explains why nerve cells are different from skin and heart cells, for example. In the early 1990s, the cDNA approach to gene isolation spawned a genetic gold rush in which scientists in the United States, England, and France used laboratory robots to race each other to rapidly identify—and then patent—human genes.

. . .

All that would come years later. In 1967, at the Salk, Baltimore and Berg would become close friends. Their friendship would have an overarching impact on biology in the decades to come. The linkage between high-powered research groups at the Salk Institute, Stanford

University, and the Massachusetts Institute of Technology would cast a long shadow over genetic research from the 1970s to the end of the century. Together, Berg and Baltimore would play central roles in the genetic engineering debates of the mid-1970s that created concern among some segments of the public about the hazards of manufacturing killer, mutant bugs in the laboratory. They would share and nurture bright young scientists, including Richard C. Mulligan, who would perform critical experiments that ultimately created the capacity to insert genes into mammalian cells with high efficiency.

And this coalition of interrelated scientists would later dismiss the importance of the work of French Anderson, an outsider from the National Institutes of Health, even though he had worked as a biochemist with Nobel-winner Marshall Nirenberg, and had been talking and writing about the potential of human gene therapy since the late 1960s. They dismissed Anderson because they felt he had not labored in the vineyard of gene transfer research; he only talked about its potential. Talk was meaningless; hard work lay ahead. Also to come was more controversy than any of them could imagine.

• • •

In 1968, after spending a year in Dulbecco's lab at the Salk, Paul Berg returned to Stanford and began phasing out the bacterial studies in his laboratory. At the same time, he started recruiting young scientists interested in growing animal cells and working with their viruses, especially animal tumor viruses. Berg chose a monkey virus called simian virus 40, a small, icosahedral-shaped viral particle that lacks an envelope. Its double-stranded DNA genome of 5,000 base pairs exists as a eukaryotic minichromosome carrying a few genes.

SV40 was first isolated in 1960 from the kidney cells of a rhesus monkey then being used to manufacture the polio vaccine. The discovery panicked the public health experts overseeing the polio vaccination program. When between 10 million and 30 million Americans, aged fifteen to thirty-five, and probably more Russians, had placed sugar cubes on their tongues to ward off the paralysis of polio, they also infected themselves with the monkey virus. Studies in 1961 and 1962 showed that a substantial portion of the exposed children made antibodies to both the polio virus and SV40. None of them seemed to be any worse off, so initially, at least, SV40 infections seemed to be innocuous.

But what about the future? Cancers, even those known to be caused by viral infections, take decades to develop. No one knew what the latency—the time between infection and the appearance of disease— might be for SV40-induced tumors. Only long-term epidemiological surveillance could tell if a rash of malignancies would crop up some- day.

As scary as the vaccine contamination had been, SV40 remained attractive as a laboratory tool because it had some of the same prop- erties as bacteriophages and Dulbecco's polyoma virus. For example, SV40 can capture bits of genetic material from the cells it infects and carry those genetic bits inside the virus. In some phases of its life cy- cle, SV40 can integrate its genes into the chromosomes of a cell it in- fects, making its genes a permanent part of the infected cell. If the virus also has hijacked cellular genes, then these too become part of the new host cell.

Berg quickly saw that SV40 might become a molecular vehicle (usually called a vector) for transplanting genes from one cell to an- other. The problem was going to be getting some interesting gene into the SV40 viruses so it could be transplanted into another cell.

The Phage Group had exploited the ability of some bacterial virus- es to steal genes from one bacterium and insert them into a different one to learn about the way genes were organized and how they were regulated. Berg hoped to repeat those successful studies in higher or- ganisms, mammalian cells and their viruses. The Phage Group work- ers, however, were not able to direct the bacteriophage to carry any specific gene. Instead, they could rely on randomness, on what the viruses accidentally picked up along the way to find a phage strain that carried interesting cellular genes. Since it only takes twenty min- utes to produce a new generation of E. coli, a plate of bacteria and their viruses could quickly produce billions of new bacteriophage, heightening the chance that an interesting rare event would occur.

Berg did the calculations for SV40 and concluded that the random events common among phage were unlikely to provide useful combi- nations of virus and cellular DNA. The life cycles of SV40 and the cells in which it grew were too slow. Besides, he wanted something better than chance. Berg wanted the ability to put any gene he chose into SV40 and then use SV40 to intentionally put the new gene into animal cells. He wanted SV40 to become his molecular delivery truck for transferring genes.

But he had a problem. These were the waning days of the 1960s, just before the discovery of unique enzymes that neatly cut and paste the long ribbons of DNA—the enzymes that would herald the dawn of genetic engineering. The first set of these enzymes would not be publicly described for another two years. In 1969, Johns Hopkins University scientists would find the first bacterial enzymes that could cut up the long strands of DNA in specific patterns, giving researchers the ability to slice up DNA and to isolate segments of genetic material that actually contained genes. But they didn't tell anyone for a year. Until those enzymes became available, Berg had to create his own system for isolating and manipulating the genes he wanted to insert in the SV40 virus.

Again, the Phage Group pointed the way. Al Hershey had shown that the DNA packed into a bacteriophage called lambda was linear, like a piece of audio tape taken off the reel. But in the test tube, Hershey found, the isolated DNA could circularize into a ring with the ends attached, like a rubber band. He discovered that the phage's linear DNA molecule had "sticky ends" that could connect if they randomly touched each other.

Hershey's sticky ends were caused by an unexpected phenomenon—little tails of single-stranded DNA hanging off the end of the virus's double-stranded, linear genome. Normally, the DNA molecule has a double-stranded structure that looks like a ladder twisted into a spiral. On each side of the ladder rises a vertical support. Between the supports are the rungs that form the steps. The DNA ladder comes apart in the middle of each rung, which is made up of two matching nucleic acids, or base pairs. Typically, these base pairs run the length of the molecule, just as the ladder's rungs reach all the way to both ends.

On the sticky ends, however, the base pairs do not reach the end of the molecule. Instead, a short, single strand extends beyond the end of the DNA molecule. If a second single strand of DNA, matching the first single strand's base sequence, comes along, the two will join together—completing the ladder, re-forming the more stable, double-stranded configuration. Sticky ends are complementary to each other the way the two different sides of Velcro match up and hang together.

If the two single strands were on opposite ends of the same linear molecule, they could join together, complete each other, and form a DNA ring. If the sticky ends were on different DNA molecules, then these two previously unconnected molecules could become connected. Conceptually, it would be possible to recombine bits of DNA into novel constructions never before seen in nature.

Berg wanted SV40 to have sticky ends like the lambda phage. If he could chemically create sticky ends, then he might be able to attach new genes to SV40's DNA, and then transfer the combined DNA molecule into cells. If this all worked—a lot of ifs—then he might be able to use SV40 to carry new genes into animal cells for the first time. Even if the experimental vision was clear, how he would do this was not.

Along with David A. Jackson and Robert H. Symons, two scientists in his group, Berg began creating a laborious technique that used existing, though expensive and rare, enzymes and chemical reactions to put sticky ends on the SV40 DNA and on a piece of bacterial gene called the galactose, or *lac,* operon—a milk gene that codes for an enzyme that metabolizes the sugar galactose—the same gene originally isolated by the Harvard team in 1969. If each bit of DNA had a sticky end, the SV40 and the *lac,* then they could be pasted together.

The galactose operon originally came from a bacterium but had been picked up by a bacteriophage sometime in the past and incorporated into the virus's own genome. The phage, however, then lost some of its essential genes and became unable to reproduce an entire virus particle. Instead of living the free and easy life of a bacteriophage, popping in and out of cells, the defective virus got stuck. Its remaining genes, with the attached *lac* operon, survived as a circular bit of DNA, called a plasmid inside a strain of bacteria. The viral plasmid reproduced itself independently inside of a bacterium but could never escape. Using a cumbersome set of enzymes, Berg, Jackson, and Symons attempted to cut the galactose operon out of the plasmid and paste it into the circularized chromosome of SV40. The work went very slowly.

. . .

As Berg and the others struggled to make the hybrid DNA through 1971 and 1972, however, trouble was brewing. The SV40 virus Berg picked to work on had a dark side. Its attractive ability to insert the genes it carried into animal cells came at a price. Stable gene transfer worked only when SV40 turned those normal cells—including human cells—into cancer cells. Although there was no evidence from the polio vaccine contamination that SV40 would cause a widespread epidemic of cancer, it remained a theoretical possibility because the incubation period between infection and tumor formation could be decades.

In the early 1970s, little was known about how viruses could turn a normal cell into a cancer cell, although researchers were focusing a growing amount of attention on tumor viruses. Still, scientists had little appreciation of the dangers they faced from working with viruses not known to cause acute human illnesses. Test tubes brimming with microscopic bugs were routinely handled on open desk tops. Mists of bacteria commonly floated in the air around high-speed centrifuges spinning bugs into pellets at the bottom of test tubes. Fluids were transferred from tube to tube with pipettes—long glass straws with measurement markings on the side—filled by technicians and researchers sucking on one end with their mouth while submerging the other end into the germy solutions. With only a cotton plug in the pipette's end to separate the scientist from the soup, a mouthful of microbes and growth media was a frequent consequence of carelessness.

Back in the Phage Group days, few scientists worried about the dangers of working with viruses, even mouth pipetting, because bacteriophage only infect bacteria, not the eukaryotic cells of higher organisms. But as more and more scientists started working with animal viruses, especially those that could cause human cells growing in culture to become cancerous, concern began to rise. By the early 1970s, an increasing number of scientists worried about the dangers of working with live, tumor-causing viruses. They also worried about the casual ways most labs operated. The word *biohazard* entered the lexicon. Scientists not working with tumor viruses questioned colleagues who were. They wanted to know whether people walking down the hallway in the same building might contract cancer from stray viruses drifting on the air currents. Despite the reassurances of some, nobody knew how dangerous things might become.

Some well-funded labs, like Berg's and Baltimore's, responded to the growing concerns by installing safety equipment, such as laminar-flow hoods and negative pressure rooms. Hoods are essentially lab benches with boxes built over them with a glass shield on the front and a space at the bottom through which the hands of the researcher can manipulate objects. A layer of air blowing across the opening prevents any viruses from floating out of the hood and into the room. Negative pressure rooms are a full-sized version of the hood; they prevent any viruses escaping the hood from leaving the lab to infect people on the outside. And most scientists stopped pipetting by mouth.

Even as the growing concern about safety slowly spread through the scientific community in 1970, Janet Mertz was packing her bags and heading west to work on her Ph.D. in Paul Berg's Stanford lab, which by then had almost completely converted over to SV40 work. Mertz had just finished her undergraduate work at the Massachusetts Institute of Technology, where she had taken David Baltimore's animal cell virology course. Her first job at Stanford was finding an interesting problem for her doctoral thesis. Berg had a suggestion: When he and the others finally made the hybrid SV40 virus that contained the *lac* operon, they intended to put the construction into animal cells to see if the newly inserted bacterial gene would function in a mammalian cell. No one knew whether genes from a primative prokaryoteslike bacteria could function in the more complex eukaryotic cell; many believed they would not.

Mertz, Berg proposed, could do the reverse experiment. She could insert the hybrid SV40 into a bacterial cell to see if the genes from the animal virus would function in the bacterium. That would test, for the first time, whether animal virus genes—and perhaps mammalian genes—could function in a lower cell, which many scientists also doubted.

Initially, Mertz resisted the idea. This was 1970. Radicalism had swept over science just as it had the society at large. Vietnam was in full swing; so were the campus protests. Just a year earlier, Beckwith and Shapiro had warned at their Harvard press conference that the isolation of the *lac* operation would lead to genetic engineering, possibly of people. Mertz, at the time, considered herself a "middle-of-the-road radical" who worried about the impact of technology on society. She struggled with the decision about whether to conduct research that might advance the technology, heightening the dangers to society.

Her desire to work with Berg, however, eventually won out: "Being able to put some DNA into *E. coli* is still a long way from being able to change people's genes. . . . It was sort of a compromise between my radical point of view and wanting to work on specific types of projects and have Paul Berg as my thesis adviser." Mertz took on the job.

By June 1971, she had worked out a technique for putting the hybrid SV40 genome into the bacteria—if Berg and the others ever succeeded in creating it. That month, she also left for a three-week course in cell culturing at the Cold Spring Harbor Laboratory on Long Island. That trip would forever change her career—and many historians mark it as a turning point for genetic science. At Cold Spring Harbor, Mertz became the center of a controversy that many say launched the public debates about gene-splicing and led to a temporary ban on recombinant DNA experiments.

Mertz attended a cell-culturing course taught by Robert Pollack, then a thirty-one-year-old Cold Spring Harbor cell biologist. He also was an SV40 expert, "and viewed it with great respect." Pollack, who was not working on SV40 at the time, had become concerned about plans by other Cold Spring scientists to grow large concentrations of SV40. They wanted to extract its DNA, and test a newly discovered class of bacterial enzymes, called restriction enzymes, that chopped up long strands of DNA in a predictable way. These were the enzymes discovered by Johns Hopkins researchers in 1969, and described publicly for the first time in 1970. Hopkins scientist Hamilton O. Smith made the initial discovery, but Daniel Nathans immediately showed how the enzymes could be put to use by creating a so-called restriction map of SV40. Restriction maps give relative positions of where the restriction enzymes cut the virus's circular chromosome and can help identify genes within the fragments.

Cold Spring Harbor researchers wanted to repeat and expand Nathans's work, including growing large vats of SV40. Pollack worried, out loud, that producing such large volumes of SV40 increased the dangers to those working in the laboratory, and that the biohazards had to be resolved before the studies were permitted.

Pollack spent an afternoon session of his cell biology course on laboratory safety, using the SV40 problem as an example. Afterward, Mertz mentioned to Pollack that the Berg team was working on a way of making a recombinant DNA molecule that contained genes from both the SV40 virus and from a bacterium. Once the technology was worked out, Mertz said, they would put the hybrid genes into both mammalian cells and bacterial cells, specifically *Escherichia coli,* a bacteria that lives in the human intestines.

The news stunned Pollack. He immediately saw a serious, potential biohazard in the idea of putting genes from a virus that normally causes cancer into a bacterium that normally lives in the human gut. What would happen if some of those genetically transformed bacteria infected one of the researchers, or somehow escaped the laboratory and infected other people not involved in the work? Conceivably, a cancer-causing bacteria—which previously did not exist—could spread through the general population, creating tumors.

"I had a fit," Pollack said. "She was bragging about how they're going to make Russian dolls: SV40 in lambda [the bacteriophage], the lambda in *coli, coli* in something else," Pollack recalled. "I said, well it's *coli* in people. She didn't know that, as far as I could see. It never occurred to her that that would be a problem."

Concerns from Mertz's proposed experiment spread quickly across the Cold Spring Harbor campus during the rest of the course, leaving the graduate student barraged by objections. Among the most aggressive critics was David Botstein, a bright and voluble yeast geneticist then at MIT, later destined for Stanford. For every possible problem posed by Botstein during impromptu debates in the lab's cafeteria, Mertz threw up a defense of Berg and the experiment. But Mertz's arguments could not eliminate the risks. Finally, she became overwhelmed. Convinced of the biohazards proposed by the others, however theoretical, Mertz decided not to put a hybrid SV40-*lac* combination into *E. coli* cells, even though it was a good thesis project and even though she already had spent months on it. Luckily for her, Mertz would soon make an important observation about sticky ends that redirected her career and had a major impact on the development of genetic engineering.

Her quitting the SV40 project, however, was not enough for Pollack. On Monday, June 28, 1971, Pollack called Paul Berg to press him on the biohazard issues of the SV40 experiment, and why it was a bad idea to put genes from a tumor-causing virus into a bacterium that lives in the human gut. Berg raised counterarguments that Pollack answered in turn. Finally, Pollack made a flat statement: The experiments should not be done. Berg was annoyed and noncommittal, saying only that he would consult with his colleagues.

To Berg's astonishment, his scientific friends also criticized the experiment, with many of them arguing that he should not do it, at least not until the dangers had been thought through. And it was a long list of colleagues, including David Baltimore and Maxine Singer. Even James Watson, by then director of Cold Spring Harbor Laboratory where the fuss kicked up, was reportedly critical.

Although this initial debate occurred only within a small circle of the scientific community—and did not involve the public at large—it would have a profound effect on both the future of genetic engineering and human gene therapy. The idea of using newly emerging techniques to make genetic recombinations both scared people and raised thorny ethical issues about the responsibilities scientists had to their colleagues and their countrymen.

While most researchers focused on the biohazards that directly confronted the scientists doing the work and, to some extent, the public at large, Leon Kass, a Harvard-trained molecular biologist with an M.D. from the University of Chicago, looked at the broader issues. Kass, who himself had evolved into a bioethicist, met Berg in the fall of 1970, before the debate about Mertz's SV40 experiments, in the home of Daniel and Maxine Singer. Maxine was Berg's friend and an NIH geneticist who had helped Marshall Nirenberg in his race to crack the genetic code. Daniel was a real-estate lawyer and former general council to the Federation of American Scientists, a liberal, Washington-based think tank founded in 1945 by members of the Manhattan Project to study how science affects society. Daniel had become sophisticated in the issues surrounding the ethical and social responsibilities of scientists.

The Singers understood that Berg's intention was to develop a technology that would allow him to move genes into and out of any cell he chose for purely scientific purposes. They, like Kass, also understood that such a technology might one day be used to move genes in and out of people as a form of medical treatment—or as a way of "improving" humanity. At a dinner at their home, which Berg and Kass attended, they discussed Berg's plans and what they might mean for society if they worked and gene transfer became possible.

After the dinner, Kass wrote down his impressions and sent them to Berg in a letter dated October 30, 1970. Kass reviewed the ethical questions about safety and efficacy of gene transfer technology, pretty much dismissing them since they were similar to existing problems posed by any new medical therapy. There were, however, thornier questions raised by the purpose for which gene transfer might be used. "Obviously, once a technique is introduced for one purpose, it can then be used for any purpose. Therapeutic use is one thing, eugenic, scientific, frivolous, and even military are quite another," Kass wrote. Then he laid out some of his concerns: "What are the biological consequences to future generations of widespread use of gene therapy on afflicted individuals? . . . Are we wise enough to be tampering with the balance of the gene pool? . . . What are our obligations to future generations?"

While he argued for "therapeutic use" only, Kass added, "I would insist that we need to foster public deliberation about this question, since I don't think this is a matter to be left to private tastes or to scientists alone." And as he considered how gene therapy might be controlled, Kass worried that "the more . . . I contemplate the possible widespread consequences of genetic manipulation, the more I believe that all decisions to employ new technologies, and even to develop them for employment in human beings, should be public decisions. "How to do this is not obvious. . . . "

It was not obvious to Berg either. And, of course, it wasn't just the Berg and Mertz experiment that would raise questions. Another involved NIH researchers who had discovered several strains of virus that had undergone a natural genetic recombination. The new virus carried genes from SV40 and genes from adenovirus, a common cause of respiratory infections in humans, that had been packed inside the adenovirus capsule. The hybrid adenovirus with the SV40 genes retained the ability to infect human cells easily and proved capable of making human cells cancerous in the laboratory. Andrew Lewis, Jr., a physician struggling to understand the new recombinants, wrote: "The pathogenicity of non-defective viruses which are hybrids between human pathogens and other viral agents such as SV40 . . . represent unknown hazards." He saw the dangers in naturally occurring recombinant viruses, but he also knew that researchers planned to make such recombinants on purpose. Lewis feared "that agents which people were creating in the laboratory could be considered as pathogens." He was among the first to raise the biohazard issue that viruses, genetically altered in the laboratory, could become new sources of human disease, and perhaps, even become more pathogenic than the natural virus from which they evolved. Lewis was not the last to raise that fear.

By tradition, scientists who publish their work are obliged to share their reagents—unique chemicals and living organisms like bacteria and viruses used in the experiment—with any other researchers who wish to confirm their work—or advance it. It's a unique social value—unlike anything in business—where competitors can ask their opponents for assistance and get it. Help must be provided for the good of all science, or the violator faces social sanctions.

When NIH began receiving requests for the SV40-adenovirus hybrids, Lewis, who had the viruses in his lab, balked, fearing that the viruses might be too dangerous to release. The requesting scientists from Cold Spring Harbor protested, and the issue of scientific etiquette became a bureaucratic nightmare as the NIH researchers kicked the problem upstairs to their administrators, who then had to grapple with the concept of biohazards. The bureaucrats took a predictable approach: They issued a policy statement that became, by the end of 1972, the first government regulation of genetic research. Researchers wanting the SV40-adenovirus recombinant had to agree to conditions that among other things limited the distribution of the virus.

. . .

Berg, increasingly persuaded by the biohazard arguments, reevaluated his plans to build a hybrid SV40 virus and inject the genes into animal and bacterial cells. He decided the chances of the experiment's success were small but the questions of safety were large, if theoretical. He decided not to put the SV40 hybrid genes into any cells, but his group continued working on techniques to make the hybrid. In an interview with MIT's Oral History Program four years after he decided to cancel the experiment, Berg minimized the risk of introducing bacterial genes into animal cells with the SV40 hybrid since mammalian cells in culture cannot escape the lab and live on their own in the wild. But he saw the risk of putting animal or SV40 cancer-causing genes into bacteria as completely different. Bacteria can live on their own outside the laboratory. What's more, the new genes they carry from the hybrid experiments, genes that can cause cancer in animals, would be reproduced and might become widely distributed in the environment. The possiblity of risk to people was real. An Andromeda strain might be made, a breed of *E. coli* that spread infectious cancer was theoretically possible, and such a bacterium might escape the laboratory, threatening the population at large.

"I realized that while I could argue that the probability [for hazard] was low, I could never say that the probability was zero," Berg admitted to former *Rolling Stone* reporter Michael Rogers. Berg, in the fall of 1971, decided not to put the hybrid DNA into any cell—eukaryotic or bacterial—at least for now.

It was a timely decision. In fall and winter of 1971–72, Berg, Jackson, and Symons finally succeeded in constructing a hybrid DNA to insert the *lac* genes into the SV40 chromosome. The work was technically difficult, involving an intricate series of enzyme steps that gave increasingly smaller yields with each step. And the piece of DNA they made turned out to be three times bigger than the SV40 chromosome; the genetic ensemble was too big to package into the SV40 capsule.

Shortly after making the recombinant DNA, in the spring of 1972, David Jackson left Berg's laboratory to set up his own lab at the University of Michigan. The work of building hybrid SV40, so difficult and now considered so dangerous, simply stopped. Berg told everyone that his lab was no longer constructing hybrid DNAs. Because the techniques were so tedious and painstaking, few other labs anywhere could pick up where Berg stopped. The issue seemed dead.

It wasn't. Even though Berg's plans to put the hybrid DNA into cells were abandoned, the biohazard questions raised by the experiment would not go away. And Berg was now fully engaged in the debate. He and others decided the best way to resolve the questions about hypothetical biohazards from experiments with the genes of SV40 and other viruses was to assemble the world's experts and hold a conference to address the issue. The experts gathered on January 22–24, 1973, at the Asilomar Conference Center at Pacific Grove, California, about three hours south of San Francisco. Only 200 yards from the white sands of the Monterey Peninsula beach, Asilomar comprises a collection of small redwood dormitories and large meeting rooms scattered through a seaside forest of redwoods and Monterey pine. Despite thorough review and heated debates about safety, not much new was learned at the conference. The only consensus was that someone should conduct epidemiology studies to assess biohazards, and that there should be a follow-up meeting in two years. Neither recommendation would be carried out. Events would soon move too fast, and the relatively small scientific community that began the discussions about genetic safety would soon lose control of the debate. Sticky ends would make everything come apart.

· · ·

Even as Berg's team began their struggle to make the SV40 hybrids, a series of critical discoveries were being made elsewhere that would accelerate the field of gene engineering. The techniques would make it easier to achieve what Berg wanted, but they would also raise concerns about genetic safety to new heights—and in a very public way.

It started when researchers Matthew Meselson and Robert Yuan, his postdoc, discovered a class of enzymes in *E. coli* K bacteria in the spring of 1968 that would allow the manipulation of genes in a new way. It turned out that bacteria were not quite defenseless against the bacteriophage that preyed on them. When the phage injects its DNA into the target cell, many bacteria launch an enzymatic counterattack with restriction enzymes made inside the bacteria. These enzymes recognize specific parts of the viral DNA as alien and grind it up into useless bits of nucleic acid. This prevents viral reproduction.

Salvador Luria had noticed these bacterial defenses in the early 1950s. He and others saw that phage that grew well on one bacterial strain often failed to grow on another related bacterial strain. And viruses from the second strain would fail to grow on the first. They were, Luria wrote, "restricted" to their own host bacteria.

Even as Meselson was discovering the actual enzymes that blocked viruses in *E. coli* K, Werner Arber, a Luria associate, simultaneously discovered restriction enzymes in the B strain of *E. coli*. Arber showed that when the viral DNA entered the cell, a bacterial enzyme destroyed it.

Yet somehow, the same enzymes did not dice up the bacteria's own genes. In 1965, Arber proposed that the destructive enzymes recognized some unique sequence base pairs along the viral genes, and then cut up the DNA at that point. What's more, Arber reasoned, there had to be some enzyme that recognized these same sequences along the bacterial chromosome and somehow modified them—perhaps by physically attacking a blocking chemical, probably a methyl group—to protect the DNA from the restriction enzyme. It would take Arber three years to find signs of the enzyme he proposed.

Once the restriction enzymes were purified, however, the floodgates to genetic engineering were opened. After reading the Meselson and Yuan paper in the spring of 1968, Hamilton O. Smith and Kent Wilcox at Johns Hopkins School of Medicine looked inside *Haemophilus influenzae,* strain Rd, which they were working with at the time, for similar restriction enzymes. They hit the jackpot by finding *Hind II,* a bacteria enzyme that differed from the ones found by Meselson and Arber. Their enzymes degraded invading DNA, but they cut it up into random pieces. *Hind II* cut the foreign DNA very precisely, every time. It apparently could recognize a specific site on the viral DNA, and cut only through that site.

This new level of precision would give scientists unprecedented control in being able to cut the long fibers of DNA down to manageable sizes. And it didn't matter whether they were being spliced up to make genetic maps, or whether the smaller DNA pieces were being prepared for recombination. Many scientists rapidly moved into the field, searching for better and more precise restriction enzymes. Soon there would be dozens, later hundreds, of different restriction enzymes, all slicing up DNA at different points.

· · ·

In spring 1972, a doctoral candidate in Herbert W. Boyer's University of California at San Francisco laboratory discovered yet another restriction enzyme from *E. coli*. The new enzyme was called *Eco R1*. Boyer shared the new enzyme with a number of laboratories, including Paul Berg's shop at Stanford. Berg had *Eco R1* tested on the circular SV40 genome, and discovered that the enzyme cut SV40 in only one place, giving a new reference point on the SV40 map. That small study, however, led to a major discovery.

Janet Mertz, who had abandoned her plans to insert a recombinant SV40 into *E. coli,* noticed that the *Eco R1* appeared to automatically leave "sticky ends" on the linear DNA that remained after enzyme treatment. Boyer's group followed up Mertz's observation, proving that any pieces of DNA cut with *Eco R1* would indeed be left with sticky ends. This meant any two pieces of DNA could be pasted together by first cutting them with *Eco R1* to make the sticky ends, and then mixing them together. The sticky ends would attract each other like magnets. The sticky ends Berg and his associates struggled for several years to construct laboriously by hand, *Eco R1* could make in minutes. The enzyme would make it a snap to construct new combinations of genes never before seen in nature.

Only one piece of technology was still needed: a way to rapidly reproduce the newly isolated fragments of DNA that could now be recombined at will. Somehow, scientists needed a way to "clone" the recombinant genes, to make billions of identical copies so they could be studied in the lab. They needed a molecular photocopying machine for genes.

Enter Stanley N. Cohen of Stanford University, an expert in a strange bit of bacterial genetics called plasmids. Plasmids are little rings of DNA that carry a few genes and seem to live an independent life within their host bacteria. Plasmids float around in the cell's sticky center but live separately from the bacteria's single chromosome. They reproduce independently of the chromosomes, and even jump between bacteria in a kind of primitive sexual exchange of genetic material.

When a bacterial strain becomes resistant to an antibiotic, it's often because it picked up a plasmid carrying a gene that inhibits the drug. Usually it's an enzyme that degrades the antibiotic. Cohen had developed a system for easily removing plasmids from bacteria and then putting them back in.

In November 1972, during a three-day, American-Japanese conference on plasmids in Hawaii, Cohen heard Boyer give a talk about the *Eco RI* restriction enzyme and the sticky ends it could make. Cohen immediately realized that if plasmids could be cut by Boyer's enzyme to make sticky ends, and then a bit of DNA—say from a virus, fruit fly, or frog—was also cut with *Eco RI* so it had sticky ends, then this foreign piece of DNA could be spliced into Cohen's plasmids. The entire genetic package could be smuggled back into the bacteria. Once inside the bacteria, the plasmid would reproduce every twenty minutes along with the dividing cell, making millions of identical copies of the isolated DNA. If the inserted DNA was a gene, then the bacteria would make millions of copies of the gene. In a flash, the basic elements needed for genetic engineering came together.

At a delicatessen across from Waikiki Beach, Cohen proposed a collaboration. Boyer accepted. By March 1973, the two laboratories had demonstrated that they could, at will, isolate pieces of DNA with the *Eco RI* enzyme, pack it into a plasmid, and insert the new recombinant DNA molecule into bacteria that would easily churn out billions of identical copies of the gene. They had, in effect, created genetic cloning.

All of the tools needed for the coming genetic revolution were now in place. The only thing that remained was a public discussion of the dangers.

• • •

In June 1973, six months after what would later be called the first Asilomar meeting that Paul Berg organized to discuss biohazards, a Gordon Conference was held on nucleic acids, the chemical building blocks of the genes. Gordon Conferences are tribal retreats for different disciplines of science held at private schools in New England. They are the ultimate in scientific exclusivity, where conference chairmen, themselves leaders in a field, invite only the best and brightest to present their hottest data, and allow only a small number of the best and brightest—usually fewer than 100—to hear about the cutting-edge work. Everything is off-the-record; outsiders—especially the press—need not apply.

NIH's Maxine Singer, Paul Berg's old friend, and Dieter Söll, a biochemist from Yale University, cochaired the 1973 nucleic acids conference. During the meeting, San Francisco's Herb Boyer—despite promising Cohen that he would not divulge their current work—described his unpublished experiments with Cohen about their new way to put DNA into bacteria. The Gordon conferees immediately grasped the importance and implications of the new technique: "Now we can combine any DNA," someone excitedly shouted at the conference.

The Boyer-Cohen technique would make it possible for any researcher, not just high-powered scientists with well-funded, technically sophisticated labs like Paul Berg's, to make hybrid DNA molecules. Scientists at the meeting told Singer they were concerned about the uncontrolled use of this new technology and the possible biohazards it posed. They wanted to talk about their anxieties at the meeting. Singer and Söll agreed to put the issue on the conference agenda: a half-hour on Friday morning, the last day.

With little time for discussion, three proposals were offered and approved, according to Sheldon Krimsky in his 1982 book, *Genetic Alchemy: The Social History of the Recombinant DNA Controversy*:

(1) a letter should be sent to the National Academy of Sciences and the National Academy of Medicine to set up an expert panel to study the biohazards; (2) each participant at the conference should do what he or she wished, but that no collective action should be taken; and (3) a group letter should be constructed and signed by all who wished to sign it.

Of the 143 who initially came to the conference, an overflow crowd, ninety were left. Seventy-eight voted that a letter expressing their concerns be sent. Then the proposal was amended to give the letter wider publicity: forty-eight voted for general distribution of the letter; forty-two voted against it. Ultimately, the letter stating the Gordon Conference concerns was published in *Science* magazine. There was little public reaction to the letter, although it did get the attention of the scientific leadership.

National Academy of Sciences president Philip Handler sent the letter to one of NAS's standing committees, the Assembly of Life Sciences. The committee members interviewed Maxine Singer about the problem, asking her who could lead the study group requested in her letter. Singer offered her friend Paul Berg. He was a credible choice. A well-respected scientist who worked in the field, Berg had shown responsibility by slowing his own work, and helping organize the biohazards meeting at Asilomar. After some thought and consultation, Berg accepted the job and assembled a committee that met on April

17, 1974. The committee read like a *Who's Who* of molecular biology: besides Berg, there was James Watson, Daniel Nathans, Sherman Weissman, Herman Lewis, Norton Zinder, bioethicist Richard Roblin, and Berg's close friend, David Baltimore.

Their first meeting concluded that enough potential dangers from the new genetic technology existed that the scientists involved should hold off on certain types of experiments until the dangers could be understood. Berg's committee, "acting on behalf" of the National Academy of Sciences, decided to publish a second letter in *Science*. *Nature* and *The Proceedings of the National Academy of Sciences* also would publish the letter. Though the letter was carefully crafted, it proved to be a bombshell: The Berg committee called on scientists to observe a voluntary moratorium for certain classes of recombinant DNA experiments, at least until an international meeting could be held to discuss the dangers of recombinant DNA.

And this time, the public heard about it, too: "Genetic Tests Renounced over Possible Hazards," announced *The New York Times;* "Halt in Genetic Work Urged, NAS Panel Warns of 'Hazards' " responded *The Washington Post.* Biohazards went public. Groups beyond the small clique of scientists who had begun the safety debate would now enter the fray.

At the academy press conference announcing the moratorium, Berg explained: "We've taken this route because we feel that the scientific community should be given a chance to regulate itself in its movement in the future. . . . I think most scientists agree very readily that the hazard is there and would like to see the hazard removed in some way. . . . "

Many who entered the debate were scientists themselves who had concerns about the social responsibilities scientists had toward the public. For example, Nobel laureate George Wald at Harvard University spoke against genetic engineering for a leftist radical group, Science for the People. Harvard Medical School's Jonathan Beckwith, who isolated the first bacterial gene, continued worrying about the development of government programs to manipulate genes in people.

Scientists working in the field of molecular genetics also were moved to action. Roy Curtiss III, a microbiologist at the University of Alabama in Birmingham, wrote his own sixteen-page, single-spaced memorandum about the potential biohazards of genetic engineering, and what ought to be done about them. In his conclusions, he urged his fellow geneticists to take these early concerns seriously, adding that they should remember how the initial warnings about the hazards of radiation had been ignored, only to be later proved correct.

Allusions to radiation would become common. The shift in public approval for nuclear power, from its early days of acceptance to its later fall from favor, helped the molecular biologist understand the issues they faced. Increasingly, writers, social commentators, even some scientists would equate the newfound ability to manipulate genes with the development of the atomic bomb. Even if gene engineering didn't explode or produce radiation, it might create a nasty life form never before envisioned in nature crawling out of the test tube to threaten life on the planet. Michael Crichton's *The Andromeda Strain* could become reality, not because some spaceship became contaminated with a cosmic pathogen, but because a careless scientist created a deadly plague and let it escape down the drain.

Deep in the back of the scientists' minds was the ultimate horror. It was not the dangers of biohazards to the laboratory workers, or even the grander science-fiction scenarios of Andromeda strains and genetic monsters. It was the genetic engineering of people. Not gene therapy to eliminate disease, but Hitlerian visions of a master race crafted at the genetic level by scientists run amok.

• • •

Paul Berg didn't put much credence in the more far-fetched notions, but he certainly took the warnings seriously. In his own lab, he already had stopped work that was considered hazardous. And now he headed the National Academy of Sciences committee set up to delineate the risks of recombinant DNA research and had led the call for a temporary moratorium. The next step would be a major conference of the world's experts to assess the known risks and to set a course of action.

This meeting too would be at Asilomar, the pleasant retreat center in the redwood forests of the Monterey Peninsula. It would be *the* Asilomar meeting—the first Asilomar meeting on biohazards in 1973 was long forgotten.

More than 150 researchers from thirteen countries—including the U.S.S.R.—and sixteen reporters converged on the quiet town of Pacific Grove, California, for the invitation-only gathering. In the silence of the retreat center's chapel, the high priests of biology assembled to contemplate biohazards and to draw up safety guidelines that would allow the experiments to proceed with miminum risk to themselves and society.

Like Woodstock, Asilomar was a defining moment. Unlike the revelers at the harmonious 1960s rock concert in a New York State pasture, however, the scientists at Asilomar seemed to end up in a free-for-all. Like squabbling schoolboys, members of the different scientific factions—there were experts from many subspecialities including plasmid designers, phage engineers, virologist, and cell biologist—feared that rules would be written to limit their work, but not that of others. The different groups jockeyed to protect their own positions, threatening to splinter the conference, preventing any chance of consensus.

Although they had been called to California to discuss risks and propose a system of self-regulation, most scientists there were more interested in gossip about the latest techniques for transferring genes hither and yon than they were in grappling with unseen dangers. Many thought the concerns about potential risks were overblown. No one could calculate what the dangers actually were in statistical terms, as Jim Watson kept pointing out. Watson, a signer of the original Berg letter calling for a temporary moratorium, had changed his mind. He, like fellow Nobel laureate Joshua Lederberg from Stanford University, considered the experiments safe and wanted the moratorium dropped. Stuck in a centrist position, Paul Berg had to balance the calls from the Watson-Lederberg faction to drop the regulations with demands from the far left—represented by groups like the Boston-based Science for the People—for even stricter controls.

When data were actually presented on risks, it tended toward the theoretical—or sometimes the theatrical. Ephraim S. Anderson, director of the enteric reference laboratory at the Public Health Laboratory Service in London, demonstrated that the laboratory workhorse, *E. coli* strain K12, was probably safe to use. Isolated in 1922 from the feces of a patient recovering from diphtheria, researchers believed that K12 had evolved over the decades of pampered existence of continuous laboratory culture conditions to be less vigorous and able to survive on its own than undomesticated bacteria. To test that, Anderson had conscripted eight volunteers to drink germy cocktails of K12 in milk. The study showed that K12 could not be recovered from the volunteers' feces after six days, and usually not after three days, indicating that it could not establish a sustained infection in people. English microbiologist H. Williams Smith from the Houghton Poultry Research Station in Huntingdon, conducted the same experiment—on himself. He had similar results. K12 actually seemed pretty safe for gene engineering experiments.

Even so, the debate raged on about whether it was possible to ensure that a genetically engineered bug would *never* escape a laboratory and wreak havoc on the public at large. That's when British biologist Sydney Brenner stepped in and saved the conference from chaos by enthusiastically pushing the notion of "biological containment." The earlier safeguards all entailed physical containment, conducting the experiments inside high-tech hoods and high security labs, where no one could just walk in, and no germs could walk out. Everyone agreed physical containment would not be enough; there would always be mistakes that allowed gene-carrying bacteria to escape.

Brenner, a brilliant, bushy-browed experimental biologist from the Medical Research Council in England, led the discussion toward designing enfeebled bacteria unable to live outside the comfy confines of a modern laboratory. If the bug couldn't grow in the wild, then it couldn't spread disease. Simple.

Roy Curtiss, the Alabama gene jockey who early on supported the research ban, predicted an enfeebled *E. coli* could be constructed in a month. Curtiss was dead wrong: it would take fourteen months, but it was doable. His cavalier conclusion, however, helped sway the day; biological containment would be the solution—both biologically and politically.

On the evening of the third day, Wednesday, February 26, the scientists received a reality check. A postdinner panel of lawyers organized by Daniel Singer gave a sobering discussion about the scientists' responsibility to society and the liabilities they might incur if an experiment went terribly wrong. Alexander Capron, then a law professor at the University of Pennsylvania, forcefully argued that "the public" had to get involved in this discussion, and that "the public" meant "the government." Capron concluded that the scientists would have to accept three inevitabilities: some regulation would be necessary, some experiments would be restricted, and the government—from Congress to the Occupational Safety and Health Administration—would have the final say. Roger Dworkin, a law professor from Indiana University, led the researchers through the frightening wilderness of liability, negligence, and multimillion-dollar lawsuits. The attorneys shook any residual beliefs among the biologists that this little biohazard problem of theirs was going to remain under their purview. Outside forces were ready to pounce, if they were not already at work.

From the first presentation on the first day, in which David Baltimore exhorted the gathered experts to find a consensus because there would be no one else to turn to, to the final half day chaired by Paul Berg, in which the final consensus document was approved, the 150 scientists struggled for an agreement. "The proceedings rapidly developed the appearance of some obscure primitive tribe eons ago, accidentally stumbling by trial and error onto the secret of parliamentary procedure," Michael Rogers wrote in his *Rolling Stone* account of the meeting. In the end, a set of guidelines did emerge, stating just which types of experiments would be allowed, and which would not.

Asilomar was a momentous event in the evolution of genetic engineering. It would cast a decade-long shadow over the fledgling field of molecular biology, and lay the foundation for regulating genetic research. It was the first time since World War II that scientists struggled to rein in a newly emerging technology. Many physicists, including many who helped make the bomb, tried to reharness the atom once it had blasted out of the barn, but to no avail. Asilomar had been intended to assess the risks of redesigning genes, and write the guidelines that would allow the experiments to proceed safely. Paul Berg and the rest of the molecular biology leadership wanted to prove that they were responsible citizens who could regulate themselves to ensure the public safety. What's more, Asilomar was a political act, an attempt, in part, to maintain scientific autonomy. If the dangers were great and the threat to the public real, political leaders were likely to weigh in with Congress writing regulations that would hamstring science. This is what the lawyers had predicted, and what the scientists feared. And that is what nearly came to pass.

Many within the scientific community saw Asilomar as a way to prevent the federal government from writing regulations to limit genetic research. That strategy failed miserably. The day after Asilomar ended, February 27, 1975, the newborn Recombinant DNA Advisory Committee (RAC) of the NIH met in an ornate conference room in a San Francisco hotel. Although few Asilomar participants knew it, Robert S. Stone, then NIH director, had empaneled the RAC on October 7, 1974, four months before the Asilomar meeting. Stone had ordered the select group of scientists and government administrators to gather immediately after the meeting and do exactly what the conference participants feared most: turn the general conclusions of Asilomar into legal regulations that would specify what constituted a genetic biohazard, how it would be controlled, and which experiments would be prohibited.

Even as the RAC began its work around the polished wooden table in relative seclusion, the sixteen science reporters left Pacific Grove, and filed their mostly laudatory stories. Even as the press praised the scientists for their social responsibility, the reporters described the potential dangers well enough to frighten readers. The genie was out of the bottle. No longer would the scientific elite control the debate about genetic engineering. Now that the public—and its elected representatives—knew about the biologists' concerns, they wanted some input. The lawyers' prophesies would come to pass.

Writing recombinant DNA guidelines turned out to be a difficult task. As the first drafts of the regulations circulated in the summer and fall of 1975, critics on both sides attacked: some said the proposed NIH rules were too strict. Others—including Paul Berg and some fifty biologists who signed a petition organized by Richard Goldstein of Harvard Medical School—complained that they were too lax. They were certainly much less restrictive than the guidelines put forward at Asilomar.

Still others said the whole process was a sham because those writing the regulations were the scientists pioneering the field. The fox had been put in charge of the henhouse. The scientists who created the biohazard had taken charge of writing the rules to limit the hazard. Consequently, they could write in the loopholes they needed to allow their own work to proceed while restricting the work of others—especially their competitors. MIT's Jonathan King, then a member of the Scientists and Engineers for Social and Political Action, charged that it appeared that the role of the RAC was "to protect geneticists, not the public."

Stung by the criticism, the RAC took another stab at writing tougher regulations in December 1975. After much debate, it proposed stricter rules that required four levels of physical containment and three levels of biological containment. Physical containment was labled P1, P2, P3, and P4 in ascending order of severity, where P4 was a high-security lab where workers wore special clothes and the lab's air flow and waste disposal was carefully regulated. Biological containment was labled EK1, EK2, and EK3, and defined the organisms—so-called safe bugs—into which recombinant DNA molecules could be placed. By the end of 1975, Roy Curtiss believed he had created an enfeebled *E. coli* strain that could not survive outside the laboratory. The Curtiss strain would be a safe place to put recombinant DNA molecules. Biological containment became a practical reality.

On its second effort, RAC actually produced tougher rules than those called for by Asilomar. For Paul Berg personally, the regulations would mean abandoning ongoing work with SV40 recombinants that his lab had been doing for five years. Berg decided to switch over to the mouse polyoma virus, a process that would cost him at least six months in a field that was rapidly becoming highly competitive.

As RAC's regulations moved toward publication in spring of 1976, the debate about the wisdom of the guidelines intensified. Two solid voices in the field of biology came forward as unlikely critics, not because they felt the public needed to be involved, but because they feared the scientists themselves had not stopped long enough in their headlong competition with one another to consider the profound impact these new techniques might have on life on Earth.

One was Erwin Chargaff, a biochemist at Columbia University who discovered that there were as many adenosines in DNA as there were thymidines, and as many cytosines as there were guanines, providing Watson and Crick with a key clue to the structure of DNA. Chargaff worried that genetic meddling might lead to unforeseeable problems. "Have we the right to counteract, irreversibly, the evolutionary wisdom of millions of years, in order to satisfy the ambition and curiosity of a few scientists?" he asked in *Science* magazine. He argued that the recombinant research should be prohibited altogether for at least two years, to give the scientists time to cool off and reflect on the risks they were about to take for all of humanity.

Chargaff's judgments, however, were not just technical, but also moral. "I am one of the few people old enough to remember that the extermination camps in Nazi Germany began as an experiment in genetics."

The other notable critic was Robert Sinsheimer, chairman of biology at the California Institute of Technology, a biophysicist and expert in the virus phi-X174. Sinsheimer criticized the entire NIH guideline-writing process because it focused too narrowly on the possiblity of health hazards without ever considering the possible evolutionary consequences of intentionally changing genes within organisms—including humans. Sinsheimer believed there was a natural barrier between prokaryotic cells like bacterial and eukaryotic cells of mammals that prevented these two different groups from interacting at the genetic level. Breeching this barrier, Sinsheimer feared, could be cataclysmic. Of the NIH guidelines, Sinsheimer would write that "they regard our ecological niche as wholly secure, deeply insulated from potential onslaught, with no chinks or unguarded section or perimeter. I cannot be so sanguine. In simple truth, just one—just one—penetration of our niche could be sufficient to produce a catastrophe."

*   *   *

Despite the ongoing objections, the first official NIH guidelines for the conduct of recombinant DNA research finally appeared in the *Federal Register,* giving them official weight, on June 23, 1976, nearly a year and half after Asilomar. Although the guidelines were stricter than the Asilomar conferees envisioned—and consequently outlawed some ongoing experiments, requiring scientists actually to destroy recombinant DNA molecules and years of work because their labs did not meet the required containment conditions—the official rules would end the moratorium and define the conditions under which the experiments could proceed.

There was also a certain genius to RAC's final pronouncement: They were guidelines, not formal regulations or law. This legal status would make it easier to change the guidelines as time went on, and—as scientists expected—the experiments would prove safe, allowing the strict standards to be relaxed. But the guidelines also had limitations: They applied only to institutions receiving grant money from the NIH, so private companies funded with their own, or investors', money could conduct whatever experiments they chose.

In any event, the guidelines provided molecular biologists with some breathing space. Research could continue—presumably safely. But the respite was short-lived. On the day the guidelines came out, a new commotion erupted in Cambridge, Massachusetts.

On June 23, 1976, the very day that the NIH guidelines were released, Mayor Alfred Vellucci rapped his gavel on the podium to open a formal hearing of the Cambridge City Council into Harvard University plans to build a gene engineering lab in the heart of the city. The lab—called a P3 lab because it is a secure facility designed to contain and isolate any organisms, dangerous or not, produced through recombinant DNA reseach—would be built on the fourth floor of the aging Harvard Biological Laboratories on Divinity Avenue. Initially designed to culture animal cells, P3 would also offer access to university gene splicers. Of the three rooms to be rebuilt, only one would get the special safety equipment needed to achieve P3 status—such as locked doors, special air flow systems, and laboratory hoods, designed to prevent engineered organisms from escaping. Only places like the U.S. Army's Fort Detrick in Maryland, the home of the military's biological weapons research where the most exotic and deadly germs could be safely handled, had the more secure facilities known as P4.

The Harvard controversy began because some scientists working in the university's bio lab worried that the genetic engineering studies might threaten them personally, and they wanted the lab project stopped. The opponents argued that the bio lab building was the worst place to put a P3 lab with its potentially dangerous genetic experiments. The old building, constructed in 1931, was in the center of a densely populated neighborhood, suffered frequent floods from broken pipes and electric failures that might help spread any engineered organism. It was intractably infested with pharaoh ants that might carry any P3-made organisms around the building and out into the community at large. It wasn't worth the risk, the opponents argued; build the P3 lab elsewhere.

"The proponents' answer, in essence, was that the risk was too slight to justify the inconvenience" of building elsewhere, wrote Nicholas Wade in *The Ultimate Experiment*.

Ruth Hubbard, a spry scientist who had left active research for reflection on the history and ethics of science, had been active in several political movements during the late 1960s and early 1970s. Now, in spring of 1976, she could see the political opposition within Harvard to the P3 facility was losing the battle. A key meeting would be held with Harvard administrators in May 1976 to debate the virtues and hazards of the new lab. Hubbard decided to call in reinforcements. She phoned Barbara Ackerman, former Cambridge mayor and then a member of the city council. Hubbard knew Ackerman from earlier Vietnam protests, and told her about the P3 lab and her concerns about it. Ackerman, once employed by a scientific publishing company, found the issue interesting, and agreed to come. For gene splicing proponents, however, the timing of late night television shows could not have been worse: The night before the hearing, Barbara Ackerman stayed up and watched the 1971 movie version of Michael Crichton's *The Andromeda Strain*.

For most of the three-hour meeting on May 28 in the Harvard Science Center, Ackerman simply listened to the debate. When Ackerman finally stood to speak, however, the Harvard deans knew their internal struggle was no longer private. The presence of Ackerman and a reporter from *The Boston Phoenix* ensured that the Harvard debate would get a larger audience.

When Cambridge Mayor Alfred Vellucci read an account of the meeting in a *Boston Phoenix* article entitled "Biohazards at Harvard," a week later, he went through the roof. The article, based on testimony from Harvard opponents during the hearing a week earlier, describes the exotic, though theoretical, dangers that faced his city, including infectious cancers and other new diseases that might spread to the surrounding neighborhoods. This, in Vellucci's mind, made the P3 lab a public health issue over which he had jurisdiction.

Less than a week after the *Phoenix* story, on June 13, 1976, Harvard announced that the P3 lab would be built. Ruth Hubbard had been preparing information packages about the lab and the potential hazards for Ackerman and the rest of the city council members. The day after the Harvard decision, she, along with her husband, George Wald, bundled off to the massive brownstone on Massachusetts Avenue that houses city hall. During the visit, Hubbard and Wald met with an angry mayor. They explained their fears and agreed that they would testify if asked. Vellucci, supported by a Nobel laureate, immediately decided that the Cambridge City Council would hold a public hearing on Harvard's P3 plans. "We want to be damned sure the people in Cambridge won't be affected by anything that would crawl out of that laboratory," Vellucci said when he announced the hearing.

The evening of June 23, 1976, found the city council chambers packed to overflowing. Floodlights from network TV crews lighted it up like Fenway Park. Signs like "No Recombination Without Representation" festooned the room. Reporters from big-league newspapers across the country squirmed for space so they could cover a local city council meeting with national implications. Whatever happened in Cambridge could affect local governments everywhere.

In a circus atmosphere, Vellucci held forth, alternately asking questions of the gathered experts and then berating them into silence for past offenses. By the end of the hearing, Vellucci proposed a resolution insisting that "no experimentation involving recombinant DNA should be done within the City of Cambridge for at least two years. . . . " The resolution brought Harvard biologist Mark Ptashne to his feet, demanding that Vellucci clarify the resolution because it would immediately shut down all of the biochemistry and half of the biology department, including experiments not considered dangerous by anyone.

Vellucci's resolution did not prevail, and after a second hearing on July 7, Cambridge City Council voted 5 to 4 to suspend P3 and P4 recombinant research within its borders for three months—the moratorium actually lasted seven—while a specially elected committee, composed principally of lay people and a physician but no scientists, evaluated the dangers of recombinant DNA.

To many, the Cambridge Experimentation Review Board, headed by former mayor and heating oil businessman Daniel Hayes, was a remarkable success of democracy at work. It held seventy-five hours' worth of meetings, enough for the lay members to actually understand the issues, and in the end came to a fairly reasonable conclusion: The P3 recombinant work would be permitted under the NIH guidelines, with a few additional precautions, such as extra monitoring for escaping engineered organisms. Experiments requiring P4 containment, which no one had proposed, were banned outright. The board made its recommendation on January 5, 1977, and it was accepted, with minor modifications, at the city council's February meeting.

The resolution, however, came too late for at least one gene engineer: Thomas Maniatis, a gene researcher, abandoned Harvard for the relative security of Cold Spring Harbor Laboratory, and then fled to the West Coast, to the California Institute of Technology, where he would later be the first to discover the gene for beta hemaglobin using a gene-engineering technique called "shotgun cloning."

The Cambridge decision unleashed the scientific community's worst nightmare: a flurry of action by local governments from coast to coast. With the media spotlight focused on Cambridge, other states and communities began to rethink recombinant research in their own backyards. Could the scientists be trusted? Should the work be regulated? The NIH guidelines clearly did not cover everyone: Private companies using their own money, from the fledgling California biotech company called Genentech, which had been founded in April 1976 and was reportedly trying to synthesize and clone the human insulin gene, to General Electric, were exempt from the federal rules.

Recombinant concerns spread like wildfire during 1976 among local municipalities. Cities across the United States set up committees and held sometimes raucous public debates about the wisdom of genetic engineering and how it should be regulated. Some of those involved included San Diego, California, Madison, Wisconsin,

Princeton, New Jersey, Bloomington, Indiana, and Ann Arbor, Michigan. State governments in New York, New Jersey, California, and Maryland also held public hearings and struggled with the regulation of genetic research. Maryland passed the first state law requiring all scientists within the state, as well as private companies, to comply with the NIH guidelines. In short order, scientific leaders faced a crazy quilt of differing rules regulating genetic research around the country.

Not only did local communities begin to take notice but so did Congress. On September 22, 1976, shortly after the Cambridge hearings and partly stimulated by a provocative article in *The New York Times Magazine* about potential global destruction crawling out of gene engineering laboratories, Sen. Edward Kennedy, chairman of the Senate health subcommittee, held a congressional hearing on recombinant DNA. He opened the hearing with an admonition: "The debate over genetic engineering must go on. Scientists must tell us what they are capable of doing, but we, as members of society, must decide how it should be, or whether it should be applied." Kennedy did not immediately follow up with proposed legislation as expected, but others did in 1976 in both the House of Representatives and the Senate.

The situation did not get any better in 1977. A three-day symposium at the National Academy of Sciences in Washington, D.C., nearly turned into a melee when protesters from the Coalition for Responsible Genetic Research demanded that genetic engineering studies be shut down, signing "We Shall Not Be Cloned," and carrying a banner quoting Adolf Hitler that said, "We will create the perfect race." George Wald and Ruth Hubbard, along with Nobel laureate Macfarlane Burnett, and protesters from the Friends of the Earth, an environmental group, showed up, objecting to recombinant DNA research.

The public brouhaha, well covered by the major media, propelled Congress into action. The technical staffers from several House and Senate committees were given marching orders to figure out what to do. Hearings were organized in the spring of 1977. Study groups of congressional staffers were organized to draft legislation that would be introduced in time.

Fearing the worst, the new NIH director, Donald S. Fredrickson, who had taken over NIH in July 1975 as the crisis was heating up, organized a committee within the federal government that included members from every federal agency that had any interest in recombinant DNA, from the National Science Foundation, which funded many of the scientists, to the Environmental Protection Agency, which would have to worry about any environmental disasters caused by microbial renegades crawling out of recombinant test tubes. After judicious posturing and sometimes internecine turf wars, the group generated its own proposed legislation for Congress to enact as a preemptive strike against something worse. The Administration's bill was duly introduced and immediately ignored because of its flaws. States' rights would become one of the central issues in gene engineering legislation: Should Congress enact a law that preempted a state or city's right to impose stricter rules on genetic research?

The scientific leadership had mixed feelings about the impending national legislation. On one hand, institutions like Harvard and MIT lobbied hard for legislation that would level the playing field by eliminating the different rules that governed research in different regions of the country. What's more, such legislation would also regulate private companies, not just federal grantees. On the down side, legislation would be more rigid and more difficult to change as knowledge gained from genetic engineering continued to expand at a rapid rate. And for the scientists, the prospect of laws governing science seemed to signify the "intrusion of Big Brother into the sanctity of the laboratory."

On April 1, 1977, Senator Kennedy finally introduced his long-awaited legislation. S. 1217 sent chills down the backs of scientists who bothered to read it. Patterned after the Atomic Energy Commission Act of 1946—with "recombinant DNA" substituted where "atomic energy" appeared in the earlier act—Kennedy's bill would have set up a national commission to regulate all recombinant DNA activity in the United States. Kennedy appeared to consider genetic engineering as dangerous as radioactive material. The provisions so frightened the scientific community that groups of scientists, and some scientific organizations, launched an intense lobbying attack that rapidly destroyed what little support Kennedy's bill enjoyed. Kennedy eventually withdrew the proposal.

Although all this activity on Capitol Hill scared the scientific community, most members of Congress were uninterested. "The recombinant DNA controversy in mid-1977, the time when it peaked, was still a minor issue in Congress," wrote Burke K. Zimmerman, for-

mer science advisory to Rep. Paul Rogers of Florida, then the powerful chairman of the House health and the environment subcommittee.
"In order for an issue to rank with the top political issues of the day,
it must have immediate consequences, it must affect some large or at
least powerful segment of the population, and it generally must deal
with one of the following: taxes, employment, war, inflation, gaso-
line, or some other issue touching either the pocketbooks or the hearts
(and in a contest between the two, the pocketbooks) of a significant
portion of a lawmaker's constituents. But among those who were concerned [with recombinant DNA], the battle raged and emotions ran
high."

NIH director Don Fredrickson was one of those who cared deeply.
Recombinant DNA now dominated nearly half his time. In addition
to appointing the official Recombinant DNA Advisory Committee to
follow the government's administrative procedures, Fredrickson assembled a kitchen RAC, genetics experts from across the NIH's 300-
acre campus, to give him up-to-the-minute advice on techniques and
political strategy. But the director also reached beyond NIH to a lifelong network of powerful friends and associates the urbane Fredrickson had cultivated. Among them was Congressman Olin "Tiger"
Teague, then chairman of the House Science and Technology Committee, through which the recombinant regulations would have to
pass. One night, in the midst of the 1977 legislative crisis over genetic engineering, Tiger Teague, who had become ill himself, summoned
Fredrickson to his Naval Medical Center bed in Bethesda, just across
Wisconsin Avenue from the NIH. Teague quizzed Fredrickson for an
hour and a half about the risks of recombinant DNA. After he left,
Teague decided to claim jurisdiction over the issue; all bills introduced into the House would indeed have to go through his committee
where he gave them all a long and thorough hearing. And from which
they never emerged.

Several bills were introduced and fought over in 1977 and 1978. After being burned by S. 1217, Kennedy decided to await House action
and offered to introduce a reasonable companion measure in the Senate. Although one bill nearly made it through the House, by the end
of 1978, all of them died when the 95th Congress expired. A few
furtive efforts would be made to revive the issue and some legislation, but generally the steam had gone out of it. As more and more
genetic engineering experiments were performed without the end of
the world occurring, the horrific scenarios woven by recombinant op-

ponents seemed to recede more and more into the theoretical. Even the radical left had to agree that the risk of putting new genes into bacteria seemed to grow vanishingly small.

The technology would continue to raise profound questions for which society would have to find answers. A scientist at General Electric, for example, would file a patent on a strain of bacteria he "created," though not with gene engineering techniques. The ensuing legal and philosophical battle reached the U.S. Supreme Court, which concluded that living organisms indeed could be patented. In the early 1980s, a University of California at Berkeley scientist developed a strain of bacteria that prevented the formation of frost on potato plants—a valuable advance for farmers. The scientist wanted to spray the genetically engineered bacteria—dubbed Ice-Minus—on a test patch of potatoes in rural northern California, but a storm of criticism enveloped the project. Environmentalists challenged the safety of intentionally releasing gene-altered organisms (which the 1970s debates fought so hard to keep in the lab) into the environment. A turf war among federal agencies like the EPA and the U.S. Department of Agriculture, not to mention the RAC, over who would regulate the intentional release of genetically engineered organisms, held up the experiment for years. Eventually, in the mid-1980s, the experiment proceeded.

What's more, behind the 1970s arguments about putting genes into bacteria hovered the notion that someday scientists would want to put genes into people. Few researchers, even in the late 1970s, could envision how doctors would put genes into people, so the issue itself mostly remained theoretical. The problems debated during the recombinant DNA wars, however, would be rehashed in the 1980s—and in many of the same forums, such as the RAC—as genetic engineering of people moved toward reality.

Even as the recombinant DNA debates laid the foundation for the social mechanisms eventually used to resolve the controversies soon to engulf human gene therapy, the gene engineering technologies developed in the early and mid-1970s were leading to the development techniques needed for injecting genes into eukaryotic cells, like those found in mammals, such as humans.

And as for Paul Berg, the Stanford scientist who in some ways started the debates, he managed to survive the disputes even as he helped influence them. Futhermore, he managed to keep his hand in the lab, contributing to the developing gene-handling technologies enough for the Nobel Committee to award him the 1980 Nobel in chemistry.

Meanwhile, French Anderson began his own campaign to find ways to put genes into cells, not just to conduct basic research but to find a way to make the new technology useful for treating human diseases.

Chapter 5
Putting Genes
in Mammals

"Biologists have long attempted by chemical means to induce in higher organisms predictable and specific changes which thereafter could be transmitted as hereditary characters."
—*Oswald Avery, 1944*

It is not easy to put genes into cells. In the more than 3 billion years that living cells have existed on Earth, a growing number of defenses have evolved to protect cells from stray DNA. The reason is simple: If information is power, then the data stored in DNA can be life-changing. Experiments in the early 1970s showed that isolating all the DNA from adenovirus, and then sticking only that DNA, the whole amount, into a cell could produce infection-competent adenovirus particles. Conceivably, absorbing a single gene in a stable way could change forever a cell's form and function.

To keep that from happening, cells acquired sophisticated methods to keep out, or destroy, random bits of DNA. The simplest barrier is the plasma membrane, the boundary that separates the cell's insides from the outside world. It's made from an outer envelope of fatty molecules that form a bi-layer of water-repelling compounds. Proteins poke through the fatty membrane to provide carefully regulated access for some chemicals. The proteins floating in the fatty membrane open pores, or channels, for water and other small atoms to enter. The long, charged molecules of DNA cannot easily pass through the pores or the membrane to reach the gooey cytoplasm within the cell.

Still, if the membrane that demarcates a cell's boundaries were totally rigid—like the wall of a castle—then nothing could enter, and the cell—or town—would die of starvation. Various mechanisms exist to allow cells to take in the raw nutrients and building chemical blocks—including DNA—that it needs to survive. With one mechanism, called endocytosis, the cell wall bulges inward, like a finger poked into the side of a toy balloon, and surrounds some juicy collection of chemicals. Once the cellular meal is surrounded, a bit of the plasma membrane pinches off, forming a bubble that can float in the cell's cytoplasm. The bubble then merges with other structures that digest the contents.

Even if unwanted DNA manages to cross the membrane hurdle, there are plenty of other destructive cellular defenses, including the grinding, chopping, shredding actions of enzymes designed to destroy any intact, incoming genes. In their native state, these enzymes, the restriction endonucleases, prevent foreign DNA from finding a safe home in the cytoplasm. Once isolated from the bacteria, the purified enzymes gave scientists a powerful molecular scissors for cutting up DNA fibers in a precise way.

These defenses can be found in both the most primitive cells, the prokaryotes such as bacteria, and the evolutionarily more advanced eukaryotes. As nature progressed, the eukaryotes developed additional armaments. The most obvious is a second membrane, the nucleus, which separates the chromosomes from the rest of the cells. Even if DNA penetrates the cell and survives the enzymatic attack, it must still pass through the nuclear membrane. And even if it gets through, for the DNA to survive for a long time in the eukaryotic cell, it must somehow be spliced into the chromosomes within the nucleus, becoming part of the cell's native DNA. So many barriers all standing in the way of foreign DNA—and of the scientists who wanted to get genes into cells to see what they would do and how they could work.

Oswald Avery and his colleagues at Rockefeller University pointed the way. Their successful transformation experiments proved in 1944 that DNA carried the genetic code. They also showed that it was possible to get bacteria to soak up functional bits of DNA that could change the characteristics of the cell. In the 1970s, the first genetic engineers showed they could easily insert their newly isolated bits of DNA into bacterial plasmids, and smuggle the plasmids back into bacteria, just as Avery had done with whole DNA. This approach propelled the development of biotechnology, and led to the first genetically engineered drug, insulin, in the late 1970s.

But this only worked in the test tube—and only in simple bacteria and later yeast. If researchers were really going to pursue a gene treatment for human ills, they would have to find ways to get DNA into the mammalian cells, the eukaryotes with the multiple layers of defense. And not just in mammalian cells, but mammalian cells in mammalian bodies where there were many other defenses outside the cells—in the gut and the bloodstream—to prevent DNA from surviving long.

• • •

In 1974, when French Anderson decided to abandon his growing empire in clinical hematology, the work on the Cooley's anemia program was a success. His lab now had nearly two dozen thalassemics in its program, mostly under the care of Art Nienhuis. As Anderson watched the genetic engineering revolution from the sidelines, he knew he had to get back in the lab and back in the hunt for a technology that would allow him to transfer genes into his patients, people like Nick and Judy Lambis. If he didn't get back into the lab and start learning genetic engineering, he would never be in a position to help launch the field of human gene therapy.

As he turned over more and more responsibility to other people in the lab, Anderson withdrew to begin contemplating what to do next. "I realized I had to make major new approaches," Anderson recalled. "So I went back to Cold Spring Harbor to take a course in tissue culture and recombinant DNA." The kind of skills that in the late 1960s could only be found in Dulbecco's Salk laboratory had become widespread. No longer would scientists like Paul Berg have to make sabbatical pilgrimages to some scientific Mecca to learn the new techniques. Places like Cold Spring Harbor now handed out prepackaged training courses, complete with hands-on experience and how-to cookbooks.

The Cold Spring Harbor course marked a transition point for Anderson: "I started to do gene therapy very actively."

Well, not exactly gene therapy. He—like so many other scientists at the time—began looking for a way to put genes into mammalian, and ultimately human, cells. Anderson picked up the first likely cellular model to test his emerging ideas about gene transfers: red cell ghosts. The body teems with red blood cells. They carry oxygen from the lungs to the rest of the body and return the spent carbon dioxide for exhalation. Of all the cells in the body, red blood cells are unique:

They have dumped all their chromosomes, an evolutionary lightening of their load. They even jettison the nucleus. At their most mature state, red blood cells are basically bags of hemoglobin with little left in terms of molecular machinery, enzymes, or nearly anything else except for the oxygen-carrying red pigment.

These characteristics made red blood cells an ideal target for inserting genes. If Anderson could find a way to introduce genes into these cells, then perhaps they could be used to ferry genes into the human body. To heighten the chances of success, Anderson and others working on these cells went even further to make room: They bled the red cells of their hemoglobin. All that remained was a colorless membrane that held the classic concave shape of a red blood cell. They were red cell ghosts.

Even if the gene insertion techniques that had worked with bacteria proved successful with red cell ghosts, it seemed unlikely that gene-charged ghosts would be of much use. After all, there was nothing left inside them to carry out the gene's instructions. All the protein making machinery—the ribosomes, the amino acids, transfer RNAs, and enzymes—had been sucked out of the cell. Worse, the eviscerated the red cell ghosts would not take up all that much DNA.

Anderson abandoned red cells as useless and turned to liposomes. Made from the same fatty material as the plasma membranes of cells, liposomes are like soap bubbles. Where the hollow red cell ghosts would have carried DNA only in the body, liposomes held the promise of smuggling genetic material actually into human cells. The approach would work like this: DNA would be loaded into liposomes. Then the liposome would be fused with the membranes of a cell—like two soap bubbles merging. As the cellular soap bubbles fused, the contents of the liposome would be dumped into the cell's cytoplasm. The approach would get the genes past the first barrier—the plasma membrane—and into the cell's cytoplasm. The genes still would have to survive enzymatic attacks inside the cell, and then migrate to the nucleus and become part of the chromosomes. A lot of ifs. Still, it would be a first step in the right direction if it worked.

It didn't. Liposome technology was too new. The genes could not be packaged in functional ways. The approach just was not working. Anderson needed something entirely different, completely new. That's when he heard about Elaine G. Diacumakos, who had just become the head of the laboratory of cytobiology at Rockefeller University in 1976.

She had studied biology at the University of Maryland, graduating in 1951, and earned a Ph.D. from New York University in 1958. A special government fellowship allowed her to move to Rockefeller University in 1962 to work in the laboratory of Edward L. Tatum, one of two researchers who won a Nobel for proving that one gene carries the information needed to make one enzyme. At least, that is how George Beadle and Ed Tatum phrased their discovery. Later studies showed that one gene specifies one polypeptide, a string of amino acids. Many proteins are made from a single polypeptide; others are made from two, three, or four polypeptides working together. Hemoglobin, for example, is made from four polypeptides, two alpha globin and two beta globin polypeptides. Alpha globin is coded for by one gene, beta globin by another.

After the fellowship expired, Diacumakos stayed in Tatum's Rockefeller lab. She began developing techniques to micromanipulate cells by probing them with microscopically thin glass needles that could penetrate a cell without destroying it. Once she had her needle inside the cell, she could either inject material or extract a cellular structure. All the manipulations were monitored under the microscope. She started in the late 1960s with Tatum, in part to test whether the organs of the cell itself, called organelles—such as mitochondria, the cell's power plant—carried any genetic material that affected the cell's inheritance.

When Tatum died in 1975, Diacumakos was orphaned, said former Rockefeller president Joshua Lederberg. "Tatum had managed the lab, and arranged financial support. He died, and she kept the lab." But times were tough for a woman at an elite, mostly male research institution, and grant money was hard to come by. Lederberg remembers her as "someone deeply dedicated to her scientific work. She was preoccupied with it. I think she was a good scientist. She hit on a technique that seemed to require extraordinary skill, rather than having a larger theoretical conception. I tried to get people interested in it, and tried to get support from commercial companies." The companies did not respond.

Others remember her as a reclusive and reticent woman who mostly worked alone in an area of science with almost no competitors. Only Adolf Graessmann of the Free University of Berlin developed a similar microinjection technique using a different kind of microscope that made it easier to inject the cells. Graessmann's approach proved simpler than Diacumakos's, and ultimately, he could inject many more cells far faster than she could.

Without much funding during the 1970s, mostly working alone, Diacumakos concentrated on honing her skills, using fine glass needles to either inject or extract what she wanted to manipulate inside the cells. She even reportedly became proficient at extracting organelles, including intact chromosomes, with the microscopic needles.

• • •

The entire field of microinjection got started when several researchers discovered that the plasma membrane of cells was self-healing. Where a balloon bursts when the bubble is pricked, the fatty membrane around the cell opens when a needle is pushed through, and closes when the needle is withdrawn. In 1960, John B. Gurdon, a zoologist at Oxford, in England, showed that microinjection could be used to learn about genetics and cell biology.

Gurdon started by injecting whole, fully differentiated nuclei from intestinal cells of a tadpole into denucleated frog eggs. To most biologists' surprise, the hybrid eggs developed into normal frogs, and they were genetically identical to each other. They were clones carrying the identical genome from the tadpole that donated the intestinal cells. The experiment proved that every cell in the body contains all the genes needed to make a complete organism. This suggested that genes were turned on and off in the different types of cells in the body to give different tissues their unique characteristics. Previous theories suggested that genes were lost in different tissues: Heart cell lost all the genes to be liver cells, for example. Later, Gurdon used microinjection to introduce many other biologically active molecules into frog's eggs, including purified genes and messenger RNA.

Compared to human cells, however, frog eggs are a massive target, the biological broad side of a barn. They are nearly a millimeter in diameter, visible to the naked eye. What's more, the volume of a frog egg is nearly 100,000 times larger than the volume of a mammalian cell. Diacumakos worked with mammalian cells.

First, she had to make her own equipment. The needle used for microinjection is made from a glass capillary tube, a glass rod really, that is rolled over the fire of a micro Bunsen burner, and slowly bent into a Z-shape. One end of the Z-shaped tube was connected by platinum tubing to a micrometer, a device that applied pressure to a column of silicone oil inside the tubing in a very controlled way. The micrometer determined how much material was pumped into the cell and how fast it flowed.

On the other end of the Z-shaped tube, which was called the pipette, Diacumakos attached the glass needle. It was made by dabbing a glob of molten glass on the end of a pipette and extruding strands of glass that grew thinner and thinner. The glass ends created an extremely fine glass skewer, thinner than hair. In scientific terms, its outer diameter was only one micrometer, one thousandth of a millimeter; the inner diameter was half that.

Such a fine tool couldn't be held in the hand and moved with any hope of hitting a microscopic target. There would be too little control. Under the magnification needed to match up needle to cell, a hand-held injector would appear to gyrate wildly with each tiny movement of the fingers, heightening the chance that the cell would be punctured or shredded in the process. The Z-shaped tube, with the glass needle attached, was held in a device called a micromanipulator that would move it slowly and precisely across the stage of a microscope until the glass spear lightly touched, and then punctured, the cell. All of this was overseen by looking through the microscope, or via a video camera that looked into a port on the microscope and was attached to a television monitor.

To hold the cells in place, Diacumakos developed a complicated system in which she stuck the cells to a microscope slide cover slip. Then she inverted the cover slip on four pieces of glass so the cover slip would not lie flat on the microscope slide. The cells to be injected (she started with mouse fibroblasts, a skin cell) hung down from the bottom of the cover slip but were well above the microscope slide. This kept the cells accessible to the fine needles. Then, she sealed the edges between the cover slip and the slide with silicone oil to form a chamber that could be filled with culture media. The culture media nourished and protected the fibroblasts.

The entire assemblage was placed on the stage of her phase-contrast microscope at 2,000-fold magnification. Using the micromanipulator, she would then maneuver the glass needle in from the side, through the silicone oil dam, through the culture media, and spear the cell hanging down from the cover slip. She could then either inject the cell or suck out whatever cellular component she was interested in. Over the early years of the 1970s, Diacumakos demonstrated that she could successfully deposit microscopic amounts of inert material—like silicone droplets—into virtually any part of the cell, including the nucleus, which holds the chromosomes and mitochondria. But Diacumakos was not injecting biologically active material—such as genes or chromosomes—though she now clearly had the ability to do that.

. . .

It was around 1976 when Anderson first heard about Elaine Diacumakos. "There was this crazy woman up at Rockefeller," Anderson recalled. "She claimed that she could microinject protein molecules not just in the nucleus, but in the nucleolus [a spherical structure filled with proteins and RNAs inside the nucleus]. And that she could inject mitochondria, and she could pull out individual chromosomes, and so on. Everyone I talked to said she was crazy." Still, Anderson was intrigued. "It made sense to me. If you could microinject genes, it might work. I called her up and said, 'People tell me it can't be done.' She said, 'Why don't you come and see for yourself?' So I said, 'All right, I will.' I went up the next day."

Anderson found Diacumakos, as she often was, alone in her tiny lab with her microscope and cells. "She was a nice little lady, a pure scientist. All she did was sit in front of the microscope all day, and play with her chromosomes and things. No one took her seriously," Anderson recalled of the initial visit. "She went in and microinjected a microdrop of silicone [oil] right into the nucleus. It absolutely amazed me. She was doing it, but nobody knew or cared."

Anderson cared. The technique developed by Diacumakos clearly could deliver material into the nucleus of cells, and that meant it could deliver genes into cells. With the recombinant DNA revolution in full swing, it would only be a matter of time before a large number of genes became available for genetic therapy. With microinjection, Anderson might be able to solve the problem of crossing the evolutionary barriers and get genes into mammalian cells. If the approach worked, he would be able to put the new genes into defective cells and cure them. He immediately went back to Bethesda and converted a table top in the back of his long, narrow office into a microinjection platform.

The NIH engineers helped him assemble the equipment he needed to make the needles, but they did their own calculations and said it wouldn't work. The sheer forces caused by pressing the long DNA fibers through the half-micron orifice was sure to rip up the genes into useless bits of nucleotides. James Watson would later say the same thing about attempts by a former student, Mario Capecchi, when he independently began using microinjection around the same time.

Anderson also ran into skepticism from those in his own lab. Art Nienhuis, his close scientific colleague to whom Anderson had given more than half the branch, called microinjection Anderson's toy. Among many biologists, the importance of microinjection was never really accepted. And neither was Diacumakos. "She was an artist," Anderson said. "But the molecular biologists said she was a mechanic, not a biologist. I aligned with her, and people put me in the same category. I was not helped by the fact that Art Nienhuis and the others thought it was a waste of time, that I was just playing. My reputation was not enhanced when the people in my own lab thought it was stupid."

Stupid or not, Anderson decided to pursue microinjection. Over the next several years, he—and later members of his lab—would become experts at extruding glass to make the fine needles needed for gene transfers. He set up his microscope and inverted sets of glass slides with silicone oil seals to make media chambers for cells that they would genetically transform with injected genes.

"So I would play back here," Anderson said. "Elaine would come up here once a month or twice a month, and I would sit here for hours and hours and break pipettes. Everything had to be made by hand—extruding the glass pipettes, everything. It took months and months and months. I finally got to where I could inject a silicone oil drop."

Now that he was getting the technique down, Anderson began to assemble the other components of the experiment he had formulated. He had begun recruiting postdoctoral researchers again, bringing in those who already had begun learning the new techniques of genetic engineering. At the same time, a number of potentially useful genes was being isolated that Anderson thought might be worth injecting into cells. At least, these genes might allow him to prove that the concept of microinjection was valid and potentially useful for gene therapy.

The first was a gene close to Anderson's heart: the beta globin gene, the one that makes half of a mature hemoglobin molecule. Defective copies of beta globin caused the thalassemia that hobbled Nick and Judy Lambis. Getting the hemoglobin genes in hand, and finding a way to inject them into patients might make a difference in the lives of Anderson's first two patients.

Human beta globin was first isolated by Thomas P. Maniatis, a molecular biologist who fled Harvard University when the Cambridge City Council hearings and moratorium shut down all gene engineering research at the university. In 1976, he escaped Cambridge for Cold Spring Harbor to complete an attempt to isolate rabbit hemoglobin. He used a technique in which messenger RNA was isolated from rabbit blood cells, and then converted into DNA, using the reverse transcriptase enzymes Howard Temin and David Baltimore discovered a few years earlier. Since RNA contains all the coding information needed to make the protein, simply converting that information back into DNA gives the bare essentials of the gene.

Once Maniatis had the rabbit beta globin gene in the DNA form, he began using it in an attempt to understand how the gene is regulated, and to determine whether it could be inserted back into cells. But by then, however, Maniatis had fled west to the California Institute of Technology, in Pasadena, because New York State was considering a gene engineering ban similar to the Cambridge moratorium. Jim Watson at Cold Spring Harbor and others were instrumental in blocking the New York action, so it never went into effect.

Once settled in California, Maniatis began to develop an approach for cloning DNA in animal and human cells. It relied on a technique called shotgun cloning, in which all of the DNA from an organism is isolated, diced up into small pieces, and packaged into bacteriophage. As the viruses go through their replication cycles in bacteria, they reproduce the passenger DNA.

If enough pieces of DNA make it into viruses, it's possible to produce a library of DNA fragments that represents nearly all of the DNA in an organism. Basically, it works like this: All the genes within a batch of identical cells are isolated and then chopped up into pieces by enzymes, producing millions of fragments of DNA. It's like ripping all the pages out of a book of blueprints and scattering them around a room. Some of the pages get torn, some shredded, some lost completely. But many pages come out intact, containing the exact instructions for constructing a building, or in this case a gene, a protein like beta globin.

Next, all of the DNA fragments are randomly stuffed in a small bacterial virus called lambda, which can manufacture millions of identical copies of a DNA fragment inside its host bacteria. It's somewhat like randomly collecting all the scattered blueprint pages—the torn and the whole—and sticking them in thousands of photocopying machines.

"You need to make 1 million independent clones of twenty kilobases to get a library [that represented all of the genes]," Maniatis said. "We managed to overcome those problems. The first library that we made was the rabbit library, which we then screened with the rabbit cDNA."

Next, Maniatis said, "We made the first human genetic library, and screened that with human beta globin cDNA, and found that there was a cluster of [globin] genes, and they were linked."

Maniatis was a basic scientist trying to understand how hemoglobin production is regulated. Now that he had clones of the beta globin gene, he could begin taking it and its regulatory structures apart to understand how they worked together to control so precisely the expression of alpha and beta globin to make intact hemoglobin. In addition, he began using his collection of clones to screen humans with various blood diseases, particularly thalassemia, to see if he could sort out which forms of the illness were caused by defects in the hemoglobin's structural genes that actually made the proteins, and which were caused by defective gene regulation elements.

More important for other scientists around the country, Maniatis believed in one of the old tenets of the Phage Group: Once a scientist makes a discovery and publishes his work, he must make his materials—reagents, viruses, bacteria—available to other workers in the field so they can confirm and advance his discoveries.

"The policy in my lab has always been that we provide everything we publish to everyone who asks for it," Maniatis said. "Many times we provided it to competitors who were doing exactly the same experiments."

For Anderson, this was a godsend. Here was one of the key human genes he had always been interested in: beta globin. It was neatly cloned and packaged so he could manipulate it in the laboratory. With microinjection coming along, it might be possible to use the tiny glass needles to inject the beta globin genes into diseased cells that lacked the gene. But first he had to prove gene transfer by microinjection was even possible.

All of the gene transfer techniques under development shared a common problem: It was hard to tell if the gene went into the cell. In the 1960s, Ted Friedmann and others at NIH had tried to use HAT medium to select, or identify, cells that had been transformed by bulk DNA carrying the HGPRT gene. The cells lacked the gene, and consequently could not live in HAT medium. If any of the normal

HGPRT genes got into the cells, they would survive. The experiment didn't work because none of the normal HGPRT genes got in, but the principle was sound.

Gene engineers working with bacteria used the same concept of selection to identify transformed bacterial cells. Scientists would insert a gene that made the bacteria resistant to a particular antibiotic, and then grow the genetically transformed cell in the presence of antibiotics. Only those that had taken up the new genes could survive the lethal drugs.

Selection became a running theme throughout the field of genetic engineering: Find a method for inserting a gene. Test it by transferring a gene that gave the cell a unique property—such as antibiotic resistance. Then, grow the cells in the presence of the antibiotic, and only those cells that had been genetically altered would survive.

Anderson, and others already in the race to find ways to transfer genes to mammalian cells, decided to go back to HAT medium, but with a twist. Instead of trying to insert the HGPRT gene as Friedmann had tried, he would use a unique gene found in the herpes simplex virus. First of all, the HGPRT gene had not yet been isolated, so Anderson could not inject it into a cell. Second, researchers at the National Cancer Institute and Yale University had recently isolated a DNA fragment containing a unique herpes gene, the thymidine kinase gene, or TK. The enzyme is essential in the biochemical pathways used for making new molecules of DNA. If a cell lacks TK, it too cannot grow in HAT medium, just as if it lacked HGPRT.

Mutant cells, fibroblasts from a mouse that lacked the TK gene, had been found in 1963. The cells needed to be grown in a special medium to keep them alive. What's more, these cells, called L minus or L$^-$ cells, could not grow in HAT media. This made for a simple, elegant experiment.

Anderson fixed the mouse L$^-$ cells on a cover slip and placed them in the growth media chamber made by turning over the cover slip and sealing the sides with silicone gel. Once they were on the microscope stage, Anderson would microinject the cells with the thymidine kinase DNA, just as Diacumakos had taught him. The microscopic needles were made by hand, connected to the micrometer for pressure control, and loaded on the micromanipulator for precision movement. The tubes were filled with silicone oil, except for the very tip of the needle, which he loaded with the TK gene.

After the cells were injected, the cover slip was taken off the microscope and the cells placed in a growth medium for a short period to nourish those that survived microinjection. Then, the cells were placed in HAT medium to kill off any cells that had not been genetically transformed. Cells that failed to take up the TK gene that protected them from the toxins in HAT, died. Those cells that survived had taken up the TK gene.

"It was a feeling of absolute joy to have the cells survive after you put in the gene," said Lillian Dean-Killos, who was a graduate student in Anderson's lab and who became one of the expert microinjectors in the experiment. She also knew the tedium and difficulty of the technique. "You learn how to do it. You explode a cell or two [by injecting too much material], or someone could distract you, or you bump something and break off the end of the pipette. It was painstaking work."

When microinjection succeeded, however, the results were dramatic. And the TK gene made all the difference. It allowed the cells to live where they should have died. The experiments proved the NIH team could get the gene into the TK-deficient L$^-$ cells and genetically heal them. Additional studies proved that it was the herpes TK gene—not a mutation that reactivated the cell's own nonfunctioning TK gene—that had restored the cells to normal.

Once it was clear he could inject a functional gene and cure it so that it could live in HAT, Anderson wanted to take the next step and show he could microinject a gene that might make a difference in human disease: the beta globin gene Maniatis had isolated.

Anderson and his team repeated the steps, except that they added both the TK and the beta globin genes. When Anderson and the others looked for signs of the beta globin gene, however, they found only disappointment. There were clear signs that the beta globin gene had gone into the L$^-$ cells, but something was wrong. They could find no evidence that the mouse cells were producing the human beta globin protein.

Normally fibroblasts do not make beta globin, so any production of the molecule would have come from the inserted gene. Members of Anderson's lab sorted through the messenger RNA molecules in the L$^-$ cells looking for signs of beta globin. They found an extremely low level of human globin mRNA: two to ten molecules per cell. Clearly the globin gene had gone into the cell, but something was preventing the protein from coming out of the cell's synthesizing machinery.

The explanation came some years later from Tom Maniatis. Although he had successfully cloned the structural portion of the human beta globin, he had not yet identified the regulatory sequences—the genetic on-and-off switches—that turned on globin mRNA production. And neither had anyone else. Simply putting the structural gene sequences into the cell was not enough for the enzymes that read the information and transcribed the DNA code into RNA.

Hemoglobin is among the most complex, and best studied, sets of genes in the body. To make a mature hemoglobin molecule, one molecule of alpha globin must be produced for every molecule of beta globin. Thus, the two genes must be turned on and turned off at the same time. If there is an excessive production of one or the other globin genes, then a disease like thalassemia results. This genetic two-step ensures that the regulation of globin is enormously complex. Simply putting in the structural portion of the gene could never be enough.

The answer was tough to find because "the expression of globin genes involves sequences that are located more than sixty kilobases away from the gene," Maniatis said. That's a gigantic distance in genetic terms considering that the regulatory sequences of the *lac* operon, for example, virtually abut the gene itself. "That problem [of controlling hemoglobin gene expression] still has not been resolved," he added nearly a decade and a half later. "That is why treating thalassemia with gene therapy is not on the horizon."

• • •

Neither Anderson nor his associates understood the problem then. They were still hoping to turn microinjection into a technique that could lead to a viable treatment for thalassemia patients like Nick and Judy Lambis, who were not doing as well in the latter part of the 1970s as they had in the beginning of the decade. Desferal treatments were helping to remove the iron overload from their bodies, but the two had been receiving blood transfusions for so long that they had accumulated enough iron to turn their skin bronze, and make their hearts beat with irregular rhythms. Nick was working at NIH as a part-time personnel clerk, but his duties were frequently interrupted by treatments to keep his condition under control.

Anderson struggled with another problem: Even if he could find a way to increase the productivity of the beta globin gene placed in cells by microinjection, it was not clear how he could use the technique to put globin genes into enough cells to make a difference biologically. To cure the Lambises, and thalassemics like them, he would have to inject millions, perhaps billions of cells. There was no physical way to do that using the Diacumakos approach. That would take a production line of microinjectors organized in teams to fix the cells to the cover slips, mount them on the slides, place them on the microscopes, microinject them, and grow out the clones that produced the right amounts of beta globin.

Worse, there was no way to get the right cells out of a patient's body to mount on the microscope stage for the microinjection. For the children to be cured, Anderson would have to inject the gene into special cells found in the bone marrow that give rise to all other blood cells in the body. These immortal cells, called stem cells, had not yet been identified. They were known to exist and were essential for the success of bone marrow transplants. But they make up a very small portion of the bone marrow—much less than 1 percent of the cells. They can't be seen directly because they are identical in shape and size to the ordinary mortal cells, which already had begun to differentiate into white blood cells or red blood cells. Inserting genes into these differentiated bone marrow cells was useless because the gene would be lost when the cell finally matured and died. Only the stem cells would provide a permanent cure. Anderson had no way to get at them.

Undaunted, Anderson presented the microinjection results with the beta globin gene at a symposium sponsored by Chicago's Comprehensive Sickle Cell Center in October 1979: His team had managed to insert the TK gene and the beta globin gene. They had gotten production of the TK enzyme, but not the beta globin gene. The announcement caused a sensation at the meeting.

Other scientists working in the field were less dazzled about the advent of microinjection. They could see the limitations as clearly as could Anderson. It was a nifty mechanical trick, and it could lead to a few interesting scientific results, but really, it was just theater: It would never lead to efficient enough gene transfers to make a difference in disease.

What the naysayers did not see was the impact microinjection would have on another field: the development of transgenic animals. Frank Ruddle, a geneticist and head of the Yale University division of biology, decided that Anderson and Diacumakos might be onto something. He also heard about work by Mario R. Capecchi at the University of Utah, who independently began working on microinjection in 1977. Capecchi got the idea from watching neurobiologists stick microelectrodes into nerve cells to measure electrical activity. The microelectrodes were made from extremely thin glass needles filled with electricity-conducting fluid. "It was not a big jump in imagination that if you could put a needle into a cell, why not put something into the needle besides salt," Capecchi said. "I copied their technique of electrical measurements, and instead of reading electrical potential, I started injecting molecules into the cell."

Capecchi, a basic scientist, was not interested in gene therapy, but he was interested in understanding how genes worked in the cell, so he began injecting different types of genetic material into cells growing in laboratory cultures. When he thought about what might be the best cells to target, Capecchi realized that it might be possible to inject genes into fertilized eggs. Though he didn't do it, he talked about it as a possibility.

Frank Ruddle decided to do it. He took the train down from New Haven to New York and spent the time to learn microinjection at the master's hand—Elaine's. Ruddle was not locked into looking for genetic cures. He was interested in using microinjection to study biological functions of genes, among other things. When he thought about which genes would be the most interesting to inject, he decided on the fertilized mammalian eggs. Any gene incorporated into the egg would likely be found in all the cells of the animal, and might even be passed on to succeeding generations.

Ruddle was not alone. A small group of researchers had been searching unsuccessfully for a way to repeat John Gurdon's experiments of the early 1960s, in which he implanted whole tadpole nuclei into frogs' eggs, with fertilized mouse cells. As news of microinjection began to spread, a growing number of researchers became interested in possibly using it to put genes into fertilized eggs. "Once you develop the technology and see how it works, you can extrapolate to see where it goes," Capecchi said. "Microinjection directly led to the development of transgenic animals."

A transgenic mouse is made by removing a fertilized egg, placing it on the microscope slide, and injecting some gene of interest directly into the nucleus, just as Diacumakos had done with her cells for more than a decade. Instead of putting the engineered cell into tissue culture, however, the egg is placed into a foster mother and allowed to develop. The technique, however, is rather inefficient: only 10 to 30 percent of the microinjected eggs survive. Among the survivors that grow up to be mature animals, somewhere between a few percent and up to 40 percent carry the gene in their bodies.

In 1980, Frank Ruddle and his Yale colleagues worked out the first techniques, and showed that transgenic animals could be made. They assembled a plasmid with bits of viral DNA that stimulate the reproduction of viral DNA and the TK gene from the herpes virus. They injected this genetic construct into the mouse eggs, and implanted them into pseudopregnant foster mothers. In three of the surviving 180 mice that were born, they were able to find signs of the gene. The TK genes did not seem to function, but they were present in all the cells of the mice.

Two years later, in December 1982, Ralph Brinster at the University of Pennsylvania, Richard Palmiter at the University of Washington in Seattle, and others dramatically demonstrated the power of transgenics—where the gene actually functions—by inserting the rat growth hormone into mouse eggs. It wasn't hard to find the mice that had been transformed: They weighed twice as much as normal.

Transgenics would go on to become a valuable tool in basic biology and biotechnology. Mario Capecchi went on to use it to insert genes in their natural place in the chromosome through homologous recombination between the old gene and the newly injected gene. A Harvard team put two cancer-causing genes, called oncogenes, into a transgenic mouse to study how they worked together to cause cancer. That resulted in the first patented animal: the Harvard Mouse. Genetically engineered bacteria had been patented in the 1970s.

Other gene engineers began using transgenics to produce animals that manufactured molecules that would be of commercial value to biotechnology companies, including a drug that dissolves the clots that cause heart attacks and pigs that make human hemoglobin.

• • •

Although all these accomplishments seemed like good spin-offs, microinjection was never going to do what French Anderson wanted it to do: cure human disease. Even transgenics wasn't a likely bet since no one was even talking about putting genes into fertilized human cells to perhaps prevent a child from developing an inherited illness. It would be technically difficult, and present thorny ethical questions, such as, What if the inserted gene caused more harm than help? What if the researchers created a human monster? Who would be responsible for it, the parents? No one wanted to take chances like that.

Clearly, some other approach than microinjection was needed to get genes into enough cells to make a difference. But what? Several techniques had been tried through the 1970s, but none of them seemed adequate. For example, researchers had learned to use virus particles to fuse different types of cells together. It was much like merging soap bubbles together. The resulting hybrid cell contained the chromosomes from both original cells. Eventually, the cells would kick out many of the extra chromosomes until it reached some stable level.

Although cell-to-cell fusion worked, the results were unstable and difficult to make. What's more, there was no way to control which chromosomes, and the genes they carried, remained in the cell. It was unsuitable for treating patients.

Another approach—called electroporation—literally shocked the cells with electric currents to blast open holes in the cell membranes to allow naked DNA molecules to slither into the cell's gooey cytoplasm. The technique was unreliable and unsuitable for treating the cells of patients, especially if the cells could not be removed from the person for treatment.

· · ·

Even as Anderson and the others worked on their own physical attacks on the cell with microinjectors, a team of scientists at Columbia University College of Physicians and Surgeons hit upon a chemical means for getting genes into cells. In 1972, Alexander J. van der Eb and Frank L. Graham from the Netherlands discovered that they could get mammalian cells to absorb viral genes if the DNA was precipitated out of solution with calcium phosphate. They discovered that the salt somehow forced open the cell's membranes to allow the long filaments of DNA to slip into the cell. Once the naked DNA molecules got inside, a few of them would survive the other cellular defenses and become a stable part of the cell's chromosomes.

These, however, were the earliest days of the gene engineering revolution; few pure genes had been isolated. Those that had been discovered did not lend themselves to studies such as van der Eb's because there was no way to identify which cells had absorbed the genes. Since van der Eb and Graham did not have any pure genes to work with, they inserted purified DNA from an adenovirus that lacked the intact protein coat normally needed to get the virus into the cell. This way, they could tell when they had successfully inserted virus genes into the cells, because the transformed cells would produce whole viruses, and the culture would become infected. Since there were no other genes to test, and since there was no other way to determine when a gene had gone into the cell, they stopped after proving the calcium-phosphate-precipitation approach worked.

When the recombinant revolution finally did make a growing number of genes available several years later, Richard Axel, the Columbia lab chief, along with Michael Wigler, Saul Silverstein, and Angel Pellicer, who were all working in his lab at the time, decided that they would try van der Eb's precipitation technique with the same herpes thymidine kinase gene that Anderson, at the time, was still trying to microinject into L^- cells.

In 1976 and 1977, using restriction enzymes, the Columbia team sliced out and purified a 3.4 kilobase fragment of herpes genome that contained the TK gene. They mixed it with a large amount of salmon sperm DNA, which acted as a carrier for the smaller herpes TK gene. The entire genetic solution was then precipitated in calcium phosphate solution (the Columbia team modified van der Eb's method) so that the long strands of DNA came out of solution and settled to the bottom in a cloudy layer that could be removed from the test tube. The researchers then sprinkled the DNA precipitate, along with the calcium phosphate on top of a layer of L^- cells growing in cell culture.

With the precipitated form of the DNA apparently blasting holes in the cell's membrane through which the herpes TK gene could pass, the transplanted DNA managed to overcome the evolutionary hurdles, and find its way into the cell's nucleus, where it randomly takes up residence in a chromosome. They then grew the cells in HAT media and waited to see if any of the cells survived, and could form colonies. The technique produced a handful of cells that could grow in the media normally poisonous to L^- cells. Clearly, the herpes TK gene had gotten in, and was producing the TK enzyme, which was protecting the cells from the toxin.

Not only did the gene get in but it got in stably and persisted for hundreds of generations, as long as the cells were growing under the selective pressure of HAT medium. When the cells were grown in normal medium, however, the TK gene would be gradually lost.

The success of the experiment caused a great deal of excitement among genetic engineers. It now looked as if it would be possible to transfect mammalian cells just as Avery and Hotchkiss had been able to transfect bacterial cells. The TK gene seemed to provide an adequate way to find those few cells that had successfully taken up the new DNA. Adding the TK gene along with whatever gene was to be transfected should allow researchers to find the few cells that carried both.

Beta globin would be one of the first nonselectable genes to be tested. A bicoastal collaboration formed between the Columbia and Cal Tech scientists. Axel and his team developed the delivery system; Thomas Maniatis had the gene, the rabbit beta globin gene. If they succeeded in introducing the rabbit beta globin gene, then they would likely be able to introduce the human beta globin gene once it had been identified—a study which Maniatis already had underway.

For Maniatis, this would be a chance to test whether he had isolated enough of the beta globin's regulatory elements to actually get gene expression, the production of the beta globin protein. For him, it was a chance to study gene regulation, not the first steps toward gene therapy.

The researchers mixed the TK gene with the rabbit beta globin gene, but did not chemically link them together. Each gene would have to find its own way into the cell during calcium phosphate precipitation. The Columbia team believed that only some cells in the population of L⁻ cells was actually susceptible to transfection, so these cells were equally likely to pick up the TK gene, beta globin, or both. And that is what they found. Using HAT medium, they identified a few cells that contained both the TK gene and the beta globin gene.

The team even detected small amounts of messenger RNA made from the transplanted rabbit beta globin, but no protein production could be found. As Anderson had discovered when he attempted to transplant the human beta globin gene through microinjection, scientists still did not understand how the on-and-off switches for the beta globin gene worked. Even if they could get the beta globin gene into a few cells, the level of protein production was so low that it would not be useful as a therapy.

The bigger problem with what became known as the Axel-Wigler technique was its relative inefficiency. Only about one in a million cells manages to take up the gene of interest during calcium phosphate precipitation. The rest of the DNA added to the culture soup just hurled itself against the cellular ramparts, and was repulsed. The scientists had to spot that one-in-a-million cell where the DNA did penetrate, using HAT medium or some other selecting medium. Then, they had to pluck those genetically transmogrified cells out of the mass of millions or billions of unchanged cells, and stimulate them to produce millions of identical copies of themselves. Each new cell would then carry a copy of the newly transplanted gene.

• • •

This approach would prove useful for getting genes into cells to study them in the laboratory, but it was not clear how it could be used as a treatment. And finding a treatment was growing increasingly urgent for Anderson and his patients. Even as Axel and Wigler, with Maniatis, were sneaking beta globin genes into a few cells in culture, and Anderson was preparing to jam genes in with microinjection, Nick Lambis's body was beginning to break down. All along, Anderson had been telling Nick and Judy Lambis, his first two patients, that their contributions to research into thalassemia over the years increased their chances that they would be among the first to be genetically cured when it became possible. In the face of astonishing progress and promise, with it now clear that genes—including the beta globin gene—could be transferred into mammalian cells, Anderson and the Lambises were running out of time.

After twenty-seven years of two to three unit transfusions every few weeks, the iron buildup in Nick Lambis's body was taking a serious toll. He had lived longer than most thalassemics, who usually die in their late teens. Now, his doctors had him wearing a drug pump that constantly infused the iron-binding drug Desferal into his ravaged body. They were doing whatever they could to drain the toxic mineral that makes red blood cells but now was killing him. The Desferal wasn't enough. His heart beat with an erratic rhythm. His organs were beginning to fail.

"Nick and Judy had so much iron overload for so many years, we could not get all the iron out," Anderson said. "They had major endocrine, liver, and cardiac problems before we could get them on chelation [the treatment that removes iron]. Basically, their bodies just wore out."

Nick died first, and it was a terrible, lingering death. With his heart beating out of control, he went into congestive heart failure. Although his heart still beat, it failed to pump blood effectively enough to sustain life. Drugs lightened some of the burden, lowering his blood pressure, reducing the amount of fluid the heart had to push. He was admitted to the NIH intensive care unit, tubes plugged into every part of his body. As he lingered, Nick became despondent, writing notes that pleaded for release, for death. Finally, on November 20, 1979, he died.

It was an emotional, draining time. Anderson had always considered the Lambises surrogate children. He spoke at Nick's funeral, helped arrange larger than usual obituaries in local papers, crediting Nick Lambis with helping advance the scientific understanding of beta thalassemia.

A year later, Judy died too. She had gone to a Greek festival after Thanksgiving, and given a demonstration of a traditional Greek dance. When it was over, she sat down, slumped over, and died of a heart attack. Anderson considered it a blessing, since she was spared the technological terror her brother faced in the ICU.

The loss, however, remained with him. "We were developing gene therapy for the Lambises," Anderson said. "They provided the emotional drive for the iron chelation work and the emotional drive for gene therapy. Nick and Judy were the first patients we were going to do gene therapy on. That was what was going to save their lives. They believed it, and I believed it."

It was not to be. More than a decade after their deaths, gene therapy researchers are no closer to treating blood disorders like thalassemia than they were in 1980. The gene transfer methods that existed then were just too primitive, and it was becoming increasingly clear that globin genes were going to be far more complex than simple genes like thymidine kinase that could be put into a cell without complex on-off switches. TK just had to be put in, and it was turned on. The expression of beta globin was so low that there was little chance of benefit.

. . .

Even as Anderson began his collaboration with Rockefeller's Elaine Diacumakos in 1976, a new graduate of the Massachusetts Institute of Technology was camped outside the office door of Stanford University's Paul Berg. Richard C. Mulligan, tall, intense with a flamboyant, dissolute air, wanted to work with Berg, and he wasn't taking no for an answer—though that is exactly what Berg had said.

Mulligan was a diamond in the rough. Born in Summit, New Jersey, in September 1954, he was the classic youngest child. Bright, carefree, a heavy partier, Mulligan was not afraid to take chances. He once pushed his two-stroke Yamaha 350 motorcycle up to 120 miles per hour as he whined along Storrow Drive, the curving four-lane that runs along the Charles River in Boston.

Tall, and skinny until later in life, Mulligan looked down on the world from his 6 foot 4 inch vantage point, where he judged who was a good scientist and who was not. "I am like my father in that I am assertive, not suffering fools gladly," Mulligan said. His father, an engineer, ran the National Academy of Engineering staff before becoming dean of the University of California at Irvine. Mulligan's mother was a teacher.

The Massachusetts Institute of Technology accepted Mulligan as an undergraduate. While there, he worked with Alexander Rich, an expert in the physical structure of DNA. Mulligan did high-level research as an undergraduate, helping learn where the genes for the different proteins of SV40 were located on its genome. The work was good enough to gain Mulligan unlimited access to Rich, who treated the younger man like a postdoctoral fellow. Mulligan produced useful information on the transcription and translation of proteins from the SV40 virus. It was good enough to impress a lot of other scientists and get him several coauthorships on papers in the scientific literature.

When he graduated from MIT and went looking for a graduate school to earn his doctorate, his advisers told him Stanford University was the most exciting place for gene engineering. Mulligan decided to cross the country.

On the basis of his work at MIT, Mulligan had little trouble getting into the university's biochemistry department. The system at Stanford tends to be pretty open. A graduate student joins the department generally, and hangs around for four or five months while deciding on a research project, and selecting a lab in which to work. Unless a lab is overcrowded, the student is accepted into the lab. Mulligan wanted to work with Berg and made it clear right from the start.

"Berg was such a big shot that everyone came to the lab to work there," Mulligan said. "Berg was incredibly good, and he was so able to motivate people. He was remarkable. You had the impression that this was just the best place." Mulligan wanted to be part of it.

"I refused to take Mulligan at first because he had already worked on SV40 as an undergraduate," Berg recalled. "I thought it was probably a bad choice for him to continue working on SV40 because that would not give him a chance to work on something new. So, I said no.

He camped on my doorstep because he refused to take no. After a couple of weeks of passing him all the time, I finally gave in and accepted him."

With his black leather jacket, a full beard, and a pony tail down the middle of his back, Mulligan settled into Paul Berg's straight-laced lab. In 1972, the early days of the gene engineering revolution, Berg and his colleagues struggled to splice and paste bits of DNA from the *lac* operon into the SV40 genome. Now, four years later, the techniques of genetic engineering had advanced so rapidly that it was relatively easy to recombine chunks of DNA from SV40 with other DNA.

Still, there were problems. Relatively few genes had been isolated to work with. Two groups of scientists—one in California and one in Massachusetts—were racing to be the first to clone the human insulin gene and make the first biotech product, but the gene would not become easily available.

What's more, while scientists could now easily insert genes into bacteria, they still had trouble getting them into mammalian cells. And lastly, none of the groups were getting the genes to function very well once they got them into mammalian cells.

Berg set Mulligan to the task of looking for a way to turn on the transplanted genes in mammalian cells. Working with the entire SV40 genome, Mulligan began to design an expression vector. A vector is any molecule that transports genes into cells in such a way that they survive. An expression vector allows the cell to use the newly transplanted genes to make the protein encoded on that stretch of DNA. For an expression vector to work, it must contain the genetic on switches that allow the necessary enzymes to physically bind to the DNA and read the encoded information needed to make the protein.

Berg and others in his lab had successfully inserted a number of genes into SV40 vectors that reproduced in mammalian cells, but the transplanted genes did not seem to make either the messenger RNA or the protein. Getting transplanted genes to function was, after all, the goal Berg had set nearly a decade earlier when he went to learn polyoma work with Dulbecco, and it was the principal focus of his group in the early 1970s when they struggled to make SV40 recombinants.

Making the recombinants was now easy, so Mulligan began to slice up the SV40 virus in different ways to search for the genetic on switches. He started by inserting into SV40 a bacterial gene called XGPRT, an evolutionary relative to the HPGRT gene found in people and associated with the human mental illness called Lesch-Nyhan syndrome.

The gene's advantage was its ability to detoxify HAT medium— just like the thymidine kinase gene. If the SV40-XGPRT construct was going into cells, they would survive growing in HAT. In short order, Mulligan had turned the SV40 genome into a delivery truck for the bacterial XGPRT gene, and deposited it into monkey cells, which then made the protein. If the cells didn't make the protein, they died when grown in HAT medium.

"It was really the first time one demonstrated you could actually express a bacterial gene in animal cells," Mulligan told author Yvonne Baskin in the early 1980s. "It blew aside all these myths that maybe there's something special about animal cell protein-coding sequences."

· · ·

But animal coding sequences were the main point. Could the same approach be used to insert mammalian genes into mammalian cells, and get them to work? Mulligan began by removing a big gene from SV40—the VP1 gene, which makes one of the three proteins used to produce a virus capsule. He left in place, however, most of VP1's regulatory information, so any gene inserted into its space would likely function. These signals would be crucial for the function of any inserted gene.

But what gene to insert? Tom Maniatis's rabbit beta globin gene, of course. It had all the advantages: It was isolated and available; it was small and could easily fit into the SV40 genome after VP1 was removed; and since beta globin protein had been so well characterized, if it was produced, Mulligan would be able to detect its presence easily. Maniatis provided the gene, and Mulligan set to work.

Mulligan had to develop many genetic tricks to get the beta globin gene prepared with just the right sticky ends to be dropped into the SV40 genome, but eventually, he worked out the details and inserted it. Then, Mulligan inserted the recombinant SV40–beta globin combination into the monkey kidney cells.

Normally, monkey kidney cells do not make beta globin, so any hemoglobin production would be easy to spot. If the genetics had worked as he hoped, the genetic on signals from SV40 would turn on the rabbit beta globin gene, and it would make protein.

To find out if the gene was functioning, Mulligan used a number of different techniques—including monoclonal antibodies that recognized beta globin protein, classic chemical extraction, and molecular probes for globin mRNA. Mulligan showed conclusively that the monkey kidney cells were producing rabbit beta globin. Not only did they make the protein but they made lots of it. "We estimate that B-globin and VP1 are synthesized at nearly equal rates," Mulligan wrote in *Nature* magazine in January 1979.

It was a major paper, a breakthrough in protein expression in mammalian cells by transplanted genes. He had not even completed his doctorate yet, but Mulligan was well on his way toward a major research career. Some say this paper helped pave the way toward his winning a MacArthur Foundation "genius" award a few years later. The genius awards are given to those believed to be making a significant contribution to their field. The MacArthur Foundation provides varying amounts, up to several hundred thousand dollars over several years, to free the genius from financial worries so he or she can pursue novel ideas.

It was the first time anyone had transferred a mammalian gene into a mammalian cell and got it to function normally. At last, it was possible to move genes into the cells that make up humans. For the first time, a discussion of human gene therapy could be based on solid science, not just dreams and speculation.

Berg, however, didn't see it that way. Every time someone in his lab announced another advance in gene transfer or gene expression, he would move in quickly to quash speculation—especially predictions in newspapers and in television news reports—that this or that new technique would quickly lead to human trials of gene treatments. It was one thing to put new genes into cells growing in the test tube, Berg would say. Putting genes into a person was entirely different.

Scientists still didn't know how to target specific human cells within the body; they couldn't insert genes into their original place in the chromosome, nor exert normal genetic control over the transplanted gene; and they still didn't have many genes that could be used therapeutically. Human gene therapy, Berg believed, was decades away.

"Berg believed science was going to make this happen, not hype," Mulligan said. "So do I. We took the attitude not to hype the concept of gene therapy. We were very aware that the approaches we were taking would be the basis for it. That was an important theme. We

were more interested in talking about model systems. We were looking at the biological issues. It's a funny thing, but there was clearly
a recognition that this was important, but that we should focus on
doing the science. Even as an assistant professor [later at MIT], I
never mentioned gene therapy even [when my] talk would reek of
gene therapy."

If gene therapy was going to become reality, however, it was not
going to be with the SV40 expression vector. Although Mulligan's
beta globin experiments did show that the problem of protein production could be overcome in mammalian cells, SV40 had too many
other problems to make it practical. First, it was too inefficient. Since
the expression vector was too large to fit into the SV40 capsule, it had
to be inserted into cells using chemical tricks like Axel and Wigler's
transfection technique. Even if the new genes could fit in the relatively small SV40 capsule, it could not fit very many. What's more,
the virus's limited host range meant it could only infect a small number of cell types, and most of them were in monkeys, not humans.
Even in monkeys, it often caused disease or other complications.

A bigger problem, however, was what happened to the expression
vector once it got into a cell. For the most part, SV40 vectors remained outside the chromosome as independent genes floating
around in the nucleus. While the enzymes required for converting genetic information in mRNA and ultimately proteins did work on these
SV40 minichromosomes, the activity of the transplanted genes did
not persist very long. Although the SV40 vectors produced large
amounts of beta globin protein, Mulligan found, they made it for only
about thirty minutes. That's plenty of time for using the system to
study the basic control systems in genes, but it is not enough to treat
a human with thalassemia.

Lastly, the fatal flaw was lethality itself. The SV40 expression vector ultimately killed transfected cells. For gene therapy to work, the
treatment must be benign so the cells can live a normal lifespan and
produce protein from the transplanted gene.

At the same time, in 1980, Mulligan was coming to the end of his
time in Berg's lab. He had completed the work needed to earn his doctorate, and he was leaving the lab with a tremendous number of papers for a student not yet on his own: nineteen published, some of
them major.

As a final part of his doctorate degree, Stanford required Mulligan
to write a description of what he intended to study once he left grad-

uate school and was on his own. Now thoroughly invested in developing genetic transfer systems that allowed proteins to be made in mammalian cells, Mulligan laid out a description of how a different virus, a retrovirus, might be used to transplant genes directly into mammalian chromosomes. "The idea was based on what had been done in the SV40 system," he said.

But the idea went well beyond what was possible with the monkey virus. Retroviruses—Mulligan talked about using mouse retroviruses—are unusual even among viruses. They carry their genes in the form of RNA, and then convert them into DNA so they can be inserted into the infected cell's chromosome. Best of all, retroviruses usually don't kill the cell they infect. Once their genes are intercalated among the host genes, they quietly crank out new virus genes and proteins that get assembled into new infectious viral particles. An infected host cell lives an essentially normal lifespan, which, depending on the cell, can be years.

Retroviruses already had proved their usefulness in pointing the way to the discovery of oncogenes, usually normal genes that cause a cell to become cancerous when they are mutated or deleted. Retroviruses also produced the reverse transcriptase enzyme, the revolutionary protein that violated the Central Dogma of biology by passing genetic information from RNA to DNA, winning a Nobel Prize for Howard Temin and David Baltimore.

By using some fancy genetic engineering, Mulligan predicted that it would be possible to create a cell line that contained a defective retrovirus genome that made all the proteins needed for producing normal virus particles but would be unable to package its own genes into the virus capsule. Although the cell was infected and producing viral proteins, it would not be able to make any infectious virus.

At the same time, Mulligan imagined, a second retrovirus genome could be inserted into the same cells. The second genome would be engineered to carry a gene that was needed for therapy, such as the beta globin gene. The rest of the genes in this second retrovirus—the ones needed to make retroviral proteins—would be removed. The second retrovirus genes would, however, retain the signals needed to package its genome into a retroviral capsule if one was present.

Placing both of these retroviral genomes into a single cell—one that made all the viral proteins but couldn't package its own genes, and one that carried a therapeutic gene but could make none of the viral proteins—would make a complementary system that produced infectious retroviruses filled with the therapeutic gene. The retro-virus could then be used like a molecular delivery truck to deliver the therapeutic gene into cells that lacked the gene. As Mulligan said, these were ideas that reeked of gene therapy.

"There already were complementation systems," Mulligan said. "There was an example of monkey cells transfected with SV40 T [gene], that made it possible to complement reproduction of SV40 constructs that did not have T. It was a sort of helper system. There were other complementation systems where you have two different viruses with a mutation in different functions. Mixing them restores the function. There was not much on that, but my interest was how do you take the concept of the SV40 stuff—making expression con-structions, using bits and pieces of sequence to make protein, and the notion of stable transformation—to stably get genes into different cell types."

Mulligan had picked the retrovirus because it infected cells as well as SV40, but entered more cell types, because it could stably integrate its genes into the chromosomes of the host cells, and because it could hold more genes than SV40. Besides, he didn't want to compete di-rectly with Paul Berg.

What's more, this was the end of the line at Stanford. Mulligan completed his Ph.D., and it was now time to move on. But he couldn't decide what he was going to do. Mulligan's old MIT mentor, Alex Rich, said he could come back to MIT and work independently in his lab, that he would give the younger man the resources he needed to pursue his ideas. It was an offer Mulligan liked and accepted.

Before he could relocate, however, Mulligan visited the Salk Insti-tute, where he ended up at dinner with Paul Berg and David Balti-more. When Baltimore heard that Mulligan was planning to go back to the Rich lab, he asked Mulligan to come and work in MIT's new cancer center. "We can set you up, and you can work independently," Baltimore offered. Mulligan accepted, ending up with space in Philip Sharpe's lab at the cancer center. He started working on the ideas for making genetic packaging systems from retroviruses that he had out-lined as part of his Ph.D. work.

After getting set up at MIT, Mulligan would become a major con-tributor to the technology needed to conduct human gene therapy. But in 1980, he was just a bright postdoctoral fellow heading back to the east coast to begin his professional career. Back at the National In-stitutes of Health, French Anderson was becoming disillusioned with his quest to treat diseases with genes. Microinjection had a high rate of successfully transferring genes into cells, but it was labor intensive, since each cell had to be injected one at a time. It would never be suitable for genetically correcting the millions or even billions of cells needed to heal a human.

The Axel-Wigler and all the other physical and chemical approaches were just as inefficient in their ability to get many genes into many cells. True, transfection by calcium phosphate precipitation would allow physicians to sneak genes into mammalian cells, but it was not well suited for human treatment. It would mean the patient's cells had to be taken out of the body, genetically manipulated, and then placed back in the body after selection. It would be laborious, difficult, and so inefficient that success seemed unlikely.

And while there was growing interest in using viruses to transfer genes into patients, none of the existing systems were even close to being testable. It just didn't look like gene therapy was going to work anytime soon.

"The technology was simply not available to do clinical gene therapy," Anderson said. "You could not microinject all those cells; calcium phosphate precipitation was no good. There was no technology here."

In the summer of 1980, Anderson helped organize a scientific meeting near Washington at a conference center called the Airlie House. He chaired a session with several presenters who reviewed the status of genetic therapies. Anderson presented the work he and Elaine Diacumakos had done on transferring genes with microinjection. As the others presented their work with the other approaches, Anderson became more and more unhappy. "It was the realization that we had a long way to go," Anderson said. "It was like running a race, and you turn the corner, and you think that is the finish line, and instead, the course goes off into the distance in the opposite direction."

That night, he lay in the conference center bed tossing and turning, running the day's session through his mind. As he searched for a new tack to take toward gene therapy, he found none. At 3 A.M., he rose from bed and went home, despairing that gene therapy was going to happen in his lifetime. The day was June 24, 1980, his nineteenth wedding anniversary.

From that moment of darkness, Anderson virtually abandoned any hope of performing gene therapy, and essentially left the field. What's more, for the next few years, he did little of anything scientifically.

His calendar system stopped showing greens and instead displayed a growing pattern of yellows and reds. He found solace in practicing the martial art of tae kwon do and became an expert in sports medicine.

Little could he know that in the same month that he virtually abandoned the dream he harbored for so many years, another scientist was about to take one of the technologies Anderson had rejected as inadequate, and make the first stab at putting genes into a human patient. That experiment would rock the field—and society.

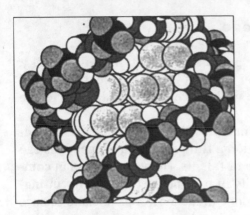

Chapter 6
The Risks of
Progress

"The code word for criticism of science and scientists these days is 'hubris.' . . . To be charged with hubris is therefore an extremely serious matter. . . ."

—*Lewis Thomas,*
"The Hazards of Science" in The Medusa and the Snail

A thin, ordinary-looking man with sandy-brown hair and glasses boarded a Rome-bound airliner at Los Angeles International Airport. He carried with him extraordinary baggage: a Styrofoam cooler stuffed with dry ice and laboratory tubes full of genes. Martin J. Cline, a physician and researcher from the University of California at Los Angeles, was heading off to conduct research overseas during the summer of 1980. His chilled tubes carried two genes in different combinations: one, a gene from the herpes virus, two, a human gene for making half of hemoglobin, the protein that carries oxygen in the body and stains the blood red, and, three, a piece of DNA in which the herpes gene was chemically attached to the human hemoglobin gene.

Cline did not know for sure what he would actually do with the genes once he reached Europe. For the last year and a half, and particularly throughout the spring of 1980, he had been laying the groundwork for a remarkable experiment. At the least, he would do some laboratory studies to solidify new collaborations he had established with researchers in Israel and Italy. But the real goal was nothing less than historic: He wanted to insert the genes in the Styrofoam cooler into humans suffering from a terrible inherited illness of the blood called thalassemia. It would be the first time anyone had put genes into people. It would be a grand experiment. It would be risky. If it worked, he'd win the Nobel Prize. If it failed, at least he would be the first to try—and failure in the face of great odds carried no shame.

All of this appealed to Martin Cline. He had a strong bias for action, for experimentation, for doing things. It fed a tremendous ego powered by a rare combination of talents. He was a physician comfortable with caring for critically ill patients, yet steeped in cutting-edge science that could only be done in the most sophisticated science lab far from the clinic. An internationally known cancer researcher trained at Harvard Medical School, he had conducted dozens of successful studies in which patients with deadly cancers were given poisonous drugs, brought to the edge of death, and then rescued from their disease and the treatment. His UCLA colleagues considered him a doctor's doctor, the kind of clinician they went to see when they or a family member were sick.

Ambitious, brilliant, energetic, organized, a gifted researcher, teacher, and administrator, Cline's presence could be overwhelming. Many considered him intense, impatient, cold, and arrogant. If someone had an appointment with Cline from 9 to 9:30, it started at 9 and ended at 9:30, whether the person was late or not. And if Cline concluded that a bit of research was poor quality and not worth supporting, he said so, strongly, without regard to personal feelings. He not only failed to suffer fools with any gladness, he would not tolerate them at all. Yet, he did not seek confrontation.

He did recognize that science was the art of asking answerable questions. "He had a knack for seeing questions that might have clinical impact," said one longtime friend. Cline had a penchant for zeroing in on the biggest questions in science. He routinely chose problems that were difficult but flashy, high profile. Succeeding with original experiments in a difficult field brought the biggest rewards— personally and professionally. Gene therapy would be the ultimate. Only one person was going to be first to transplant genes in humans. If it worked, it would revolutionize medicine. He wanted to launch the revolution.

A slight, bespectacled man—a marathoner—Martin J. Cline was born in Philadelphia, Pennsylvania, on January 12, 1934. He graduated from the University of Pennsylvania in 1954, at age twenty, with honors in chemistry, mathematics, and general academic subjects. Four years later, the Harvard Medical School conferred his M.D., cum laude general, and cum laude for thesis in a special field. Cline interned from 1958 to 1960 under George W. Thorn, physician-in-chief at the Peter Bent Brigham Hospital in Boston and a founding visionary of the Howard Hughes Medical Institute; worked as a clinical associate from 1960 to 1962 at the National Cancer Institute under clinical director Nathaniel I. Berlin; and had research appointments at the University of Glasgow, Scotland, and the University of Utah College of Medicine in Salt Lake City before settling into the hematology unit at the University of California at San Francisco in 1964.

Cline's aggressive style propelled his career like a rocket motor, bringing him all of the rewards academic medicine had to offer. In 1973, Cline won one of the few, well-endowed chairs of medicine at the University of California at Los Angeles, becoming the Bowyer Professor of Medical Oncology at age thirty-nine. He had been recruited from UCSF to create and direct UCLA's combined department of hematology and oncology, where he established the school's first successful cancer research center. His group made critical contributions to clinical medicine, such as helping develop the techniques used in bone marrow transplants in the treatment of cancer. It also helped sort out the use of growth factors in humans to speed up recovery of bone marrow after a transplant, and made advances in the treatment of leukemia.

Cline's department also was a hotbed of controversy. Its aggressive staff contained some of the most audacious figures in medicine, including David Golde, a researcher later to become embroiled in a bitter, precedent-setting legal battle over who owned the product rights to a patient's own cells after they had yielded a medically useful substance. Another was Robert Peter Gale, the bone marrow expert who, in early 1979, was turned in by his own nurses. They accused him of conducting experimental bone marrow transplants without getting proper administrative approval or patient consent. Gale also performed what many considered dubious transplants of fetal liver cells (in place of bone marrow) on Soviet rescue workers lethally exposed to radiation after the nuclear power plant accident at Chernobyl in April 1986. None of those Soviet workers survived.

One colleague called them "the three joggers"—Cline, Gale, and Golde. Three brilliant minds. Three driven researchers. Three men destined to make major strides—and enemies—in science. They even fought among themselves. In the mid-1980s, Cline sued Golde over royalty rights for the discovery of one of biotech's most valuable drugs: granulocyte-colony stimulating factor, G-CSF. It was identified at UCLA and made Amgen one the world's wealthiest biotech companies.

As a scientist, Cline was unsurpassed. He had more than 200 scientific publications in the world's leading medical journals by 1980. His research proposals and grant requests routinely sailed through the study sections at the National Institutes of Health. Study sections made up of scientific peers expert in a given field are generally considered the ultimate judges of a researcher's success as a scientist. The study sections regularly gave Cline high marks for the quality of his ideas, followed by grants that generously supported his research. In 1980, his large lab had some twenty postdoctoral researchers and technicians funded by nearly $1 million in federal grants each year.

Simply put, Cline was a star. He was a high flier whose achievements brought him international notoriety and acceptance by scientists and physicians everywhere—many of whom wanted to work with him. One scientist called Cline "the most glamorous personality in hematology." He had just spent a pleasant sabbatical year conducting research in Australia. Transcontinental travel had become an interesting way to do science and see fascinating parts of the world simultaneously. Life was good.

Now, on June 8, 1980, Cline headed for Europe and the Mideast for more international adventures. He was excited. The success of recent animal studies made Cline itch to put genes into people. He believed he had found a way to treat, maybe even cure, thalassemia. Whether he could test his ideas on this trip depended on his collaborators getting the necessary approvals at their institutions. Even as his plane rose into shimmering California sunshine, Cline did not know what awaited him. Whatever happened, he reasoned, it would be an interesting, perhaps even momentous, summer.

· · ·

Genetic engineering had exploded in the last decade. Scientists found they could put genes into nearly any cell they chose, though not very easily or in very many cells. Most of the time, the genes failed to function once they got into the cell. Genetic engineers mostly tested their techniques in bacteria growing in glass dishes in the laboratory. Increasingly, however, researchers were thinking of turning to cells from higher organisms, the lab rats and mice, and even cells from humans.

During the great genetic engineering debates of the mid- to late-1970s, scientists, their leaders, Congress, and the public argued about the chances that the genetic engineers might goof and accidently create an Andromeda strain that would escape from the laboratory and spread a lethal infection worldwide. By 1980, after seven years of debate, fear of a genetic holocaust seemed to be receding from the public consciousness. Congress, after investigating the new science, abandoned legislation introduced to regulate genetic research. Scientists themselves, working through the National Institutes of Health, set up guidelines and procedures to regulate the field and limit further the apparently diminishing perceptions of danger.

Cline, however, not a player in the field of basic genetic research nor a participant in this national debate, was uninterested in monkeying around with genes in bacteria. He wanted to put new genes into sick people. He believed he was one of the first to see the new paradigm, that genes could deliver cures for fatal diseases. He could see it so clearly that he feared other scientists also could see it and were going to beat him. Cline wanted to be the first to take his gene transfer experiments and leap ahead to humans. Although his studies had provided controversial and, to many, brought inconclusive results, Cline convinced himself that his approach worked well enough to justify moving to human patients. Many others disagreed. That would lead to a classic power struggle between a headstrong scientist convinced that he is right and institutional leaders responsible for ensuring their scientists did not stray too far from acceptable behavior.

In the struggle, Cline could see only his righteousness, not his own arrogance—the one personality trait that could undo his brilliant career. He, like many doctors before him, believed that only scientists doing clinical research could really understand the risks and the benefits of untried treatments. Only they could decide when a new treatment, however risky or unlikely to work, was ready for testing in people. For all of his accomplishments, intelligence, and hard work, Cline's hauteur and aggressiveness were about to undo his life.

By the end of this trip, at age forty-six, Cline's lifetime of accomplishments would become meaningless. His reputation would change from "the most glamorous hematologist" to that of a maverick out of control, operating beyond the rules. As a result, he would be abandoned by his UCLA colleagues and international collaborators, vilified by his pioneering patients, and punished by the scientific establishment. Many researchers would write off his actions as scientific hubris. Others more profoundly disturbed by what they considered a serious ethical breech would accuse Cline of violating the Helsinki agreements on the protection of human research subjects, and compare him to the World War II German doctors who—in the name of science—tortured prisoners at the Nazi death camps.

Cline himself found the ensuing brouhaha baffling. To him, he was just doing what he always did: testing radical new treatments on desperately ill individuals nearing the end of their lives. He was not laying the groundwork for eugenics nor creating a genetic Frankenstein's monster in a secret laboratory. The goal was to heal, not harm. The result was a disaster.

. . .

Hours after it left Los Angeles, Cline's plane bounced onto the tarmac of Leonardo da Vinci airport outside Rome. Cline had come to Italy to set up a collaboration with Dr. Cesare Peschle, chief of hematology at the Istituto Patologia Medica at the University of Naples, and an expert in thalassemia and other blood disorders. Born in Pula, Italy, Peschle was more than a decade younger than Cline. Though Peschle had trained as a physician at the University of Rome, he spent his time conducting basic research, not taking care of patients. He had set up a laboratory to study defective blood cells in patients with thalassemia and other inherited blood disorders. The lab had a reputation as being one of the few anywhere in the world that could culture human blood cells in clonal assays—at the time, a difficult technique for growing identical types of cells from the blood. When Cline arrived, Peschle already had screened blood from more than sixty thalassemia patients and had some of their cells growing in laboratory cultures.

Tall and slim, with a quick manner, flashing gestures, and expressive dark eyes peering out from under a headful of thick dark hair, Peschle came to Cline's attention during an international hematology meeting when Peschle sharply questioned David Golde after he gave a lecture. Golde, Peschle, and Cline talked about the research after the session. Golde and Cline recognized familiar traits in Peschle: He was bright, tough, controlling, and ambitious—very much like them. The Americans remained in contact with the Italian after that.

In early 1980, Cline invited Peschle to UCLA to give a seminar on his work. During the visit, Cline asked Peschle if he was interested in collaborating on an experiment that Cline had in mind. It would be a revolutionary experiment that might cure children with sickle cell anemia by putting new genes into their bone marrow. Cline showed Peschle a sketch of the experiment, called a protocol, that laid out how it would work. Ten years later, Peschle recalled being intrigued, but skeptical. But Cline was Cline; it was hard to just say no. He was one of the best in the world. Peschle agreed that he would think about it.

Now months later, in early summer 1980, Cline and Peschle were together again, this time in Naples. At the minimum, they planned to use the genes Cline carried for laboratory studies on the bone marrow cells of patients with beta-zero thalassemia. At the maximum, Cline thought, he would use Peschle's connections to transplant genes into the bone marrow of thalassemia patients.

What Cline had in mind was pretty simple. He wanted to try to put a normal copy of the hemoglobin gene into the blood cells that carried the abnormal gene. Scientists had dreamed the dream of fixing diseases at the gene level ever since the 1940s when genes were shown to be made from DNA. As the techniques of molecular biology exploded in the early 1970s, it looked as if it would soon be possible to make that dream reality.

Cline, however, was no molecular biologist. He was a physician who conducted clinical trials in desperately ill patients with leukemia, lymphoma, and other blood disorders. To anyone working with cellular therapies such as bone marrow transplants, gene therapy was the next, most obvious step. To flesh out his evolving ideas, Cline decided to learn some molecular biology and gene manipulation.

Across the street from the UCLA medical center lies the collection of buildings that houses the university's biology department. In it, Winston Salser, a professor of biology, had established a reputation in the mid-1970s for his recombinant DNA research. In 1980, he would go on to help found what became one of the country's most successful biotech companies, Amgen, Inc., but at this time, he was just another academic biology professor.

Salser cuts an eccentric figure. He is well over 6 feet tall, and his wiry brown-and-gray hair tends to float above his head as though it were electrified. An equally wiry black-and-gray beard obscures his shirt collar, even as his moustache camouflages his mouth. Piercing blue eyes under thin eyebrows gaze out mercurially. In 1976, Cline decided this would be the man to teach him the new biology he needed to do gene therapy.

Salser was unimpressed at being picked. Despite his public involvement in local Los Angeles conservation movements, Salser did not much follow university politics, disdaining the petty battles, ignoring who was up and who was down, and who held power on UCLA's campus. When Cline first asked to work at his lab bench to learn genetic engineering, Salser didn't know who Cline was. "He had to tell me he was a big deal at the medical center," Salser remembered of that first encounter. Salser worried that this *doctor* would not stick it out. Molecular genetics is tedious work. Experiments usually go in small, delicate steps. Often things go wrong, for inexplicable reasons. When Cline learned how hard it was, and how unconnected to the clinic, he would quit, Salser believed. Initially, Salser tried to discourage Cline from starting.

When Cline told Salser that he wanted to learn genetic engineering so he could develop genetic therapies for patients, Salser's negative inclination strengthened. "Oh my God, this is not going to work," Salser thought. But after the third time the two discussed Cline's ideas, Salser became intrigued. Finally, worn down by a man who doesn't like to take no for an answer, Salser accepted Cline into his lab and started teaching him molecular biology. Cline learned quickly. The mentorship rapidly became a collaboration with Salser overseeing the molecular biology while Cline dreamed up novel experiments that mixed genetics and bone marrow transplantation in mice.

Around the same time, in the mid-1970s, a series of developments occurred in several laboratories around the country that helped advance Cline's ideas of using gene therapy for the treatment of human disease. The first was the isolation of the human beta globin gene, one of the few human genes that had been purified up to then. Thomas Maniatis, the molecular biologist at the California Institute of Technology in nearby Pasadena, had isolated the beta globin gene.

The news about the isolated gene spread quickly. Cline asked Maniatis for a copy so he and Salser could try putting it into cells to see if they could get the gene to make the hemoglobin protein. Maniatis, a believer in scientific openness, sent Cline the gene. It was a generous act. Maniatis's team also was trying to put the globin gene into cells growing in the laboratory. But the Cal Tech scientist—like a number of other research teams—was not having much luck getting the gene to function. Although the globin gene was going into the cells, there was no evidence that it was producing the blood protein. Unless the gene turned on the production of hemoglobin, inserting it into a cell would be meaningless for a patient.

It may be one of those odd turns in history. If the Cambridge City Council had not temporarily shut down recombinant research at Harvard, Maniatis may never have come to California to conduct his landmark isolation of hemoglobin. And if Maniatis had not been so close, Cline might not have enjoyed such easy access to the beta globin gene that occurred when his UCLA colleague David Golde took a sabbatical in Maniatis's Cal Tech lab. It's true that Cline probably could have gotten the gene even across the country after Maniatis published his results, but proximity made it easier for Golde to work with Maniatis, and connect Cline to the transplanted Boston biologist.

The isolation of the beta globin gene gave Cline a specific disease to target: beta thalassemia, or perhaps the beta form of sickle cell anemia, another disease of the red blood cell typically found in African-Americans. With the hemoglobin gene in hand, Cline fixed his sights on genetic therapy even more firmly.

Cline and Salser now faced another hurdle: getting the gene inside blood cells. All of the existing techniques used to transfer genes into cells worked, but none of them worked very well. They simply were not efficient. Scientists had a difficult time getting just a few cells among the millions or billions swirling around in the test tube to take up a new gene. They decided that the best existing gene transfer technique came out of the labs of Richard Axel and Michael Wigler at Columbia University in New York City in 1978. They had shown they could open holes in cells by mixing them with calcium phosphate. Once the holes were formed, DNA could slip inside, and become a permanent part of the cell.

But the New York scientists had worked only with cells in tissue culture. Cline needed to demonstrate that it would work in whole animals if he ever hoped to test the techniques in people. Cline then came up with his own remarkable idea: He would use Darwinian selection to find genetically altered cells, not in a petri dish, but in a whole animal.

For decades, researchers have used mice as living test tubes. They've been poked, prodded, infected, irradiated, and pumped full of chemicals, toxic and benign, to wrest more information from nature about the biology of humans. Animals have been at once incredibly powerful models for human ills and simultaneously so complex that scientists sometimes cannot sort out the results of the experiment.

Despite the complexity, Cline and Salser turned to the animals to show they could use Axel and Wigler's calcium precipitation technique to insert the gene for the herpes simplex virus's thymidine kinase (TK) or the gene for dihydrofolate reductase (DHFR) into mouse bone marrow cells growing in a laboratory culture.

To test their gene-transferring abilities, they mixed the genes—either the TK gene or the total DNA that contained many DHFR genes—with the mouse bone marrow cells in the proper lab conditions to get the genes into cells growing in culture. Next, they grew the gene-treated cells in a growth medium laced with toxins that killed normal cells. Only cells transformed by the gene treatment grew unharmed in the lethal broth. The experiment succeeded, confirming for Cline and Salser that they could transfer genes into some cells, and that their Darwinian selection with toxic drugs could identify the individual cells that had taken up the new genes in the lab. This, however, merely repeated what the Columbia scientists already had accomplished. Cline wanted to go further; he wanted to know whether it would work in an intact animal. That would be new.

To test that, the UCLA scientists mapped out a fairly complicated experiment. In late 1978, they placed bone marrow from two different types of immunologically related mice in separate laboratory dishes. One of the cell types had a distinct chromosomal marker; the other did not. (It would be like using red cells and green cells. The chromosomal marker just allowed the UCLA researchers to tell the cells apart.) To the marked cells, they added the TK gene with the calcium phosphate so some cells could absorb it. The marked bone marrow cells (the red cells), some of which now carried the TK gene, were mixed with unmarked cells (green cells) that had not received any new genes.

At the same time, the test mice were treated with full-body radiation. The radiation treatment kills the animals' bone marrow, making room in the bone marrow for the cells Cline and Salser were about to transplant. Radiation treatment is a normal part of bone marrow transplants used to treat human patients. Because radiation kills most of the cells in the bone marrow, the animal—or patient—would die without a follow-up marrow transplant.

After the radiation, the mice received the mixture of treated and untreated bone marrow cells (red and green) as though it were a normal bone marrow transplant. This would "rescue" the mice from the lethal effects of the radiation and restore the animals' blood-producing ability.

To see if the newly transplanted genes were working, however, Cline added a twist to the normal bone marrow transplant procedure in the form of natural selection: He treated the transplanted animals with the toxic anticancer drug Methotrexate. Normally, Methotrexate kills bone marrow cells when enough is given, and it can prevent a bone marrow transplant from succeeding. Because the TK gene makes an enzyme that blocks the effects of Methotrexate, however, any cell carrying the newly transplanted TK gene would survive. If enough Methotrexate was given, only the gene-altered cells could live in the mice, so only gene-altered cells would restore the bone marrow. All the other bone marrow cells that had *not* been genetically altered would be killed by the drug.

That is just what Cline and Salser claimed they found. The UCLA scientists reported they could detect Methotrexate-resistant cells in the animals. Of the mice treated with the TK gene, two of six, and eight of fourteen animals in two experiments showed signs of genetic alteration. What's more, the genetic integration tended to be unstable: "less than 15 percent of animals . . . show retention of the viral gene for several months."

In the DHFR tests, the numbers of mice showing resistance to the effects of the poisonous drug were equally small. Those cells that did take up the DHFR gene, however, tended to remain stably transformed for up to eighteen months.

In the mouse bone marrow, the cell type (red or green) carrying the transforming genes seemed to predominate, an indication that the genes were integrated, working, and conferring a selective advantage to those cells. There were no signs of toxicity throughout the experiments.

But, Cline acknowledged, the genes entered few stem cells, the most primitive type of bone marrow cell that, through a series of transformations, can convert itself into red cells and any of the dozen different types of white blood cells. Unless a large percentage of the stem cells were genetically altered, the effect of the gene transplant would fade as the more mature cells that had absorbed the genes died off. Stem cells are the wellspring for the blood. Only the stem cells are immortal and able to reproduce all other blood cells. If the stem cells were genetically changed, then *all* subsequent blood cells they produced would be changed as well. Since Cline and Salser seemed to have fixed so few stem cells, the effects of the gene transfer should be transient. It was, after all, the first time that anyone had put genes into a whole animal, and demonstrated that the genes functioned. Cline was encouraged.

When he reported the results in 1980, a year and a half after they began, in *Nature* magazine in April, and *Science* magazine in May, Cline wrote that these experiments directly pointed the way toward human trials. "Two obvious potential uses of this gene-insertion technique are apparent for use in man," Cline wrote in *Science;* the first could be the "induction of a higher degree of bone marrow resistance to the toxicity of anticancer drugs like Methotrexate." In *Nature,* he wrote that a second use would be the "insertion of drug-resistance genes coupled to other genes to treat human genetic diseases such as the haemoglobinopathies," diseases like sickle cell anemia and thalassemia.

Not everyone agreed. "This approach remains firmly in the realm of speculation," Robert Williamson, a molecular genetics expert at St. Mary's Hospital Medical School, University of London, wrote in *Nature* when Cline published the first mouse data. Williamson reviewed the difficulties and inefficiencies with the existing gene transfer techniques and, for the first time, raised the critical question of controlling protein production directed by a newly inserted gene. "Expression in inappropriate tissue or abnormal control in the cell in which they are normally expressed could be disastrous. A surplus of one globin chain would be as harmful physiologically as a deficit, substituting an iatrogenic [doctor-caused] illness for a genetic one."

Williamson raised questions for which no one, not even Cline, had answers. For example, no one had any idea how to regulate the amount of protein made by a transplanted gene in the recipient cell. For a complex structure like the hemoglobin molecule, where two separate genes must be turned on simultaneously to make exactly the same amount of proteins, making too much of either alpha globin or beta globin could conceivably cause disease.

• • •

In the spring of 1979, a year before the *Nature* and *Science* papers were published, Williamson's questions had not yet been raised. What's more, Cline had six months of his own preliminary mouse data to consider. To him, the initial results were compelling: The experiment was working. The genes were getting into the mouse bone marrow, and they seemed to be functioning. The time had come to try gene transplants in people. Besides, his competitive instincts had been aroused. The idea of putting genes into people was so clear and elegant that he became convinced that other research groups, in Boston or Bethesda perhaps, would try it soon. Certainly French Anderson at the NIH and Ted Friedmann at the University of California at San Diego, just down the road from UCLA, had been talking about gene therapy for years. If he didn't hurry to launch his own human trial, he would miss the boat. Nobels do not go to second place.

Cline, with Winston Salser and two other colleagues, Howard D. Stang and Karen E. Mercola, formulated a plan that laid out how they would repeat the animal experiment in people. On May 30, 1979, they submitted a study outline, called a protocol, to the UCLA Human Subjects Protection Committee, also generically known as an institutional review board or IRB. The committee had ultimate jurisdiction over all human experiments that took place on the UCLA campus. It was the committee's job to make sure that all clinical trials were conducted ethically, that human patients were not exposed to undue risk, and that there was some evidence that the volunteers, often dying patients, had some chance of being helped by the treatment. It wasn't ethical for doctors to take a shot in the dark just because the patient was going to die anyway.

The Cline-Salser protocol said they would put the beta globin gene that Tom Maniatis had given them into the bone marrow of humans who were sick because they did not make enough, or make normal, hemoglobin. They gave the protocol a broad title, "Autologous Bone Marrow Transplantation in Sickle Cell Disease and Other Life-Threatening Disorders of Hemoglobin Synthesis," and sent it to the committee.

After seven years of national, public, and sometimes messy battles about the dangers of putting recombinant DNA in bacteria, a proposal to put genes in people was stunning. The board members immediately recognized the novelty and the gravity of the proposal. It was such cutting-edge science that some board members feared they lacked the expertise to evaluate the plan properly. Cline, however, wanted his proposal kept secret, with the discussion confined to the UCLA committees with jurisdiction. He feared news of his plans would leak to the competition, stimulating them into a race to be first. This demand hampered the committee's ability to review Cline's proposal, and may have slowed down the process. Gene therapy had never been done before. The committee members had no precedent to guide them, and they needed help. Yet Cline would resist the committee's attempt to get outside assistance.

This contributed to a fourteen-month battle between Cline and the committee. Their struggle over the appropriateness of the experiment foreshadowed, in most details, the difficulty French Anderson would face persuading the National Institutes of Health to give him permission to launch his own human gene therapy test more than a decade later. The discussion turned on many of the same technical issues: How many cells would be genetically altered? How would it be done? Which patients would be candidates? How safe was the procedure? And most important, what animal studies had been done to show that the proposed technique would actually work?

Two serious differences, however, separated French Anderson from Martin Cline. Anderson better understood the molecular biology, which had advanced considerably in the time between the two experiments, and Anderson respected the public review process and the ethical theories on which it stood. Worse, Cline held the members of the UCLA Human Subject Protection Committee in contempt.

"They were not very impressive," Cline said later. "I felt the members of that committee were ill qualified to judge human experimentation because many of them had never done it. The people who constituted that committee were not stupid, but they were not intellectual giants, and they certainly had no experience in the field of human clinical research."

To make things worse, there was a simmering struggle between Cline's department of hematology and oncology, which conducted some of the most difficult and painful clinical trials at the university, and the human subjects committee sworn to protect patients from needless suffering. With the local politics as a backdrop, something as spectacular as a gene therapy review could not go smoothly.

What's more, the committee knew of Cline's contempt. By July 1979, when the UCLA Human Subjects Protection Committee sat down to quiz Cline about the gene therapy protocol, the battle lines were drawn. Everyone knew it was going to be a difficult review.

Because Cline's proposed experiment went way beyond the leading edge of science, the committee barely understood what would be done. "This [protocol], as you can imagine, is probably one of the most difficult that we've had to consider because of its newness," Esther F. Hays, a physician and acting chairman of UCLA's Human Subject Protection Committee, said at that first meeting. "We don't have any other background to go on. No other institutions have done this, or no other people have done it."

Robert Lehrer, vice chairman of the UCLA School of Medicine, whom Cline himself had recruited to the university, put it more bluntly: "We are in the uncomfortable position of trying to balance risks to subjects which are unknown versus benefits which are potentially enormous, on the basis of information we cannot evaluate."

They had no trouble understanding the proposed experiment's significance, nor that they needed to worry about Cline's motives. "There's a spot in history reserved for the first person to put recombinant DNA into people," Lehrer said. "It's like when the first person climbed Everest, and if he had found 'Kilroy was here' on a rock, it would have diluted the experience. This is the medicine of the future and the question is, is it being done now because it's time to do it, or is it being done now to get that first footprint on this new surface. I don't know."

Cline too worried about finding someone else's footprints when he walked into the new world of human gene therapy. He had declined to publish his preliminary mouse studies until the committee gave him permission to proceed with human patients. "I would prefer not to submit it [to a journal] until we could go ahead and do our human studies, because as you see from the protocol itself, it sounds so simple in essence that any of a number of groups around the world are quite capable of doing it. And this is why I am so concerned about confidentiality on this report," Cline said.

The UCLA committee pushed again to get outside consultants to help them review the study. Cline resisted. David D. Porter, chairman of the university's separate biosafety committee, a second committee that mostly worried about accidental infections from germs used experimentally and an IRB member, pressed Cline, saying that many experts from a special Cooley's Anemia Foundation committee would be perfect to review this since they know the science. Cline again objected, saying, "French Anderson is on that committee, as I recall. No, I would say no. Certainly not." The last thing Cline wanted to do was tip off Anderson to his plans. Years later, Cline would dismiss Anderson, saying, "I didn't consider him a serious intellectual competitor, and I didn't take him very seriously." Still, Anderson had declared himself in the field, and Cline was not going to give him any help.

Among the most perplexing technical questions were: Which cells would be targeted? Was there really any hope that the treatment would help the patient? How would Cline know that the treatment was working? Cline described how his team would remove some of the patient's bone marrow and mix in the genes, which could become part of any of the cells. Clearly, many of the genes would go into mature white blood cells that would be of no benefit to the patient. Most of the mature white blood cells would die off in a few weeks. The true goal would be the stem cells, those bone marrow cells that produce all the red and white cells in the blood.

But stem cells are exceedingly rare. Cline estimated that there would be some 10 billion cells in the 10 milliliters of bone marrow he expected to remove. Of those cells, only 1 million would be the pluripotent stem cells (that can become any other blood cell) and another 10 million to 100 million committed stem cells (which only give rise to one type of blood cell). But given the inefficiency of gene transfer with calcium precipitation, "there would be between 10 and 100 [pluripotent and committed stem] cells that would pick it [the new gene] up," Cline said. So 10 to 100 stem cells that give rise to all other blood cells would be corrected out of the total of a trillion (10^{12}) cells in the bone marrow. A pretty small and probably clinically insignificant number of cells, even by Cline's estimation.

Worse, it was not clear how Cline would know that the genes were even going into the cells. Simply measuring for the presence of beta globin would not be good enough because the patients also would receive blood transfusions to keep them alive. The beta globin from the transfusions would contaminate and confuse the results. If Cline measured for beta globin, he would not be able to tell whether it came from the gene transplant or the blood transfusions.

In addition, Cline faced one more perplexing, potentially explosive complication: race. Dave Porter brought it up. Since Cline initially planned to attack sickle cell anemia—not thalassemia—the experiment would be carried out in an African-American patient. "It begins to look just a little funny when you're carrying out something superexperimental on a single minority population group," Porter said. That happened in the past. In the 1950s, United States government doctors caused an ethical and political furor when they allowed African-American men in Tuskegee, Alabama, infected with syphilis to go untreated so the natural history of the disease could be studied. And earlier in the 1970s, the federal government launched a campaign to screen for sickle cell anemia, including individuals who carried one of the two defective genes needed to produce the disease. When two carriers were married to each other, counselors often advised them not to have children because the risk of an afflicted child was high, one in four. The African-American community saw this as a subtle form of genocide—using science as an excuse—and harking back to Hitler's final solution for the Jews. Porter wanted to know if Cline had talked over these concerns with any African-American leaders.

Cline bristled at the suggestion. He argued that race had nothing to do with science and clinical trials. If it was a green person's disease, Cline argued, it would be done in green patients. "I think the ethics are exactly the same as doing it in a white person. . . ."

Porter agreed, but added, "I'm not black, and black people have been extremely sensitive about the screening programs for sickle cell. . . ." Cline failed to appreciate the warning about cultural appearances and the way people react to perceived transgressions. It reflected his ability to dismiss concerns that were not his own.

After struggling over the experiment's technical details, the human subjects committee turned its attention to the patient consent form, a document intended to ensure that the patient understood the potential risks and benefits from the experiment—the committee's specialty. But by now, the patience of some committee members began to fray. In one challenge to the wording, Dave Porter blurted out, "Maybe we just ought to put in a statement that says something like 'This is an outrageously experimental piece of work. . . . ' "

The meeting ended without resolution, any decision on the proposed study deferred until additional questions could be answered, and until a second committee, the UCLA recombinant DNA subcommittee, had gathered two weeks later. Things did not get much better for Cline during that recombinant DNA morning meeting later that same month. Much of the discussion turned technical, centering on whether Cline would be using recombinant DNA or not, and whether the experiment needed to be approved by the federal government's Recombinant DNA Advisory Committee, the RAC. But Dr. Daniel Ray, a UCLA microbiologist, went further, saying "the hazards [of the experiment] appear so high as it stands that one need not invoke recombinant DNA to suggest that it may be too hazardous to do."

With the proposal under attack from several quarters, T. Randolph Wall, a microbiologist-immunologist in the medical school, complained that "this experiment is dancing on the front edge of several different kinds of very extreme technology," with many unanswered questions about its use and effectiveness.

Dr. Richard Barnes, another UCLA scientist, went further, calling the protocol "very, very disturbing." He went on to complain that the "scientific aspect of this research . . . seems to be rather fuzzy and hazy . . . [and] the investigators don't say specifically what they are going to do." He worried that the committee would tell Cline and Salser how the experiment should go, and then the committee, not Cline, would become responsible for its outcome.

The recombinant DNA bunch too declined to make a decision, leaving Cline suspended in a limbo of unanswered questions. His experiment hung between two committees, each of which could only make up its collective mind to defer to the other committee. The situation made Cline angry.

The use of recombinant DNA was a key problem. The term itself had become a lightning rod, a red flag, a code word for the gene debates throughout the 1970s. It had brought genetic research at Harvard and MIT to a halt. State and local governments across the country had threatened to regulate it. This was not merely science, but the interaction between science and society, and the interaction scared a goodly portion of the society that did not understand it.

Recognizing the difficulties they would face in trying to get permission for a recombinant DNA experiment in people, Cline and Salser changed the experiment, three months after submitting the proposal, to eliminate the recombinant DNA component. In a letter on September 18, 1979, they wrote:

> Some of the modalities of treatment suggested involved the use of recombinant DNAs. We wish once again to stress the recognition of our obligation to obtain separate approvals from both the Human Subjects Protection Committee and the Institutional Biosafety Committee before starting any experiments which involve the use of recombinant DNAs in patients. . . . Therefore, we wish to modify our original proposal . . . *to include only those modalities of treatment which are specifically and clearly exempt from the NIH Recombinant DNA Guidelines.* [Cline's emphasis.]

Cline sought to use a subtle difference between what was and what was not considered a recombinant gene experiment to avoid some of the rules, and move the approval process along more rapidly. Instead of using a recombinant form of the gene, in which genetic elements are physically pasted together—such as the cancer-drug-resistance gene linked to a carrier DNA molecule called a plasmid that allowed the gene to replicate in bacteria—Cline would snip all the genes apart from each other with enzymes just before putting the genes in the patients. The genes would still be recombined in the lab. They would still be highly engineered DNA molecules. But technically speaking, these naked genes would not be recombinant DNA when they went into patients, according to the technical, legalistic definitions used by the National Institutes of Health, the arbiter of such scientific uncertainties.

Cline and Salser would spend much of that fall lobbying NIH and UCLA officials for a formal exemption of their experiment from the national guidelines established to regulate recombinant DNA research. By year's end, they would win on that front. Their experiment would no longer be considered a recombinant DNA trial, as NIH confirmed in writing. That victory, however, proved meaningless.

A day after Cline's letter to redefine the experiment, on September 19, 1979, the Human Subject Protection Committee met again, and refused to act. The protocol had become mired in cross-committee communication and decision making, with each committee wanting something from the other committee before it would act. What's more, the human subjects committee had acquired on-campus consultants, and they too had a list of technical questions they wanted answered. The technical questions foreshadowed issues that would face a similar NIH committee when it considered French Anderson's protocol a decade later.

One UCLA consultant wanted to know whether enough of the hemoglobin gene would get into the patient's blood cells to be expressed to make a difference in the disease. Cline replied: "I don't know the answer. . . . It will not be possible to answer this question until we do studies in man." It was the ultimate defense from a clinician, a defense rejected by basic scientists who believed any biological questions could be answered in animals. Anderson would later use similar arguments about the need to answer questions in human patients.

Because Cline was not able to answer questions to the satisfaction of either UCLA committee, a protracted correspondence ensued, with the human subjects committee demanding more and more information and clarification. From Cline's viewpoint, the situation deteriorated further when Jeremy H. Thompson, a tough, London-trained pharmacologist from Ireland who ruled the human subjects protection committee with an iron hand, returned from sabbatical that fall of 1979 to retake the committee's reins. Thompson had been at UCLA since September 1962, as a full professor in the pharmacology department. By his own description, he "did some research, did some teaching, published a book or two; rather mediocre research work." After joining the committee around 1970, Thompson became its chairman in the late 1970s, having become a leading expert from a major institution on the evolving guidelines for protecting human subjects, and for running institutional review boards, IRBs, as his committee was called in federal government regulations. Thompson and Cline had crossed sword points many times before; no love was lost.

But now Thompson was back, and he had power over Cline's hottest idea. There were repeated rejections by the committee and a seemingly endless list of technical questions. Each time, Cline responded in writing, laying out the arguments, reviewing the animal data. Each time, Thompson would write back, saying, "The Human Subject Protection Committee has carefully evaluated the above study, but unfortunately cannot approve it in its present form."

Finally, on February 29, 1980, Cline wrote to Thompson:

> I would like to . . . request of Vice Chancellor [Albert A.] Barber that he convene a special committee comprised of experienced hematologists and of molecular biologists in order to review these protocols. . . . I feel the Human Subject Protection Committee is thoughtfully and honestly attempting to grapple with the issues contained within these protocols, but that the Committee is struggling in water that is deep and perhaps beyond its depth. This is not altogether surprising, in view of the unusual nature of the protocols and the fact that the Committee has few experienced clinical investigators in the area of hematology, and as far as I am aware, no molecular biologists included amongst its membership.

Cline was becoming frustrated. He had expected the review to take a couple of months, but now it was coming up on a year. The animal studies were virtually completed, and draft reports had been written for *Science* and *Nature*. Convinced the results were genuine, Cline urgently wanted to start the human trials. More than a month later, April 3, 1980, Thompson agreed "that expert review of your proposal by 'experienced hematologists and molecular biologists' will be necessary. . . ."

Another review, another delay. Cline accepted the review, but asked for face-to-face meetings over future questions, complaining that it was too difficult to answer complex questions through the UCLA mail system. And since the turnaround time between questions and answers was a month, "committee members must surely lose their grasp of the essential components . . . ," Cline wrote.

By May 1980, Thompson began to assemble a panel of outside experts and allowed Cline to reject anyone as a consultant who posed a competitive threat. Cline constantly insisted on confidentiality by the committee because he still feared some other scientist would attempt gene therapy once they heard of his ideas or, more important, read the details of his proposal. Cline listed seven scientists for rejection: Tom Maniatis from Cal Tech, the researcher who gave Cline the hemoglobin gene; Richard Axel from Columbia, the developer of the technique Cline would use to put the gene into the bone marrow cells; French Anderson and Art Nienhuis at the National Institutes of Health; David Nathan at Harvard University, who was to some the father of modern hematology and a collaborator with several emerging gene therapy experts, including Richard Mulligan; and George Stamatoyannopoulos, a bone marrow expert at the University of Washington in Seattle, whom Cline considered a threat.

. . .

Even as the torturous review process ground on through the fall of 1979 and the next spring, Cline worried about another problem. He needed an ample supply of human volunteers deathly ill from sickle cell anemia or thalassemia to test his ideas about gene therapy. His survey of the Los Angeles area found no patients with beta thalassemia at UCLA and only five patients at Children's Hospital of Los Angeles, a part of the University of Southern California, too few families with children seriously ill from the disease to launch the kind of clinical study that would probably take years of early failures and subsequent refinements to achieve an ultimate success. There did not appear to be enough patients in Los Angeles to conduct a major trial of human gene therapy. He needed a better supply.

The best source for patients would be in Africa, where sickle cell disease is common but research facilities are scarce, or in Mediterranean countries where thalassemia is frequent and facilities adequate. Cline chose the Mediterranean and set about establishing relationships with doctors he knew on either side of that sea; they could provide him a steady stock. Besides patients, Cline also needed a more benign political climate in which to make history, rather than face the relentless delays by the UCLA committees.

Israel was the best hope because Cline had his strongest connections there. He had become friends with Eliezer A. Rachmilewitz (*rock-ma-lay-vitz*), a physician and hematologist who ran the blood department at the Mount Scopus Hospital, a part of the Hadassah–Hebrew University hospital system in Jerusalem. Cline had visited there before, and had a special fondness for Israel. Rachmilewitz even had tried to help Cline's son become involved in an archaeological dig in the ancient city.

Early in 1980, Cline had written to Rachmilewitz to tell him he was thinking about gene therapy for thalassemia and about coming to Israel that summer. Cline wanted to know if Rachmilewitz was interested in participating. He was. Rachmilewitz wrote a one-page letter that he put it into the hands of his father, Moshe Rachmilewitz. The senior Rachmilewitz, a towering figure in Israeli medicine, was on his way to Los Angeles to be a visiting professor in Cline's department. About gene therapy, the younger Rachmilewitz wrote, "We are thinking along similar lines and if you would like to work with us, it will be an honor and pleasure to have you." Cline had his first invitation.

Rachmilewitz went on to describe the work underway in his Mount Scopus lab, including cultures of hard-to-grow thalassemic cells from peripheral blood that survived fourteen days, quite a feat at that time. This gave Rachmilewitz enough blood cells to measure messenger RNA and protein synthesis in diseased cells.

Then Rachmilewitz wrote the magic words: "This is our 'ace' because to my knowledge not many places have so many available patients at any time for growing the cells." "Available patients." That's what Cline needed. The rest of the technical descriptions in Rachmilewitz's letter, dated February 27, 1980, were comparatively inconsequential.

The letter arrived in Los Angeles a few days later with Moshe Rachmilewitz. Cline acted immediately. On March 5, 1980, Cline sent Eliezer Rachmilewitz the current version of the proposed study before the UCLA committees and a preprint of the *Nature* paper. "I believe we are light years ahead of any possible competitors . . . ," Cline said. "We are now ready to conduct these trials in patients. If you can provide one or two patients with severe beta-thalassemia, then I would undertake the research outline in the protocol with you during my visit to Israel."

Then Cline turned up the pressure. "Obviously, the potential rewards of such research are enormous—the first demonstration of the correction of a genetic defect in man. If you feel that you do not have access to such patients or cannot undertake the research, please let me know as soon as possible in order that I can organize alternate collaborative arrangements in Greece or Thailand. Your reply will therefore determine how much time I spend in Israel."

Rachmilewitz felt the pressure and leaped at the chance to have history tumble into his lap. "I would like you to know that we are looking forward to work [sic] with you and are willing to do everything to have this project get started," he wrote back on April 28, 1980. "The major obstacle that will have to be solved is the permission from the Human Experiments Committee, fortunately headed by my father."

Rachmilewitz went on to discuss the pros and cons of different strategies in approaching the Israeli committee. Then, he raised the most troubling point: "One of the bothering issues which I have heard already is why does Marti Cline have to come all the way to Jerusalem to do it, probably because nobody in the U.S. is willing to cooperate with him in this fantastic venture." Rachmilewitz worried that the committee too would find this strange, and urged Cline to provide a good explanation.

Although both Israel and Italy took great pride in the quality of their medical systems, neither was considered a leader in many fields of research. The hottest biomedical research tended to come from the United States, though sometimes also from other countries like England or France. Israel, despite good connections to many leading universities, tended to be a user of medical technology; Italy too was inclined to be a research backwater. Why, indeed, did the great Martin Cline need these scientifically small countries if he wanted to make history, and could do it just as easily at his own lauded institution where he would get all the credit?

Cline responded debonairly. On May 20, 1980, he wrote to Rachmilewitz, "I would greatly like to do the experiments in Israel but it is not absolutely necessary. We are planning to do our first patient within the next two weeks." He already knew that the UCLA human subjects committee had just acquired outside consultants to begin a new round of reviews, and that he would never start the experiment that soon. "And we might even be able to begin a second patient before I leave for the East Coast, Western Europe, and Israel in early

June," he went on. "If we can get through the political intricacies in the [Israeli] human experiments committee then we shall do an experiment together on a thalassemic patient; if not, then we shall merely restrict ourselves to some *in vitro* studies. . . ."

Rachmilewitz reacted with horror. "We are *most* interested that the first experiments will be done *here* and nowhere else [his emphasis]," Rachmilewitz wrote on June 2, 1980, a few days before Cline left Los Angeles. Rachmilewitz went on to promise that his family would do everything possible to make the trip a success.

A similar, though less warm and personal, exchange of letters occurred over the spring with Cesare Peschle, head of hematology at the University of Naples in Italy. Cline sent Peschle a preprint of the *Science* article describing some of the animal experiments. Although Cline discussed the protocol in general details with Peschle during the Italian's visit to UCLA, Cline did not send a copy of the written protocol under review by the UCLA committees that he had sent to Rachmilewitz.

Cline had pinned his hopes on Rachmilewitz; Peschle was the backup. Yet Cline was a cautious and pragmatic man. Something could always go wrong. He still did not have approval from the Israeli human subjects committee, and after more than a year's delay by the UCLA committee, over which he had some influence, he was not going to take any chances. Italy, where the rules about testing experimental treatments in people were much looser than in the United States or Israel, would be a good alternative if Rachmilewitz failed to come through.

When Cline's plane lifted off for Italy that summer, all of the experiments hung in the air: UCLA's human subjects committee had yet to rule on the protocol he had first proposed more than thirteen months before. The Israelis had not given the green light, and the Italians wanted to talk about it before going to their superiors.

In Naples, Karen Mercola, a researcher in Cline's UCLA group, was already at the Istituto Superiore di Sanità setting up a series of experiments to demonstrate that they could transfer genes into blood cells growing in test tubes when Cline arrived. In these experiments, it was a true collaboration. Mercola came with the genes Cline and Salzer had prepared in California; Peschle had cells, unique cells. His lab was one of the few worldwide with good success growing cells from beta thalassemia patients in the test tube.

With the laboratory experiments launched in Naples, talk turned to Cline's first priority: treating patients. Although Peschle was trained as a physician, he did not have patients of his own. He had confined his work to the laboratory. But Peschle knew where to get patients: from Dr. Velma Gabutti of the Second Pediatric Clinic at the University of Turin. Velma Gabutti was a foot soldier in the medical trenches battling blood disorders common in Italy and throughout countries confining the Mediterranean Sea. Her team cared for an enormous number of thalassemia patients, around 250 in all. And her patients, it was said by those who knew her, were treated as though they were her own children.

Gabutti flew into Naples to meet Cline, and to hear about his ideas for using genes to treat patients with thalassemia. She was intrigued and agreed that it might be possible to make a patient available. But, before anything could be arranged, time ran out for Cline. His wife Evie and son already had arrived in Jerusalem two days earlier and were waiting for him. Eliezer Rachmilewitz too was expecting him to arrive from Italy. As head of hematology at Mount Scopus Hospital, Rachmilewitz knew firsthand the pain of thalassemia; his small group of doctors managed about 150 patients with the illness, Jews and Arabs alike.

Initially, Israel disappointed Cline. Mount Scopus Hospital was a new, beautiful, and somewhat underused facility. Despite its association with Hadassah Medical Center and Hebrew University, Mount Scopus was not really a research facility, just a hospital. Cline scrapped the lab studies. Instead, he and Rachmilewitz decided to approach Hadassah's own human subjects protection committee: He wanted approval to try human gene therapy.

The Israeli human subjects protection committee had been set up similar to the American system. A few years earlier, Hadassah had received research funds from the U.S. National Institutes of Health for a small clinical study, and Hadassah and Hebrew University were due to receive an additional $1.5 million from NIH that year. NIH rules require any institution using its money to establish a human subjects protection committee, and the Israelis took that charge seriously. In every hospital that did research, they already had set up ethics review groups, called Helsinki committees, in honor of the international agreement signed in Finland that protects patients from unethical experimentation, and ensures that the patient has the opportunity to give informed consent before any procedure is performed.

Rachmilewitz had decided in advance that no proposal would be made to any of the Israeli committees until Cline arrived. And Cline would make the pitch himself, alone, without Rachmilewitz. But the Israeli scientist would provide some key assistance: his influential father. Moshe Rachmilewitz, the retired dean of the medical school at Hebrew University, the friend and personal physician to the country's prime ministers, and a man who had been a visiting fellow in Cline's UCLA department of hematology-oncology, chaired the Helsinki committee that Cline would have to convince. That, however, was no guarantee of success, just assurance that Cline would get a friendly hearing.

Still, the younger Rachmilewitz had his own concerns about the experiment. The whole idea seemed so preposterous. Yet, he didn't really understand the molecular biology enough to evaluate it, so he was forced to trust Cline's judgment that it would work. At the same time, he knew these patients were desperately ill, and the one he had in mind for the initial test was dying. To Rachmilewitz, there did not seem to be any serious risk to the patient—at least that is what Cline's animal studies showed—even though some pretty invasive procedures would be used. If there was little chance it would hurt and some chance that it might help—no matter how small—then it could be worth it. Besides, if it worked, his career would soar, and even his father would have to admire the experiment's boldness.

But not everyone on the Helsinki committee admired it. Cline's proposal was right on the edge of acceptability in the eyes of many at Hadassah. They *were* skeptical about why this famous researcher wanted to conduct such a flamboyant experiment in such a tiny place and not at his own major institution. Cline turned on his persuasive charm. He described his recently published animal experiments and laid out his hopes for a revolutionary genetic treatment. He also admitted that his proposal to do the same thing at UCLA had languished for more than a year before the university's human subjects committee, and that, as far as he knew, no decision had been made.

Despite their misgivings, the Israelis also understood the suffering of thalassemia and their own limitations in treating it all too well. Like the Italians, they were somewhat dazzled to have a major scientist from a large American university offer to bring them a treatment that might cure the illness, a treatment that would change the future of medicine, and perhaps some share of a Nobel Prize.

Cline made his push in Israel, believing he could get the permission he needed if he just applied enough persuasion and pressure. Cline lobbied the Hadassah Helsinki committee for its consent. It was a small committee, only four members, chaired by the senior Rachmilewitz.

Hadassah officials also called Dr. Leo Sacks of the Weizmann Institute in Rehovot, who chaired the Israeli Academy of Science's committee on recombinant DNA experimentation. The committee mirrored the American Recombinant DNA Advisory Committee, reviewing plans by any Israeli scientist who wanted to do genetic engineering. After the experiment was explained to Sacks on the phone, his response was terse: "No, don't do it."

It was not the answer Cline wanted to hear. He countered by explaining that he had managed to avoid a federal review by the NIH RAC by redefining the experiment so that recombinant DNA would not be used. Helsinki committee members pressed Cline to assure them that the experiment followed all of the American guidelines, especially regarding recombinant DNA. Cline assured the Israelis that the experiment could be done without recombinant DNA, only with naked genes. He offered the same version of the study that had been revised to skirt the NIH rules about recombinant DNA experiments.

Cline offered the Israelis something he could not deliver, at least not immediately. The human and viral genes Cline carried with him were all spliced into plasmids, the rings of bacterial DNA used to reproduce large amounts of the genes. They were recombinant in form. Cline told the Israeli committee, however, that he could do the experiment without recombinant DNA. In California, Cline had planned to use enzymes to cut away all of the plasmid DNA and use the naked gene alone. When Cline traveled to Israel, according to to his partner Winston Salser, he possessed neither the reagents required, nor the technical expertise needed, to create naked genes from the recombinant forms he had in the Styrofoam container.

Nonetheless, to buttress the claim that the experiment could be done without recombinant DNA, Cline had Salser send a telex from the UCLA chancellor's office to Yochanan Benbassat, associate professor of medicine at Hadassah University Hospital, confirming that the experiment (at least as envisioned in California) would not use recombinant DNA material. Salser liberally quoted from a November 7, 1979, letter from UCLA biosafety chairman David Porter, concluding that after reviewing the Cline proposals, the UCLA committee agreed that the study did not involve the use of recombinant DNA and was not covered by the NIH guidelines.

Still the Israelis were cautious; the human subjects committee showed the telex to Dr. Ezekiel Halpern of the Hadassah–Hebrew University Medical School. The telex didn't convince Halpern that the experiments were exempt. All of these assurances had come from UCLA. So, Halpern called Dr. William Gartland, then director of the NIH Office of Recombinant DNA Activities (ORDA) to ask if it was indeed true. Gartland confirmed that there had been many letters between UCLA and his office, and that the UCLA experiment, as Cline and Salser had modified it, would not be defined as a recombinant DNA. The UCLA experiment could proceed in the United States without first getting approval from the Recombinant DNA Advisory Committee.

Where the breakdown in communication occurred is not clear. Gartland merely acknowledged that the Cline experiment was possible without using recombinant DNA and that the NIH was aware of the current plans being reviewed in California—plans that did not require NIH approval. The Jewish scientists interpreted this to mean that none of the material was considered recombinant DNA, and that they could go ahead and still comply with NIH guidelines. Cline did not resolve the ambiguity.

Leo Sacks, the head of the Israeli recombinant DNA committee, was called back and informed that the experiment did not use recombinant DNA, according to the NIH guidelines. His committee's consent was no longer needed. Sacks recalled telling the Hadassah official that they should not do the experiment anyway. He thought it was premature, that the science was not well enough developed. Sacks was ignored.

At the same time, the Hadassah committee called in two outside advisers, both experts in genetic engineering, to review the proposal. At least one of these consultants had concerns about the experiment, the unnamed scientist said to reporter Paul Jacobs of the *Los Angeles Times* in October 1980, well after the experiment was publicly criticized. Jacobs had gone to Jerusalem to interview the participants in the study; one of the outside experts, who refused to be named in print, said he believed that "the gene transfer was premature." Apparently, his concerns also went unheeded.

Compared to the months-long delay at UCLA, Cline's proposal sailed through the Israeli committee: It only took them a week to make a decision. According to a report in the French newspaper *Le*

Monde, it had even been authorized by the Israeli Minister of Health. But to get that decision, Cline again had to turn up the pressure.

Even as he negotiated with the Israelis, he continued telephone discussions with Peschle, Gabutti, and their superiors in Italy. By Wednesday, July 9, 1980, no Israeli decision had been made, and Cline was preparing to leave Israel and return to Italy. He indicated that he had gotten a preliminary indication that he would be allowed to proceed with the experiment in Italy. The Israelis, apparently seeing their chance to participate in history slipping away, made a quick decision.

The night before Cline was to leave, Rachmilewitz called Cline at the Jerusalem apartment where he was staying, and told him the Helsinki committee had decided to give their blessing to the experiment.

Months later, Kalman Mann, director-general of Hadassah, would have second thoughts about how the review of Cline's experiment proceeded, and whether Moshe Rachmilewitz should have been allowed to review a controversial research proposal on which his son was cosponsor. "In looking back, it was not an aesthetic thing," Mann told the *Los Angeles Times.* "Maybe he should have disqualified himself."

Still, Mann argued, the Israeli review was thorough. Although the possible benefits were not clear, there seemed to be little danger to the patient. "If there is the slightest chance that [a treatment] is helpful, then it is worth trying," Mann told Jacobs. Besides, the Israelis asserted, the pioneering patient was so sick from her disease that she was not expected to live long. On that basis, the decision to try the experiment was ethically justified. Martin Cline was relieved. Finally, after more than a year and half, he would be able to get on with it.

• • •

The patient would be a twenty-one-year-old Kurdish Jew with severe, lifelong beta-zero thalassemia major. Her name was Ora Morduch, and she lived in Jerusalem, not far from the Mount Scopus Hospital. Her family had been among the legions of Kurdish Jews who emigrated from Iraq to Israel in the 1950s—when whole villages were packed into planes and transplanted in the newly formed Jewish state, spreading across the desert, claiming the sands for Israel. They had come from a part of the world that had long suffered the sting of malaria-carrying mosquitoes. Iraqi Kurdistan was heaven for *Plasmodium vivax,* the most common type, and *Plasmodium falciparum,* the mostly deadly form of the malarial organisms, and the female, blood-sucking *Anopheles* that carry them. As in any area with endemic malaria, thalassemia and other genetic blood disorders that conveyed some protection against *Plasmodium* also were common. This proved especially true among tribal Kurds where inbreeding within isolated villages was common.

Ora Morduch's parents were Kurdish carriers. They each had one defective beta globin gene and one normal beta globin gene. This genetic shuffle gave them some protection against malaria without other serious symptoms—just mild anemia as is common among people with this genetic mix. For Ora, however, the genetic role of the dice turned out differently. She received a defective beta globin gene from her mother and a defective hemoglobin gene from her father. Consequently, she could not make any globin, and she was without the ability to make normal red blood cells.

The girl had suffered severe anemia. Because she was born before transfusion treatments were widely available in Israel, she suffered deforming disease in her bones as her marrow expanded within the rigid matrix to produce more and more hopelessly flawed red blood cells. Consequently, she was short, almost dwarflike. The long bones of her legs had changed so much that she limped when she walked, and she suffered repeated fractures from bony deformities and chronic pain. Even the bones in her skull and face had tried to grow more bone marrow, expanding to open lacunae, creating in the process typical "gargoyle" features on her disfigured head and face that typify the disease. It even affected her skin, which had become darkly stained from the deposits of iron caused by the constant transfusions.

Despite the lifelong struggle with physical defects, Ora Morduch was a remarkable person. Highly intelligent, she attended Hebrew University, earning a master's degree in social work. She was ambitious, spoke fluent English, traveled widely around the world to attend scientific meetings on thalassemia, and had become an activist leader of the local patients' organization for thalassemics. To Rachmilewitz, she was both a favorite patient and a troublemaker, always looking out for the other thalassemics, complaining when she didn't feel the care measured up to what they all needed. She also tried to ignore her own disease; this made her difficult to treat because she was often late, by days, for her scheduled transfusions and follow-up visits. Yet, he had grown fond of her, and was amazed how well she survived. "Everything [in her body] was rotten except her brain," Rachmilewitz said.

But now the effects of the constant blood transfusions were catching up with her. Each transfusion deposited more iron in her body. Her overloaded liver dumped the excess back into the bloodstream, where it accumulated in other organs, especially her heart. She suffered intermittent cardiac arrhythmias, galloping, uncontrolled heart contractions that fail to pump the blood adequately and threatened heart failure. Ora had to be hospitalized frequently to stave off sudden death from a heart attack. Unless some of the iron could be removed from her body, she was going to die.

For her, Rachmilewitz wanted Marti Cline's "fantastic venture" to work. Besides, she was an ideal candidate who fit Cline's criteria: He would only do the experiment on someone in danger of imminent death and someone intelligent enough to understand the treatment. Rachmilewitz explained the procedure to her, told her what he knew of the details of how it was supposed to work, and described the risks, which he considered small. She decided to try it.

When Cline went to see the girl for the first time, he was impressed. She had a clear sense of what was going to happen, and that the chances of her actually being cured on the first try were slight. For half an hour, Cline sat and talked with Ora, going over the procedure step by step. First he and Rachmilewitz would remove some bone marrow from her hip. The marrow would be placed in a laboratory flask and treated with the gene. Meanwhile, the girl would be sent off to the radiation therapy department where a fifteen-centimeter section of her thigh bone in one leg would be doused with 300 RADs of radiation. That, Cline believed, would kill off the bone marrow in that part of the leg, creating a space for the gene-treated bone marrow cells to settle and grow after being injected back into her body. This "making room" hypothesis, Cline admitted later, was the weakest part of the experiment. Only suggestive studies had been done to test this hypothesis in animals. No data existed to suggest it would work in humans.

But Cline was on a roll. The experiment was finally going to happen. He would be the first to perform human gene therapy, a historical breakthrough. Yet, he told Ora, he did not expect it to work—in her case, putting the beta globin gene into her bone marrow cells probably would not make any difference in her physical condition. He knew that he would not get the gene into many cells. He knew that he could not even tell if the globin gene would be turned on. He also knew that there was little chance that it would get into enough stem cells, those bone marrow cells that give rise to all other blood cells, to make any long-term clinical difference in her disease. Still, he said, it was worth a shot.

Like many patients with an ultimately fatal illness, Ora Morduch had only two defenses left: denial and hope. Denial she used regularly and effectively, ignoring her illness as much as it would let her. She also had hope and the capacity to hear what she wanted to hear, that this experimental work might make her well for the first time in her life. And if it didn't work now, perhaps it would work someday, and because she was a pioneer, she would be alive long enough to benefit. Despite Cline's caveats, she agreed to go ahead with the procedure, expecting the best.

Cline and Rachmilewitz received official permission from Hadassah officials at 7 A.M., Thursday, July 10, 1980. Events moved with lightning speed. Rachmilewitz ordered Ora Morduch to come to the hospital immediately. By 9 A.M., she had arrived at the Mount Scopus Hospital and was assigned a bed. Tension began to build. Surgical equipment was brought to a small outpatient examining room, not a sterile operating room but sufficient.

By 11 A.M., Ora had been transferred to the procedure table in the examining room. Rachmilewitz numbed her skin with a local anesthetic. Using a small scalpel, he nicked the skin to make a small opening. Then Rachmilewitz pushed a long, shiny needle—strong enough to burrow into bone—through the slice in the skin, through the fatty tissue above the iliac crest, the back of the hip area just above the buttocks, and down to the bone on the back of the pelvic girdle, just above the hips. Once the cutting edge of the thick needle rested on the bone, Rachmilewitz twisted it as he pressed with his weight so it could crunch through the bony surface to the spongy marrow below. Despite anesthetic drugs, the procedure is unpleasant and moderately painful.

With a vacuum bulb on the needle, Rachmilewitz sucked out a few drops of bone marrow and placed it in a sterile container. This process was repeated. Nick the skin. Drill through the bone. Aspirate the marrow. Move on to another chunk of bone. Again and again until he had collected 15 milliliters of bone marrow, a little more than an ounce, or 0.1 to 1.5 percent of her bone marrow cells. After it was over, Ora Morduch went off to the radiation therapy department where her leg would receive 300 RADs of radiation.

Meanwhile, Rachmilewitz gave Cline the precious cargo to be carried back to the lab. He placed the cells under a microscope to take a look; the cells glistened in the microscope light. The bone marrow cells from thalassemics have too many red blood cell precursors—the cells that produce mature red blood cells—and they usually look somewhat abnormal because of the disease. Cline counted them—between 4 and 6 x 10^8 cells in the total sample. Cline ensured that anticoagulant drugs kept the cells from clumping together. Everything went routinely.

Next came the genes, just a speck of material really. They sat in labeled capped tubes, jammed upright in an ice bucket. Cline towered over them, thoughts of the protocol, about what to do next, racing through his mind. He had agreed to use the naked DNA, the genes sliced free from each other, and from their bacterial carrier plasmids. He didn't have any in that form. All he had were the recombinant forms of the genes. The Israelis had worried about recombinant DNA. Cline had reassured them it would not be used. The fact was, he had no choice. It was all he had. All the genes in the ice bucket were recombinant genes. He lacked the material he needed to cut the individual genes into naked beta globin, naked thymidine kinase from the herpes virus.

Cline felt the frustration of the limitations that had been placed on him. They were political not scientific. The limits were not even rational. The naked DNA, when it was cut out of the plasmids, had sticky ends. These sticky ends would find each other in the test tube, or inside the cells, and reattach to each other in random combinations. Even if he put in naked DNA, Cline reasoned, the genes would recombine anyway, and he would end up with recombinant DNA inside the patients. That's what nature did; no one could stop it. So scientifically, the legal requirement to use naked DNA instead of recombinant DNA did not make sense. The definition he had wangled from NIH to free him from the additional federal review was illogical and technically unimportant.

What's more, Cline had limited, preliminary evidence from his own animal studies that there was a better chance of getting the new genes to integrate into the target cells' chromosomes if they were linked together in the recombinant form. If this experiment was going to work, if this girl was going to be helped, then the only rational thing was to use the recombinant DNA. This ultimately proved wrong—scientifically and politically.

Cline reached into the ice bucket and pulled out the tubes containing the recombinant DNA. Just a sprinkle really, 4 micrograms of thymidine kinase linked to its plasmid; 4 micrograms of human beta globin linked to its plasmid; and 12 micrograms of thymidine kinase and human beta globin linked together into a single plasmid. He used the recombinant DNA, the stuff he promised not to use. This was the better experiment, Cline rationalized. He didn't tell Rachmilewitz, who was standing nearby. Cline did not believe the Israeli doctor would understand the difference. Cline did not tell Ora Morduch either. She had agreed to one experimental protocol, the one without recombinant DNA, but Cline gave her another. A technical difference, but it breached the trust between doctor and patient.

Next, Cline added the calcium phosphate, the holy water in the reaction that mysteriously opened channels in the membranes of cells so strands of DNA could slither their way inside, and become a permanent part of that cell's genetic inheritance. For four hours, Cline and the others watched over the bone marrow cells while they incubated at 37 degrees centigrade, body temperature, with the genes.

After the genes had gotten into what cells they could, Cline looked at the cells one last time under the microscope to ensure that they still looked okay. There they were, as normal as thalassemic bone marrow cells could look. The first genetically engineered cells intended to change the course of a human disease had just been made.

This was the defining moment of Cline's career. He had spent his life working on the cutting edge of clinical research. He had tested toxic therapies on dying people without flinching. Now, he was ushering in the new era of human gene therapy. Instead of continually taking pills and potions for their chronic illnesses, patients would be cured from the inside out. The genes in their cells would be changed, their diseases banished from their bodies.

And here was the courageous doctor willing to risk his career and his patient in the name of medical progress, an effort to save the lives of many thousands of people yet to become ill. Only Cline had the unique combination of skills to make this happen. No one but him, he believed, was more qualified to make the decision to carry out this experiment. Perhaps it would bring him the Nobel Prize and the gratefulness of patients worldwide. If it worked.

The genetically altered bone marrow cells were taken out of the culture material, and washed and centrifuged to remove the free-floating genes. Then, about 800 million of the bone marrow cells were placed in an ordinary blood transfusion bag. A small sample was set aside to test for infectious materials. Another small sample was tested for viable cells: 99 percent were alive.

Ora Morduch had returned from the radiation therapy department, the bone marrow cells in her leg dying from the radiation treatment. Rachmilewitz pushed a catheter needle into her left arm and connected it to the transfusion bag full of cells. Anxiety rose in the room. Genetically altering the cells in the laboratory was one thing, but now they were putting them back into their patient. Cline was tense. None of the mice had ever reacted unfavorably to the calcium phosphate or anything else in the culture medium. But this time, they were going into a human. Anything could happen.

Five hours after they had been removed, the girl's bone marrow cells slowly dripped into her vein. Her doctors hovered, watching for some adverse reaction: a sudden change in pulse, sudden sweating or rapid breathing. Nothing happened. Her pulse and blood pressure remained steady. They all kept watch the rest of the day, Cline and Rachmilewitz and the others. But she was fine.

The next day, Cline and Rachmilewitz repeated the procedure, removing another 15 milliliters of the girl's bone marrow, treating it with genes, and returning it to her body five hours later. The radiation treatment to her legs was omitted.

· · ·

Of the 800 million bone marrow cells Cline treated with the genes, much less than 1 percent would be the self-renewing stem cells that gave rise to all other blood cells. Unless Cline got the gene into a substantial number of them—at best that would only be a fraction of the less than 1 percent—any effect from the gene transfer in Ora's body would be temporary. It would fade as those cells that were genetically altered died out.

Even though Cline had used recombinant genes—the thymidine kinase gene and the beta globin gene—that would make the cells resistant to toxic drugs, his first patients would not receive the selecting drugs as had the experimental animals. Although this decision made life easier for the patient, it may have reduced the chance that the experiment would work. Without selection by the Darwinian drugs, there was little likelihood that the handful of genetically transformed cells would have any selective growth advantage over the defective cells. Even if a few cells were transformed, they would be overcrowded and outcompeted by defective cells, and would die off before they could grow and produce the billions of healthy red blood cells that would be needed to make a difference in her disease.

Still, despite juggling different components of the experiment, despite a nagging concern about the DNA switch he made at the last minute, Cline was satisfied. The experiment was started; the field of human gene therapy was launched. He had done what he had set out to do.

He remained in Israel for a few more days, checking on the patient, making plans for the follow-up studies. Her blood would be withdrawn on a regular basis, frozen, and sent to Los Angeles. Cline's team would use the most advanced molecular tests available to look for the presence of the newly installed genes. Some blood, and even bone marrow, would be frozen and stored in Israel.

With the arrangements for continued collaboration in Israel completed, Cline boarded a jet and flew to Naples. Now that the Israelis had opened the door, Cline was sure the Italians would go along, and make their patient available. Cline wanted to do a second patient, and Italy seemed to provide the best chance. The single Israeli patient would prove nothing by herself—successful or not. He would need a couple of patients initially to show there was some biological activity from the gene transplants, to prove the concept. Then he could recruit more patients to refine the techniques, probably even use patients in Los Angeles.

● ● ●

When Cline landed in Italy, nothing was resolved. The laboratory studies they had started in Naples to show they could transfer genes were not complete. Although Cesare Peschle had considered a human trial from the moment Cline had proposed one, he preferred to get some solid test tube evidence that it worked first. Cline, however, was not going to cool his heels in Italy long enough for those results to come in. He wanted to move to the next stage: a patient.

Dr. Velma Gabutti and a handful of her young physician-collaborators arrived from Torino. They had brought along a patient, just in case. Maybe they would treat the patient with Cline's genes, or maybe would just take out some of her bone marrow and see if they could insert the genes into her cells in the test tube. They remained undecided.

Over lunch, Cline launched into a description of his positive animal studies that showed he could insert the gene, and how the selection techniques seemed to show that the genes were functioning. He also talked about what he had just done in Israel, how they had reviewed his ideas and approved them. And lastly, he reviewed the steps they would take in Italy, if they agreed. Remove the bone marrow, expose it to the genes and calcium phosphate precipitation, return the cells to the patient.

It was a heady meal for the Italians. Here was Cline, brilliant, articulate, an internationally renowned researcher, leader from one of the world's most productive departments of hematology and oncology. He was eloquent, visionary. He gave them a crash course in the future of medicine. The young researchers with Gabutti were dazzled. Peschle was not. He still wanted to see the results from the earlier studies first.

Cline pushed harder. Peschle raised more questions, but he was not a molecular biologist and only dimly understood the details of the DNA treatment itself. Certainly, it would be difficult to differentiate what was and what was not recombinant DNA, especially if Cline benefited from the ambiguity. Peschle was more of a cell biologist, a blood expert struggling to grow cells in the laboratory that didn't want to grow. His intuition told him something was wrong, but Peschle lacked the knowledge to find out what. What's more, he had his own internal conflicts over this proposed experiment. He could recognize its potential importance. If it worked, his career would take off. He would be able to escape the scientific backwater of Naples, and select virtually any job he wanted in Italy. Still, Peschle wavered. There was something still not quite right. It would be better to complete the *in vitro* studies first, to show that it worked in animals before moving to patients.

The discussion at lunch wavered back and forth, pro and con. Yes, the Israelis had done it, and after having it reviewed by their own committees. Yes, there seemed to be little danger to the patient, at least according to the animal studies. The arguments repeated themselves, both with those debated in Israel and with earlier conversations in Naples. Toward the end of lunch, Peschle felt more relaxed as he perceived the consensus leaning toward waiting. Gabutti, who had said little and listened long, made no commitment about moving the experiment to the patient. Peschle finally felt that his caution was being heard. The human trial was not going to be done—at least not now. Lunch ended.

Later that afternoon, Cline and Gabutti gathered in Peschle's office. It was decision time. Cline asked Peschle what he thought. Believing that the experiment would not be done, Peschle took a magnanimous position. He didn't see any reason to rush it, to move to the patient prematurely. There was still no evidence that the gene went into the cells and made large quantities of beta globin, either in the test tube or in Cline's animals. It was better to complete the test tube studies first. But, Peschle recalled saying later, "The patient is the patient of Dr. Gabutti and Dr. Cline. I am willing to listen to what she says."

Cline went next, countering that the only way to prove the system worked was to test it in patients. He strongly stated that he wanted to do trial and ran through the lists of reasons why it was a justifiable thing to do.

Cline then turned to Gabutti who said simply, "Let's do it."

· · ·

Peschle was stunned. Something had changed Gabutti's mind, or at least that is what he thought. Apparently, Gabutti's young physicians, excited by Cline's proposal which dripped of historical importance, jumped in on his side. They wanted Gabutti to do it. There was a rumor later that they had even threatened to drop out of their ongoing collaboration with Peschle if he blocked the way. Gabutti was swayed. She would provide a patient for Martin Cline's experiment if the appropriate permissions could be obtained, and it could be done at the University Polyclinic in Naples.

Peschle was now trapped. Just five minutes before, he had staked out neutral ground. If Gabutti, who was the patient's physician, agreed to do it, then he would not stand in the way. Still Peschle continued to wrestle with the decision to do the experiment. "It was a fantastic experiment, but there was no fantastic support to indicate that it could really have a very high chance of working," he felt. "You could not exclude it [that it might work], and the toxicity was good [there wasn't any], and the Israeli committee had said yes." The next morning, Peschle called six scientists he knew to find out if they thought the experiment might work and whether he should participate. None of them could say whether it would work or not. No one said not to do it. Peschle succumbed.

Cline had crossed the first hurdle. He had won his collaborators' support. Now all they had to do was get permission from the appropriate Italian authorities—which turned out just to be Peschle's boss,

Prof. Mario Condorelli, director of the hematology unit. At the time, the Italians lacked a rigid legal framework for approving the human tests. Condorelli knew Cline's reputation. When the American scientist arrived in Naples, Peschle presented the visitor to his boss, who was impressed, and immediately indicated he saw no problems with doing the experiment in Italy. The die was cast. The second patient would be done.

• • •

As the drama over her future played out Peschle's office, Maria Addolorata of Torino waited. She, like Ora Morduch, had been born with beta-zero thalassemia, the worst kind. Her disease was so severe that her body made little beta globin, an estimated 2.5 percent of normal. Her parents were Sicilians, dwellers along the Mediterranean where malaria was once endemic and thalassemia common. Now, however, they lived in the north of Italy, having migrated to the factories around Torino and Milano.

Though sixteen years old in 1980, Maria looked as if she were twelve. Since early childhood, she had received regular blood transfusions to keep her thalassemia symptoms at bay. The transfusions contained her anemia and blocked the uncontrolled growth of her bone marrow, preventing the physical deformities of the bones and face that had disfigured her Israeli counterpart.

The transfusions, however, had shut down her brain's pituitary gland—one of the many possible consequences of iron overload. The pituitary is a master control gland in the brain. Its signals regulate many functions, including growth and sexual development. Consequently, Maria never entered puberty, nor did she reach her normal stature and development. As in Ora Morduch, the iron overload was becoming so bad that Maria had to be admitted to the hospital again and again to control her racing, arrhythmic heart. Eventually, if not controlled, these outbursts of rapid rhythms would kill her.

Gabutti had talked to Maria and her parents about Cline's proposal. Cline himself did not participate in the discussion because the Addolorata family did not speak English, and he did not speak Italian. In contemporaneous notes, however, Cline wrote, "Both the patient and the family were intelligent and understood the implications of the procedure." What's more, the family signed a simple, four-paragraph consent form, which was later made part of the U.S. Congressional record during a hearing. The first paragraph explained who proposed the treatment and where it would be done. The second described the experiment in indecipherable terms. The third acknowledged that these experiments have been tried in animals, but never in humans.

Lastly, it says: "Fully aware of the importance this new method may have for the treatment of our children and of all others suffering from thalassemia and having heard Maria Addolorata's favorable consciously and freely expressed opinion, we agree of our own will and without any duress that our daughter undergo the treatment." It was signed by her parents on July 8, two days *before* Cline performed the gene experiment on Ora Morduch in Israel.

A week after the parents signed the informed consent form, July 15, 1980, everything had been set up for the second gene procedure. Peschle and Gabutti gathered the necessary equipment and the patient at the University Polyclinic at Naples. Cline brought the genes. Again, the team removed a small sample of bone marrow cells from the girl under local anesthesia, and gave her a 300 RAD dose of radiation to her leg bones to clear a space for the bone marrow cells once they were returned to her body.

Cline exposed the 700 million bone marrow cells sitting in the laboratory dish to a concoction of genes: the herpes thymidine gene linked to its plasmid, and the herpes gene linked to the beta globin gene—both recombinant DNA. Maria received two treatments with gene-treated bone marrow, but did not get any doses of the toxic drug Methotrexate that Cline had used to give the bone marrow cells a selective advantage in mice. The genetically engineered bone marrow would just have to fend for itself in the body's bloodstream and bone marrow. Like Ora, Maria tolerated the treatment without difficulty. And again, Cline did not tell his collaborators that he was using recombinant DNA, the kind of genes he had indicated he would not use. This would be revealed only later as the case against Cline unfolded.

One day after Maria's treatment, and half a world away, the Human Subjects Protection Committee of the University of California at Los Angeles, after fourteen months of debate and deliberation, finally came to a decision on Martin Cline's proposal to put genes into the bone marrow of patients. The committee said no.

On July 16, 1980, the UCLA committee's chairman, Jeremy Thompson, wrote Cline a short note that said in part, "I regret to inform you that the [committee] has voted to disapprove the . . . study." Thompson went on to include the questions and objections from four anonymous outside reviewers working in the field of gene engineering, reportedly including two Nobel Prize winners, but excluding those Cline had blacklisted. All four rejected the experiment as premature, saying that much more animal work was needed before the approach should be tested in humans. All four raised numerous technical objections, citing various failures to demonstrate that transplanted genes had entered a significant number of the bone marrow cells in the mice, or that they functioned if they were inserted. Several questioned the reliability of the data published in the *Science* and *Nature* papers.

"I believe that the unresolved questions . . . make it highly improbable that any patient benefits will be realized," said Consultant I. After raising concerns about the cost and risk of the human trials, he worried about "the negative societal results of a failed experiment. It is my belief that there exists a substantial segment of the population that is inclined to oppose the manipulation of a human genetic constitution. Experiments of this type will tend to mobilize this potential group of critics. An ill conceived, poorly executed experiment will do so all the more."

Consultant IV agreed with the concerns of public response to Cline's proposed experiment. "I cannot overemphasize that the stakes in this experiment are too high to gamble with the low odds for a positive outcome," he said. "Nor can I justify its performance on the basis of the argument that the patients would die anyway."

Cline found these comments in his pile of mail back at UCLA. By then, he already had put the globin genes into the two girls in Italy and Israel. He had left Italy and Israel behind, thinking of future trials and long-term follow-up for the children already treated. This was just research, another step in his career. Cline never dreamed of the furor to follow, or how it would change his life and the future of human gene therapy. Or that he would never see the two patients again.

Chapter 7
The Cost of
Getting Caught

"A study is ethical or not at its inception; it does not become ethical because it succeeds in providing valuable data."
—*Henry K. Beecher, Dorr Professor of Research in Anesthesia, Harvard Medical School, 1966*

French Anderson sat bolt upright in the lounge chair next to the pool in his Bethesda backyard. It was late August 1980, a Saturday, with the kind of hot, humid, stultifying Washington weather that makes Englishmen seek shelter. Art Nienhuis, Anderson's longtime colleague and friend, had just come over with his wife to join the annual summer party of the Molecular Hematology Branch already in progress.

"I just heard a rumor that Marty Cline has gone into patients in Israel," said Nienhuis, who had just come back from the International Society of Hematology meeting in Montreal, Canada. "This is terrible. It means we have been beaten."

Conversation paused as the news settled in.

"It can't work," Anderson stated, exhaling as he leaned back in his chair. Only two months earlier, on his nineteenth wedding anniversary, Anderson had abandoned all hope of using current techniques to perform human gene therapy. The existing gene insertion strategies were just too inefficient and inadequate to get enough genes into the right cells to make a difference for the patients. Anderson had concluded that gene therapy needed a technological breakthrough to be feasible, and none seemed to be on the horizon. If Cline succeeded, it meant he had figured out some trick Anderson and everyone else in the field had missed. That seemed unlikely. Cline was a well-known clinical researcher with a great reputation in hematology, but he wasn't a molecular biologist. His partner, Winston Salser, probably hadn't solved the problems either. But they couldn't be sure.

"I agree, it can't work," Nienhuis said.

"But if he has gone in, he is first," Anderson continued. "We have to find out if it is true." Doubt nagged him, however. "All the theories say that it won't work, but it might anyway." Later, Anderson recalled that most troubling aspect of receiving scientific rumors: "We couldn't know because Cline always worked in secret. We didn't know if he had found some trick that had made it work. Winston Salser is a creative guy, and Cline is very bright. There was a real adrenaline surge."

The first, best chance to find out would come the following week. Anderson would go to San Francisco to chair a meeting on iron chelation, the drug therapy that stops iron accumulation in thalassemia patients undergoing repeated blood transfusions. A large portion of the hematology community's leaders would be at the meeting. Surely, someone would have heard something.

After arriving, Anderson began asking around to see if anyone else had heard the rumors about Cline. In the conference hotel hallway, Anderson confronted a doctor from the Hadassah University. The man, taken aback by Anderson's direct interrogation, refused to answer, saying it was not anything he would know anything about. But, he went on, if things blew up, then it might be something he knew something about. A hedge and self-protection. It strongly suggested that there might be something to the rumors.

Anderson then ran into David J. Weatherall, head of the Medical Research Council's Molecular Hematology Unit and the Institute of Molecular Medicine at Oxford University, England. A major figure in genetics, Weatherall had the clout to ferret out the truth. Anderson told Weatherall what he had heard, and the English geneticist agreed to look into it.

"I can't confirm it, but my impression is that it is true," Weatherall told Anderson later that day after talking to some of his friends. Distressed by the accumulating evidence that Cline had launched the era of human gene therapy, Anderson flew back to Washington, and decided to tell Donald S. Fredrickson, director of the National Institutes of Health.

At 8:30 in the morning, Anderson crossed the parking lot between his office in the clinical center and Don Fredrickson's office, Room 124 in Building 1. Tall and thin, Don Fredrickson had always been one of the more unusual fixtures on the scientific landscape. Owlish behind his thick glasses, Fredrickson possessed a brilliant scientific mind. He made significant and lasting contributions to understanding the biology of blood cholesterol and its relationship to heart disease. Besides having scientific acumen, Fredrickson was a master manipulator on the political scene. He came to NIH in 1953, to the heart institute as a lowly clinical associate. He steadily moved up through the ranks, taking on successively larger and more complex administrative tasks. By 1966, he directed the heart institute. Years later, after a one-year stint as president of the Institute of Medicine, a division within the National Academy of Sciences, Fredrickson became NIH director in 1975, and immediately inherited the ultimate hot potato: the recombinant DNA debates. It immediately plunged him into the world of congressional politics that threatened to shut down the nation's genetics labs. Fredrickson thrived on the pressure, skillfully piloting the NIH through touchy congressional hearings, during which he was occasionally, in his words, "boiled in oil," and satisfactorily deflected many criticisms that the NIH was trying to protect just its prerogatives, not the public.

Despite the debate's intensity and the complaints about NIH's conflict of interest—since it funded most of the research it sought to regulate—Fredrickson managed to create an atmosphere of trust. He established mechanisms in which NIH distanced itself from blatant self-dealing. The strategy seemed to be working. By the late 1970s, congressional talk of laws to regulate recombinant DNA diminished. Fear of a legislative stranglehold on technological innovation subsided. Fredrickson had safely steered the nascent science of genetic engineering into calm waters.

Fredrickson and Anderson had a strong personal relationship that started when Anderson came to NIH in 1965. Fredrickson was the heart institute's clinical director then. Anderson had immediately attracted his attention because Anderson was bright, competitive, and able to pick good projects. Fredrickson watched while Anderson became an independent investigator only two years after arriving at the NIH—an unusually quick advancement. Throughout that time, Anderson had sought Fredrickson's counsel, showing him the latest data, discussing his theories and interpretations. Anderson's attentiveness pleased Fredrickson.

So it was not unusual for Anderson to stop by Fredrickson's long, narrow office in Building 1. But this visit would be different. After arriving, Anderson laid out the rumors about the Cline experiment. The UCLA scientist apparently had gone overseas to put genes in patients; the recombinant DNA guidelines were likely flouted.

Fredrickson listened in disbelief. Cline's actions threatened to invalidate five years of struggling to establish the NIH's Recombinant DNA Advisory Committee. Delicate negotiations had turned aside sixteen pieces of legislation introduced to regulate genetic experimentation. The concerns about sticking bits of DNA in bacteria growing in laboratory test tubes would pale next to putting genes in people. Genetically engineering humans always had been in the background of the recombinant DNA debates. Most believed it was so far in the future that no one much talked about programming people's genes. To have Cline blithely blunder ahead and secretly stick genes into people was outrageous. It would be the scientific community's worst nightmare. It threatened the freedom of scientific inquiry about genetics that Fredrickson had struggled to protect. If this was true, and it became public, and NIH was not on top of it, Congress would react very badly. Fredrickson had promised Congress that he would keep a lid on genetic engineering, that the system would be open and the public informed. This would prevent anything that even hinted at the eugenic horrors of Nazi Germany.

Something had to be done about Anderson's report, and it had to be done quickly. Anderson, however, was clearly a scientific competitor of Cline's; he could not be a part of it. Fredrickson ordered Anderson to write a memo. Dated August 25, 1980, it contained a single, simple paragraph:

> This past weekend I was at a meeting on iron chelation therapy in Cooley's anemia [thalassemia]. Several participants at this meeting, who had just come from the International Hematology Meetings in Montreal, told me of a rumor that Dr. Martin Cline, of UCLA, had administered recombinant DNA to human subjects in Israel in an attempt at genetic engineering. Based on the state of the art at this time, I believe that such an experiment, if it took place, was exceedingly premature. Could your office determine the truth of this allegation?

Fredrickson assembled the kitchen RAC—his version of Pres. Andrew Jackson's Kitchen Cabinet. Fredrickson's committee was composed of some of NIH's most able geneticists, such as Maxine Singer and Susan Gottesman, Fredrickson's special administrative assistant, Bernie Talbot, and the NIH lawyer and other advisers. Fredrickson relied on the kitchen RAC to guide him through the recombinant DNA storms of the 1970s. All their hard work could be threatened if what Anderson reported about Cline proved correct.

They gathered around the long conference table in Fredrickson's office, which had—as was the director's custom— an ornate rug on it. Anderson repeated the stories he had heard, telling the cabinet of the rumors from Nienhuis and from the San Francisco meeting. Among those called in was Charles McCarthy, director of the NIH Office for Protection from Research Risks. His department oversaw the system of university committees, called IRBs or institutional review boards, that protected patients from dangerous and unwarranted experiments. If what Anderson said were true, McCarthy realized, then the Cline case would not only test the recombinant DNA guidelines, but it would become "the first major test of whether the IRB system could be really effective."

Obviously, Anderson's rumors were not enough to go on. Fredrickson ordered Anderson to get lost. If there was going to be an official investigation, Anderson could not be involved. It would look like NIH was trying to "get" a competitor to help one of its own. Anderson dropped out of the investigation, and kept his mouth shut. "I never said it to anyone: 'I am the one who blew the whistle on Marty Cline,' " Anderson said. "I never said anything one way or the other. I wasn't proud of it; I wasn't ashamed of it."

The first steps were obvious to everyone there. NIH had to find out what occurred. Fredrickson and the kitchen RAC chose a modest course of action: McCarthy, as head of the protection office, would write to the University of California at Los Angeles administrators and ask them what they knew. Their response would decide what happened next. On September 8, 1980, McCarthy sent a letter to UCLA medical school Chancellor Charles E. Young. The letter began:

This Office has received information suggesting that Dr. Martin J. Cline of the University of California at Los Angeles (UCLA), Department of Medicine, might have been involved in some research activities without satisfying Department of Health and Human Services Regulations for the Protection of Human Research Subjects (45 CFR 46). Under the UCLA General Assurance (GO 238) on file in this office, UCLA is responsible for compliance with these regulations by all University personnel who carry out research activities involving human subjects.

The obscure citation, 45 CFR 46, stands for the 45th volume of the Code of Federal Regulations, part 46—the part of United States law that regulates institutional review board rules and the protection of human subjects. A General Assurance document essentially says that the university guarantees that all of its employees will obey this law. Even if Young did not know Cline was going to go to Israel and Italy to break the law, Young and UCLA would still be responsible for it. This was going to be serious.

The letter went on to demand that UCLA look into the matter and answer in writing by October 15. Then McCarthy laid out what he wanted to know: "Has Dr. Cline been associated with research which involved inserting DNA into human beings? If so, was the material 'recombinant DNA'?" And, if Cline was involved, "were such experiments reviewed by a UCLA Institutional Review Board? If so, what decisions were reached by the Board . . . ?"

• • •

Anderson wanted to know the same thing. After failing to find a conclusive answer in San Francisco, but before going to Fredrickson, Anderson decided to discuss this with his friend, John C. Fletcher, an NIH bioethicist. Gene therapy had long fascinated Fletcher; it raised so many interesting ethical questions. The field was going to be important someday, but for now, most of the ethics discussions were academic, anticipating the issues that might come up someday. Anderson gave Fletcher entrée to the science; Fletcher gave Anderson someone to talk to about his ethics ideas. With time, the two had become friends. Earlier that year, they had decided to coauthor a paper on the technical and ethical criteria that would have to be satisfied before a scientist could put genes in people. Anderson contributed the science, Fletcher the bioethics. Now, in the summer of 1980, their article was nearly done.

They made a good team. Fletcher was an ethics expert. He had come to NIH in 1977 to set up a bioethics program at the Warren Grant Magnuson Clinical Center, the 500-plus–bed hospital on the NIH campus. It is a remarkable place where experimental ideas leap from the laboratory bench to the bedside and into the patient's body at a ferocious rate. This close link between bench and bedside is at once a crucial strength of the NIH's effort to speed up the application of medical discoveries, and a nightmare for people like Fletcher who worry that scientists can be ethically myopic, seeing the promise of a potential new treatment and the advance of their own ideas and careers, rather than the possible dangers and the rights and needs of their patients.

Fletcher, who had been a bioethicist in residence at the NIH from 1966 to 1969, understood those conflicting ambitions and the ethical quagmires they can create. During the 1960s, research ethics were being reshaped and reformed—mostly because of outrageous medical experiments in the past—and he was right in the middle of it. Since 1950, after the Nuremberg war crimes trials had ended, the ethical concerns raised by the gruesome Nazi experiments in the prison camps finally began to register with scientific and political leaders in the United States. The Nuremberg Code, as the rules for ethical medical research came to be called, required several things: A human subject must give voluntary consent, or agreement, to take part in the study; the experiment should produce some useful results for the good of society; animal studies should be done first to show the experimental approach is reasonable; unnecessary physical or mental suffering or injury should be avoided; it should be done only by qualified researchers; and the patient or doctor can stop it at any time.

Many rules in the Nuremberg Code and the subsequent Declaration of Helsinki, first made by the 18th World Medical Assembly in Helsinki in 1964, seemed to rest on common sense. Yet, they caused a tremendous debate among resistant physician-scientists unwilling to give up control over who decided when a new treatment could be tried in people. From the early 1950s to 1966, the debate raged mostly behind the closed doors of academic medicine. Most physicians believed that only they could adequately analyze a study's risk/benefit ratio because *they* knew more than anyone else about medical technology. For most of those fifteen years, that position held firm until a series of morally bankrupt biomedical experiments came to light.

The ethical logjam started to come apart in July 1964, when Dr. James Shannon picked up *The Washington Post* one morning and found a front-page story describing how Jackson, Mississippi, surgeons transplanted a baboon heart into a human. It had never been done before. Shannon, the colorful, successful, and sometimes volatile NIH director, hit the roof. NIH had funded the study, yet there had not been one shred of public debate about the ethics of such an outrageous study. The human patient, of course, died.

Then, in June 1966, Henry K. Beecher, a Harvard Medical School researcher, published, in *The New England Journal of Medicine,* a list of twenty-two studies he considered unethical. Among them was a study at the Willowbrook State School, an institution crammed with 6,000 severely mentally retarded patients on 375 acres of Staten Island in New York. Launched in the mid-1950s, the study sought to understand a chronic, widespread hepatitis infection among the hospital's residents. To learn how long a person infected with the hepatitis virus could infect others, doctors intentionally infected all incoming children with live virus. Doctors told parents that the children would be infected, but didn't tell them the risks. Beecher was outraged. The Helsinki declaration said that a physician could not "do anything which would weaken the physical or mental resistance of a human being. . . ." Beecher added: "There is no right to risk an injury to one person for the benefit of others."

Even before the abuses of Auschwitz, Dachau, and Ravensbrück, the United States Public Health Service, or PHS, a branch of the federal government, had launched its own unethical-research study that was to last for more than four decades. Starting in 1932, the PHS recruited 412 black men in Macon County, Alabama, to study the natural course of untreated syphilis. The men, already infected when they entered the study, were compared to 200 uninfected black men. The first findings from the Tuskegee syphilis study, named for the town that served as the Macon County seat, were reported in the medical literature in 1936, with updates every four years through the 1960s.

No one considered the study unethical when it started because doctors could do little to combat the spirochete that causes syphilis. By the early 1950s, however, penicillin was shown to be an effective treatment, yet the men did not receive it. On several occasions, PHS officials fought to prevent antibiotic therapy, and as late as 1969, a committee overseeing the study at the Centers for Disease Control in Atlanta, Georgia, voted to continue it. It took the pressure of national media exposure in 1972 to force the Department of Health, Education, and Welfare to end the experiment. Of the original 400 men with syphilis, only seventy-four were still alive. At least twenty-eight, but perhaps more than 100, had died of advanced, untreated syphilis.

Even in the face of obvious outrages, change would come slowly. After the baboon heart transplant in 1964, Shannon ordered the Livingston Committee, an ad hoc NIH group then still arguing the issues of the Nuremberg Code, to take up the issue of protecting human research subjects. Scientists could not always be trusted with their patients' lives, Shannon concluded. The only solution was prior group review, a process in which experiments on humans would be approved by some committee at the institution doing the study. The committee would protect patients by balancing the research risks with the benefits that the subject might get from the experimental treatment. Prior group review and approval would become the norm for all research involving humans.

The Livingston Committee's work led to a ruling by the U.S. Surgeon General in 1966 that any Public Health Service supported research—mostly funded by NIH—had to be approved by a committee of experts at the researcher's university or institution. Those human subjects committees would come to be called institutional review boards or IRBs. Five years later, the Department of Health, Education and Welfare applied those rules to everything it funded.

Meanwhile, when the Tuskegee story broke, Sen. Edward Kennedy was in the middle of four years of hearings on biomedical ethics issues primarily related to the delivery of health care. At best, biomedical ethics was a peripheral issue in the U.S. Senate. After Tuskegee, Kennedy connected bioethics to the civil rights movement. "All of a sudden, you had civil rights lawyers pushing for better protections for people at the margins of biomedical research," said NIH's Charles McCarthy. This included protecting "those perceived to be not able to protect themselves, such as the poor, the mentally disabled, the elderly, pregnant women, fetuses, children, prisoners."

Kennedy's Tuskegee hearings led to the National Research Act of 1974. It broadened HEW's human subjects protection rules to the entire federal government. It also set up the National Commission for the Protection of Human Subjects of Biomedical and Behavioral Research. From 1974 to 1978, the national commission studied the dangers of experimental medicine, issued more than a half dozen reports and more than 125 recommendations for protecting human subjects. One recommendation led to setting up the President's Commission for the Study of Ethical Problems in Medicine and Biomedical and Behavioral Research in 1978.

Although protecting research subjects had become a national law, with regulations published in the *Federal Register* (the daily document of governmental activities read by few scientists), biomedical ethicists and research leaders had to struggle throughout the 1970s to spread the word about the new rules for IRBs and the need for prior approval of human experimentation. To help make the emerging bioethical standards part of every researcher's instinctive behavior, the NIH decided its own hospital needed to set a high standard. John Fletcher came to NIH in 1977 to help set that high standard.

• • •

At 8:30 A.M., the day before he told Fredrickson about Cline, Anderson whirled into Fletcher's clinical center office to tell him what he had heard: Gene therapy had begun. As Anderson blurted out the details of the Cline rumor—that the UCLA scientist had fled California with DNA in his pocket, and put the genes into thalassemia patients in Italy and Israel—Fletcher's unhappiness increased. It sounded very much like the violations of old—a cocky researcher calling his own shots instead of building a community consensus. If this were true, it belittled the years of work by Fletcher and others to raise the ethical standards. Clearly, Fletcher could see, if Cline had received permission from his own institution, he would not have gone abroad. Clearer still, his experiment had not been approved by the NIH, or Fletcher would have heard about something this extraordinary. It made Fletcher angry.

"Let's call Cline, and ask him what's going on," Fletcher said. Anderson balked. As badly as he wanted to know, Anderson knew he was Cline's competitor because he was widely perceived as an active worker in the gene therapy field. Yet, here he was raising questions about Cline's morality and ethics. That could be interpreted many ways, Anderson said—all of them bad. One interpretation was that if he could not beat Cline scientifically, he was going to get him in trouble with the government's science administrators. Anderson, the competitor, the long-distance runner, feared that it would look as if he had turned in Cline to get him disqualified, like an athlete revealing another's illegal use of performance-enhancing drugs. Anderson did not want to seem indelicate. He decided not to call Cline.

Fletcher felt no such inhibitions. He tracked Cline to a Montreal hotel. Cline was taking a vacation after the joint meeting of the International Society of Hematology–International Society of Blood Transfusion where he had given a paper. Fletcher had never met Martin Cline. He had no official responsibility to call him or question him. Fletcher said later his purpose was noble: "I was trying to be a friend, to help this guy and the cause [of human gene therapy] out. I knew enough about the evolution of the ethics of experimental human gene therapy to know that if this were true, this was going to set things back. I wanted to warn Cline that there were these rumors, and that he should come to the NIH and clear the air. Whatever happened in the future, it would be better if he came in, but that things looked ominous."

• • •

The phone on the hotel nightstand next to Cline's bed rang while he was getting ready to go out for a run in the warm autumn air. Cline picked up the receiver. Fletcher introduced himself.

"Dr. Cline," Fletcher said, "there are rumors that you did human gene therapy in Israel." Fletcher went on to describe what he had heard about the girl with thalassemia and Cline's attempt to use a new method to transfer genes into the bone marrow cells from the patients.

Cline listened silently, his mind racing. A call from the NIH was stunning. Damn it! This was not supposed to be widely known. Cline instinctively did not discuss ongoing experiments until he had some results to talk about. The gene therapy experiments had only begun little more than a month ago. The cells sent to UCLA had yet to be analyzed. The girls seemed fine, but he couldn't say anything more than that. And, there was something in this guy's tone—something threatening, an implication that he should not have done what he had not yet even admitted to doing.

"Dr. Fletcher," Cline responded after a pause, "I don't deal in rumors."

"Dr. Cline, you better listen to me," Fletcher continued with more urgency. "I am telling you that you need to pay attention to this, and you need to come to the NIH in person, and respond to these things yourself."

"I am sorry," Cline repeated, "I don't deal in rumors."

Fletcher selected the most direct approach he could think of: "Dr. Cline, did you put genes into patients?" he asked.

"No," Cline said. "I didn't. I didn't do that. I did *in vitro* studies, but I didn't do any gene therapy experiments in people."

The flat denial left Fletcher with nowhere else to go. He thanked

Cline for taking the time to talk and hung up. There was nothing else he could do, he told Anderson later. If Cline denied any wrongdoing, if he would not talk about what was going on, Fletcher couldn't touch him. He lacked the authority to investigate. It would be up to someone else, in Building 1 perhaps, the office of the NIH director. Of course, that investigation would soon be underway. In a month, everyone would know the story.

• • •

A crisis struck in Israel. Ora Morduch's family rushed her to the emergency room at Hadassah's Mount Scopus Hospital, faint and with a galloping heartbeat. Arrhythmia. It's a deadly condition in which the heart beats out of control, trying to shake itself apart. The heart fails to pump adequately, depriving the brain and kidneys and rest of the body of blood. Uncontrolled arrhythmias are lethal. Ora had been suffering arrhythmias for some time—a result of the iron overload caused by the repeated transfusions needed to fight her thalassemia. The iron was shorting out her heart's electrical activity. Unless something was done to reduce the amount of iron in her body, she would die.

Emergency room physicians worked to stabilize her heart with drugs, and admitted her into the Hadassah Hospital. The event horrified Eliezer Rachmilewitz. He realized that he was in serious political danger. If this patient, who had undergone an experimental gene therapy treatment, died, the gene therapy trial might be blamed. He doubted that the arrhythmia had anything to do with the gene treatment, but it would be difficult to prove it had not, especially once the study became public. Throughout Ora's hospitalization, Rachmilewitz worried. Gradually, her heart rhythm stabilized, and the girl went home. Rachmilewitz relaxed.

• • •

Martin Cline was having his own crisis. After he had hung up the Montreal hotel phone after talking to John Fletcher, Cline had a vague feeling of uneasiness. Fletcher's direct questioning had a tone that suggested the experiment might be misinterpreted, that others would not understand why Cline wanted to try gene therapy. For the first time, Cline worried that "people might react negatively to the experiment." He put on his running shoes for a long, slow jog in the lovely hills over Montreal. He had not intended to offend anyone. He merely proceeded as he always had—aggressively. He knew that he pushed in the direction he wanted to go, that he had a tendency to steamroll the opposition, even among his colleagues. It was justified in this case. It was a fatal disease, and gene therapy ultimately would help. He was convinced. If he had known what was coming, Cline would never have gone abroad.

Once Cline returned to California from the Montreal meeting, all hell broke loose. By now, UCLA's administration had received a call or two about the NIH's interest in what Cline had been doing that summer. On the same day that McCarthy mailed his letter to Chancellor Young, September 8, 1980, Martin Cline drafted a five-page confidential memo to David H. Soloman, his friend and his boss. As chairman of medicine, Soloman had recruited Cline to come to Los Angeles. Cline had told Soloman about his plans in Israel and Italy before he left, about the possibility of putting genes into the patients. Soloman apparently did not object, since Cline said he was going to where the patients were, not to avoid the UCLA regulations. Besides, Soloman understood Cline's problems with Jeremy Thompson and the review committee. When he returned to California, Cline described the summer's activities to Soloman.

Now, however, things were heating up. Cline wrote Soloman a memo, struggling to put the best face on the experiments. He laid out the details in clinical terms with frequent justifications for each step along the way. He argued that the animal studies pointed the way toward human trials in patients with limited life expectancy, that the experiments in foreign countries were governed by local regulations, that he went to Israel and Italy only because he couldn't get enough patients in the United States, and that he had refused to respond to media calls until he had some results that were peer reviewed and published in a science journal. Cline closed by pleading for confidentiality.

He wouldn't get it. In the beginning of October, Paul Jacobs from the *Los Angeles Times* called. Cline had never gotten a call from a reporter before, although he had held a press conference earlier in the year to discuss the gene transfer experiments into mice. Jacob's phone call was unsettling. Another *Times* reporter had passed along a tip about Cline's activities: Somebody who didn't like Cline, or what he was perceived to have done, was trying to get him in trouble. Jacobs, a medical reporter at the time, followed up the tip by interviewing other UCLA researchers to find out what they knew about Cline's summer experiments. Rumors were widespread on campus, so Jacobs had no trouble confirming that Cline had gone abroad to do the experiments he could not get permission to do in America.

With the story essentially confirmed, Jacobs called Cline, amazed that the scientist had made the leap from animals to humans so quickly. According to Jacobs, Cline confirmed virtually everything. Cline seemed willing to take credit for the pioneering work, laid out the sequence of the foreign experiments, and talked freely about his frustration with the UCLA review process. But he withheld one thing. He did not tell Jacobs that he had used recombinant DNA, though by then he knew that UCLA was investigating that very issue in response to NIH questions.

Jacobs's story ran on the front page of the *Los Angeles Times* on Wednesday, October 8, 1980, with the headline: "Pioneer Genetic Implants Revealed," under a kicker head, "Human Engineering." In it, the reporter announced that a UCLA doctor "has become the first known scientist to use the new techniques of genetic engineering in human subjects." The story went on to describe how Cline had performed the experiment in Israel and Italy after unsuccessfully seeking permission to do it at UCLA. In the story, Cline denied that he went abroad to avoid the UCLA rules. But Jacobs covered all the bases in his first report, laying out Cline's activities and the emerging controversy springing from his study.

Jacobs quoted several scientists who said the experiments were premature and that genetic engineering had not progressed far enough to justify such trials in people. The article also reported that the National Institutes of Health had started an investigation into whether Cline had violated "federal guidelines for protecting human subjects from possible harm." Deep in the story, on the second jump, Jacobs noted that "Cline's . . . experiment . . . technically did not involve recombinant DNA."

Winston Salser, the UCLA biologist who taught Martin Cline the molecular technologies he needed to conduct the human experiments, looked at the *Times*'s article in disbelief. It was clearly wrong. Salser knew that the bits of genes Cline transplanted into the two girls were in a form that, under the NIH guidelines, would be considered recombinant DNA. Cline had taken the genetic material from Salser himself. And Salser, Cline's teacher, knew that his powerful student still lacked the technical skills needed to purify a naked gene from the recombinant plasmids Cline had packed away in his Styrofoam cooler. They were all recombinant, even the simplest that only consisted of the hemoglobin gene, or the herpes thymidine kinase gene, linked to a carrier plasmid's DNA.

This was a serious error, but Salser believed that the *Los Angeles Times* merely misquoted Cline. Salser called Cline the next day, a Thursday. Cline never called him back. The following Sunday, however, Cline came to Salser's comfortable home tucked into the winding streets on the hills of Santa Monica overlooking the Pacific. After going over data they needed to review, Salser leaned his tall, heavy frame against the front door jamb as Cline prepared to leave, and asked him whether he was going to have the misquote about the use of recombinant DNA corrected in the newspaper. Cline brushed off his collaborator, dismissing the newspaper story as unimportant.

Salser became agitated. He understood the history of the recombinant DNA wars all too well. While Cline had been caring for cancer patients, Salser, a basic scientist, had gotten enmeshed in the genetic engineering conflict. He had been at the June 1973 Gordon Research Conference on Nucleic Acids that led to the first moratorium. Salser appeared as an expert before the Cambridge City Council when it was considering its moratorium on genetic engineering. He had attended meetings of the NIH's nascent Recombinant DNA Advisory Committee while it struggled to produce guidelines that would protect the public and forestall congressional regulation of the new science.

"The guidelines may be too restrictive, and they may be stupid," Salser said to Cline, "but if we violated them, it would be a breach of faith with the public. We can say the rules are not necessary, but this is a social contract. People fear genetic engineering. We have to send a clarification to the *Los Angeles Times* because maybe the reporter does not understand that the misquote is a big deal."

Marty Cline just looked at Salser and paused. "There is a problem," Cline admitted. "The Israelis don't know that there was recombinant DNA." After Cline's trip abroad, he told his UCLA colleagues that he had used recombinant material. They, after all, would be doing the analysis and would easily detect that that type of genetic material was present. They just assumed Cline had used recombinant material, and that the Israelis and Italians had approved it.

Salser was floored. Certainly it was possible to conduct the experiment without recombinant DNA. That is what Salser and Cline had argued before the UCLA committee, and even had gotten support from the NIH RAC. And, yes, it was true that Salser had sent the telex to Israel stating that such nonrecombinant experiments were possible under the rules and that experimentation with nonrecombinant DNA would not require NIH approval. Salser, however, had assumed that Cline would wait until he had nonrecombinant material if that was what he got permission to use. It now dawned on Salser that when Cline was unable to sell the Israelis and the Italians on recombinant forms of the genes, he simply told them he would use a nonrecombinant form, and then didn't. He used what he had available, the recombinant genes in his Styrofoam cooler.

Well, there was only one solution, Salser said. They had to go to the UCLA administrators and tell them what had happened. Cline resisted. The next day, the UCLA scientists involved in the study gathered to discuss the situation. Cline and Karen Mercola came from his lab, along with Salser and his postdoc, Howard Stang, a physician-researcher who had poured his heart and soul into the experiment. According to Winston Salser, Cline instructed the group to stonewall. "If we don't cover this up, you guys will lose your jobs along with me," Salser said Cline told the group. After a vigorous debate, the consensus among the three was that they had done nothing wrong. "If we start covering it up," Salser recalled, "then we would be doing something that is wrong. And there was no way to cover it up. The best course was to be forthcoming." Then they voted: Salser, Stang, and Mercola for telling university officials everything; Cline against.

Obstruction proved to be an unwise strategy. Cline couldn't hold back the tide with his own California collaborators. He only made them angry. It cost him their support. In the end, Cline would have to tell UCLA authorities something, and a week after the *Times* story appeared, two days after the blowout with his colleagues, Cline caved in. He sent UCLA administrators a seven-page memo explaining his side of the story—including the recombinant DNA.

The letter expanded on his initial report to Soloman, but Cline tried to put on an even better face than in the earlier memo. First he described the positive results from the animals studies from 1979 and how he tried to follow up by establishing standard clinical studies as he had done previously with cancer drugs. He described it as a Phase I study, a test for toxicity from the gene transfer. Cline also would look for some indication that the genes went into the patient and maybe even made the hemoglobin. Instead of dwelling on problems with the UCLA human subjects committee—he did mention the committee rejected his request—he described setting up collaborations with researchers who had thalassemia patients. He also talked about the committee review in Israel and how he had intended to leave that tiny country and return later to do the experiment with nonrecombinant DNA if the Israelis gave him permission. When he got permission to do the experiment before he left, "I altered my travel plans and promptly began preparations to carry out the study."

Then Cline stopped stonewalling: "I decided to use the recombinant genes because I believed that they would increase the possibility of introducing beta globin genes that would be functionally effective, and would impose no additional risk to the patient. . . . I made this decision on medical grounds," he wrote to the UCLA hierarchy.

To Chancellor Young, Cline was admitting it all. He secretly used recombinant genes, obviously went overseas to avoid the conflict with UCLA's own review committee, and used federal government money to make the materials used to treat the patients. This would be a disaster of Californian proportions—like the 1906 San Francisco earthquake. Cline's medical judgment may have been fine, but his political and ethical sensibilities were way off. He offended everyone in the process, broke the spirit and letter of rules regulating the use of recombinant DNA and the protection of human subjects, and ultimately undermined his own career.

Even as the UCLA officials pondered what to do next, Paul Jacobs's *Los Angeles Times* scoop launched a torrent of media coverage, from *The New York Times* to *Science* and *Nature*. Reporters deluged Cline and university officials with calls. Inexperienced in dealing with the media, fearing the consequences of public disclosure of an experiment he now wished he hadn't done, Cline obfuscated. As a scientist, he usually refused to talk about an experiment until he had data. Since he had no data, he said little. That only seemed to raise suspicions, compounding his public troubles. He quickly lost control of the story. Unable to explain his actions or talk about his motives, Cline became increasingly defensive. Negative statements from colleagues and molecular biologists about the prematurity of the experiment were crushing. Baffled by scientific judgments that lacked any data, Cline angrily rejected the criticism.

In desperation, Cline turned to the public affairs office at UCLA. Their answer: Hold a press conference. It was a fiasco. The whole world turned out. In the glare of the TV spotlights, Cline tried to give his defense, that this was the beginning of a long series of experiments that would incrementally advance the genetic treatment toward a cure, but the reporters didn't care about his plans. They wanted to know about the experiments in Israel and Italy, and why he tried to get around the rules in the United States. They wanted to know what happened to the patients. When Cline said he wasn't sure, the reporters concluded that the experiment failed. "I didn't enjoy my fifteen minutes of fame," Cline would say later.

His university wasn't enjoying it either. Chief of Medicine David Soloman called up Cline to talk to him about taking a temporary leave of absence as chief of hematology-oncology until this all cleared up. The general disaster of the experiment was becoming personal; Cline had begun his long steady professional decline. What's more, Cline began to question himself. His previously unshakable view that he was a good, honest doctor struggling to push back the walls of ignorance began to fail. He feared he may have erred, that maybe he did step over the line. The private pressure from his university and the public pressure from the unrelenting news coverage began to take its toll on Martin Cline.

In a remarkable personal appeal to Soloman, Cline's wife Evie pleaded her husband's case:

> Things are looking rather bleak for Marty. He is already condemned by some ambitious reporters fed, sadly, by the all too human tendency toward malice by "colleagues" in a very competitive field. I address myself to you as a friend who knows Marty and as a representative of what I hope is the majority of reasonable and objective persons who will want to wait and judge by the facts. . . .
>
> His decision, in the end, to use recombinant genes was based purely on the knowledge that this technique was more apt to work and to benefit the patients. . . . Assuredly, this was poor judgment in terms of his career; I do not think it will be considered poor judgment in the hindsight of medical history.
>
> For now, it may be easiest for people to jump on a falsely moralistic bandwagon and condemn the man. . . . To take away his job or to stop his research would be to destroy this man; it would be a punishment far in excess of the crime. . . . He is thoroughly chastised; I urge compassion.

Cline would not get that, either. Two days after Evie's letter, on October 20, Soloman formally asked Cline to take a temporary leave as chief of hematology-oncology. Soloman, who had recruited Cline to UCLA seven years earlier and helped him build his successful department, concluded that Cline needed to address the questions raised about his overseas experiments. Two days after that, UCLA Chancellor Young sent NIH the university's response to Charles McCarthy's original inquiry. On the basis of Cline's reports and extensive conversations with local authorities in Israel and Italy, UCLA acknowledged that Cline appeared to have violated several rules, including putting recombinant DNA in people without permission. The university launched an investigation to decide whether Cline's "activities constituted a culpable violation" of UCLA policies, "and for consideration of imposition of sanctions for any such violations."

Three days later, on October 23, Cline acknowledged the demand that he temporarily resign as head of the division of hematology-oncology. Still, he didn't send the the formal letter of resignation for two months. It arrived in medical school Dean Sherman Mellinkoff's office three days before Christmas. It would be the beginning of the end of a glorious clinical career, leaving in its wake a talented clinician confined to treating lab animals.

. . .

The October 22, 1980, UCLA report confirmed Don Fredrickson's worst fears. Unbelievably, Martin Cline had put recombinant DNA into patients without getting permission. If NIH did not get on top of this quickly, it was going to have terrible consequences for the institution and perhaps for much of human genetic research. Congress had finally quieted down about recombinant DNA experiments in bacteria. Recombinant experiments in people could cause a fire storm. Already, questions and gossip about Cline deluged Fredrickson during scientific gatherings. Scientific leaders demanded to know what the NIH response would be. This was different, they said. This was a big deal; it involved DNA put into people. Fredrickson had to do something.

The NIH director deferred answering "until all the facts were in." He now knew that NIH money was used, at least in the preparation of the genes. NIH would have jurisdiction, and it would have to act decisively. "The Cline case is a blow to all of the order we've *all* been trying to maintain [his emphasis]," Fredrickson wrote in his diary. "There is no difficulty judging the ethics of such an incident; the questions are how to set the proper precedent in dealing with them to avoid dangerous cascades of administrative excess."

Two days after the UCLA report arrived, Fredrickson wrestled to choose the best course:

> Yesterday morning, as I rode through the mixture of greens, golds, cold, reds and grays of October on the West NIH bike path, I had time to ponder the affair of Martin J. Cline, an NIH grantee [he wrote in his diary]. He has caused me to lean even further than I otherwise might have done to emphasize the horror I have—we all must have—of blacklists in science, arbitrary administrative actions, gestures of zeal to protect the "public interest," when no harm to that interest created by a rogue could exceed the damage of denying an innocent scientist the tools of inquiry.
>
> The news also reassures us. The Pope has decided to reopen the case of Galileo. It's but 300 years since the Inquisition purified this semi-rogue of his foolishness about the sun being more central to things than the Earth. "Still, it moves." Will another century honor Cline as the first to have courage to replace a cruelly defunct gene and give an extra length of life to a condemned child? Will some of us wear the black robes in reenactment of the inquisition?

Fredrickson decided that the case was important enough to be investigated out of the NIH director's office, but not by him personally. Fredrickson appointed a small, seven-person ad hoc committee, carefully chosen and carefully charged, to investigate Cline's activities, with Richard M. Krause as chairman. Fredrickson characterized Krause as "Institute Director, NAS, Young Turk emeritus, Chm of the rDNA Executive Committee . . . our organ for considering guideline violations . . . but too unwieldy a body for this particular case." Others of the committee were described as "Chas. [Charles R.] McCarthy, head of OPRR, a former priest, an ethical authority; Mort [Mortimer B.] Lipsett, director of the [NIH] Clinical Center, Associate Director for Clinical Care, an experienced IRB man, balancing some of McCarthy's outrage with a sense of what it is to work in the trenches; Sue [Susan] Gottesman, on the RAC and as close to the ultimate expert on the rDNA Guidelines as there is in the world. For the fifth member, one desires a clinical investigator, dedicated to 'keeping the flame of inquiry bright and free of contaminating smoke'. . . ." Fredrickson chose Harry R. Keiser, a physician in the heart institute. Bernard Talbot, Fredrickson's close aide, served as executive secretary, and NIH's general counsel, Attorney Richard J. Riseberg, rounded out the committee.

Fredrickson empaneled them on October 27, 1980. They faced a daunting task: investigating one of the country's best known hematology-oncology investigators to see whether he violated guidelines governing experimentation in human subjects and the use of recombinant DNA. It would be the first major test of NIH's regulations and of the institutes' ability to self-regulate.

It also would be the first test of the institutional review board system, the rules under which human subject protection committees operated. Charles McCarthy felt the IRB rules themselves were on trial. If Cline prevailed, the guidelines protecting patients from undue research hazards—guidelines on which McCarthy and others had spent the years of their lives—might be watered down. The rules had never been tested because no one had ever been known to violate them before.

When they started, they confronted issues as simple as they were profound. First, did the federal government have jurisdiction? Since the recombinant DNA inserted into the two girls was made with NIH funding, the government clearly had an interest. "If we have one dollar in there, it was subject to our regulations," McCarthy said. "That was a landmark decision."

Next, the NIH committee ruled that the institution, UCLA—not Cline alone—would be held responsible for any breach in the rules. A UCLA attorney initially argued that it was not their concern what their faculty did on vacation. The committee insisted that the university was "responsible for what [its] agents do to human subjects anywhere in the world if they are doing it with NIH funds," McCarthy said. "That was a bombshell in the research community."

The crux of the investigation turned on the "assurance document," an agreement between UCLA and the federal government. Because UCLA was a major institution receiving large amounts of taxpayers' money, it had agreed that all of its human experiments would conform to the evolving IRB rules, whether the government funded a particular study or not. That agreement ensured that NIH had jurisdiction and that Cline had to obey the IRB rules, even if he went overseas.

Compared to science fraud cases that cropped up later, and dragged on and on in the 1980s, justice in the Cline case was swift. In the end, Cline simply admitted it all. "Cline just said straight out, 'I did it,' " recalled Bernard Talbot.

. . .

Even as the NIH investigation proceeded in private, the public uproar within the medical community reached its zenith. French Anderson and John Fletcher published an opinion piece in *The New England Journal of Medicine.* The article they had been playing around with earlier in the year—called "Gene Therapy in Human Beings: When Is It Ethical to Begin?"—had taken on a certain urgency. Sensing history, the *New England Journal* editors rushed it into print on November 27, 1980, only seven weeks after the *Los Angeles Times* broke the story.

Anderson and Fletcher set out to review the state of the art of gene therapy and the evolving ethical underpinnings that now guided human experimentation. Then they laid out the technical criteria that needed to be successfully tested in animals before proceeding to humans: "The new gene should be put into the proper target cells and should remain there. . . . The new gene should be regulated appropriately. . . . The presence of the new gene . . . should not harm the cell." The authors added, "Once all three are satisfied, attempts to cure human genetic diseases by treatment with gene therapy will be ethical."

Anderson had to walk a fine line in laying down these standards, for ultimately he would be judged by them. He had to find criteria that Cline had violated, yet set benchmarks that could be met. If the scientific and ethical standards were too high, no one could proceed. For example, Anderson worried that studies for long-term side effects might be required in larger animals, say primates, which have life spans nearly as long as the researcher's. A scientist could die before an animal study was completed, and the field would grind to a halt while waiting for the safety data. "It would be just as inappropriate to delay treatment of patients while we are awaiting long-term results in primates as it would be to rush into experimentation with patients before studies of small animals have been completed," Anderson and Fletcher wrote.

In closing, Anderson and Fletcher addressed the Cline case directly:

> We must await a full scientific report of the facts of the recent reported attempt to treat beta-thalassemia with gene therapy before judgment can be made about this experiment. We recognize that the urgent desire to treat patients having lethal

or extremely serious genetic diseases may lead the attending physician to attempt promising, though not fully tested, new regimens. . . . We hope that this laudable goal [of helping patients] will not be jeopardized by premature experiments that could prove needlessly ineffective or hazardous.

To be fair, the *New England Journal* editors gave Cline a chance to explain his side of the story in the same issue of the journal. Cline was torn by how much to reveal, since he still did not have any information on how the children were doing, or whether the genes went into their bone marrow. Cline began by reviewing the gene transfer techniques and then his own animal studies. When it came time to talk about human trials, he never admitted that he had launched the trials that were, by then, well known through the lay press. Nonetheless, he offered a series of justifications for the experiments:

Clinicians treating patients with advanced lethal diseases must frequently exercise their best judgment about when to begin . . . studies. . . . The decision about timing was until recently an individual one, made by clinicians of proved knowledge and skills. However, such decisions increasingly are made by committees. . . . One of the undesirable consequences is that important studies are sometimes unnecessarily delayed.

He then offered his criteria for launching a gene therapy trial:

Initial trials should be conducted in patients with limited alternative options . . . ; the trials should be of potential benefit to the patient . . . ; and the study should be designed so that useful information will be obtained to aid in design of future trials.

None of his arguments and rationalizations were going to save him, either from the condemnation of his colleagues who already had judged him or from the NIH panel that was now looking into the righteousness of his activities.

Even if the NIH committee was going to judge Cline on legal procedural issues, his colleagues were going to judge him on scientific grounds. They found Cline wanting. "Cline's experiments were *fundamentally* unethical [his emphasis]," Robert Williamson, the molecular genetics expert at St. Mary's Hospital Medical School,

University of London, later told a presidential commission investigating the broader issues of gene therapy. "His own work in mice shows that there was no basis for hope that globin gene insertion into marrow cells could give clinical benefit at that time. [The subjects'] families were given hope that the gene therapy might help them in their fight for survival." Williamson had raised, in *Nature* magazine, many unanswerable questions about the application of these techniques to humans. They still had not been answered when Cline went ahead with his human studies, leaving the British scientist outraged.

Despite condemnation by many scientists, Cline received a torrent of letters of support, many from the families of patients with a wide range of diseases that they hoped could be cured through gene therapy. Others were from respected scientists who offered either private or even public support. Richard J. Wurtman, a physician-researcher at the MIT Laboratory of Neuroendocrine Regulation, wrote to Cline: "I thought it might be refreshing for you to receive a letter from someone who thinks that your genetic engineering studies were at least as ethical as the norm of clinical investigations—probably more so, inasmuch as there was (and is) the possibility that you might actually have helped your subjects."

Waclaw Szybalski, the inventor of HAT medium, who had moved from NIH to the McArdle Laboratory for Cancer Research at the University of Wisconsin, wrote a scathing critique of the NIH investigation report to ad hoc committee chairman Richard Krause, complimenting the committee on "a very legalistic report, worthy of a prosecutor office or a law firm," adding later, "this report left me with a very unpleasant feeling that this is the 1981 version of some medieval legalistic proceedings, and that the only features which were missing were the rack (not RAC) and burning on the stake." Szybalski defended Cline's experiment, concluding, "Dr. Cline's decision was a scientifically and medically correct one, under the circumstances."

. . .

The NIH ad hoc committee completed its investigation and wrote a legalistic and critical report in April 1981. On April 22, the committee sent Cline a copy for comment. On May 15, he sent back a single-sentence answer: "I have reviewed the draft of the report prepared by your Committee, and I have no comments to make at this time." Four months later, Cline would mount a detailed, spirited defense, putting much of the blame on the UCLA review committee, but by then it was too late.

On May 21, 1981, seven months after it was set up, the NIH Ad Hoc Committee on the UCLA Report Concerning Certain Research Activities of Dr. Martin J. Cline handed down its judgment and the punishment: Cline was guilty of violating the rules "and warrants disciplinary action." The punishment came in four parts: For the next three years, Cline would be required to get prior NIH approval for any experiments he conducted in human patients; he would require prior NIH approval on any experiments using recombinant DNA, whether or not they involved human subjects; individual institutes would review existing federal support for Cline's ongoing work; and in the future, whenever Cline submitted a request for NIH funding, he would have to send along the committee's unflattering report. The judgment would take effect when Fredrickson accepted the committee's report, and would remain in effect for three years.

Three days later, Fredrickson publicly accepted the report, saying it "leads me inexorably to agree with the conclusions that Dr. Cline has violated both the letter and the spirit of proper safeguards to biomedical research. I therefore accept all the committee's recommendations. . . ." He ordered them implemented.

• • •

For Martin J. Cline, the research superstar, the outcome was devastating. His temporary resignation as hematology-oncology chief in December 1980 had become permanent in February 1981. "My reasons for resigning related to the circumstances surrounding clinical trials of gene therapy conducted in Israel and in Italy . . . ," Cline wrote in December. "The initiation of these studies abroad and the subsequent publicity surrounding them have proven to be embarrassing to me and to the University. . . . I deeply regret that I conducted these studies without adhering strictly to written guidelines."

Cline remained a professor at UCLA, and retained the title of Bowyer Professor of Medical Oncology, but lost access to nearly $70,000 a year, interest earned on the $1 million endowment that supported his chair. He also lost much of his NIH funding. After reviews in 1981, Cline was removed as principal investigator on a large National Cancer Institute program grant for conducting cancer research at UCLA, which was worth $3.3 million over the next four years, though a smaller NCI grant of nearly $50,000 continued through its scheduled termination six months later. The National Heart, Lung and Blood Institute, which was funding a three-year, $243,980 grant entitled "Treatment of Hemoglobinopathies by Gene Insertion," canceled its financial support for Cline's work at the end of the first year. The National Institute of Arthritis, Diabetes, and Digestive and Kidney Diseases allowed Cline to receive the last three years of a grant, worth some $118,000, provided a UCLA official certified that "the researcher performance under this award has been carried out in keeping with the intent and conditions for which the award was made."

. . .

Cline was not the only person to suffer from the fiasco. His colleagues in Israel and Italy paid, too, though the price was far lower. In Jerusalem, the chief of hematology for all of the Hadassah Medical Center system died suddenly in the spring of 1980. Eliezer Rachmilewitz was the natural successor, but as his ascension was being reviewed, the controversy over the first gene transplant experiment blew up, giving his opponents the ammunition they needed to block the appointment for nearly a year. Officially, the Hadassah leadership said, it wanted to await the conclusions from NIH's official investigation.

Although the Cline affair slowed Rachmilewitz's promotion, he suffered little else, except gossip. The review and approval of the experiment by a committee of his peers protected Rachmilewitz. In the end, however, it turned out that Hadassah's review was in technical violation of American guidelines. From the American viewpoint, Hadassah lacked the authority to conduct the review of Cline's experiment. Where UCLA had a General Assurance agreement that covered any human experiment, foreign institutions like Hadassah had only single project assurance agreements. In other words, they could conduct reviews only for individual NIH-funded projects, not any American-related project they felt like reviewing. "They needed to ask for NIH permission to review the Cline proposal," said NIH's McCarthy. "They never sought that. . . . But we never faulted them for that because they made every effort to be in compliance."

What's more, Cline lied to the Israelis. He didn't do the nonrecombinant experiment that he said he would. When the NIH report came out, virtually blaming everything on Cline, Hadassah was off the hook and so was Rachmilewitz. His promotion went through, and he continued his international collaborations in hematology and oncology without difficulty for the next decade.

Like Rachmilewitz, Cesare Peschle, the Italian collaborator, was in the process of taking a promotion when the bad news hit. Even in Italy, the University of Naples is a bit of a backwater. Peschle, a Roman by birth, had been angling for a job at the Istituto Superiore di Sanità, a kind of Italian combination of the NIH and the FDA. He came in as a research director in hematology. After he accepted the job, but before he moved to Rome, the story of the Cline experiment hit the American press and spread worldwide. Peschle came under withering attack in the Italian press, which compared him to Josef Mengele, the Nazi Angel of Death at the Auschwitz extermination camp during World War II. Mengele conducted pseudoscientific experiments on inmates, usually resulting in their deaths.

The attacks on Peschle intensified when thirteen senior Italian scientists, including Sergio Ottolenghi, who discovered that a gene deletion causes alpha thalassemia, and Lucio Luzzatto, known internationally for his work on malaria in Africa, released a public statement repudiating the human gene therapy study. They distanced themselves from the experiment, saying that such radical gene treatments should be tested in animals, not humans. The Italian researchers also worried that the initial news reports would give thalassemia families false hope, and that it could cause a public backlash against scientists if something went wrong. While acknowledging that gene therapy had promise, the group condemned the experiment and called for its halt. The view quickly spread in Italy that there was no scientific basis for the human experiment and that Peschle and Cline had done it only to get famous.

In defense, Peschle and Cline, along with their collaborators, issued their own press release, explaining the goals of the study, that they intended to treat only one patient for now, and that the patient was fine, though they didn't know if the treatment would help. Peschle couldn't win the public debate. Because there had been no review committee in Italy, Peschle lacked the defense that protected Rachmilewitz. For a while, Peschle became a scientific pariah. Former colleagues scorned him, either attacking Peschle at scientific meetings, or no longer inviting him to speak at them. There was a near rebellion when he walked through the doors of the Istituto Superiore di Sanità. Nobody wanted to work for him.

"What saved him really was a typical Italian situation," said Fulvio Mavilio, an Italian researcher who started working with Peschle after he moved to Rome. "In Italy, you can't fire anybody. Even in government they can't say, 'We were wrong; just get out.' Once you are approved for a government position, you are there for life."

Not only wouldn't Peschle be fired but he was about to become one of the most important hematologists in Italy, because he was going to the Istituto, which reviews all experimental protocols. Many Italians were horrified, said one of Peschle's former colleagues. "The very guy who did things in an unethical way was going into the administration whose business it is to approve these things."

Once he got there, Peschle persevered. He explained his position to the staff, that he thought there was an international consensus to do this trial, that Cline, the superstar, dazzled him. He admitted the mistake of allowing himself to be pressured by Cline when he was unsure of the experiment's value. Peschle, bright and energetic, quickly won over the younger scientists, and outlasted those who continued to harbor disdain for the Italian. In the end, there was no official condemnation, Peschle continued to conduct good basic research, and finally, over the years, the story faded. He was promoted to run the entire department of hematology and oncology with nearly 100 scientists and technicians answering to him. He continued his interest in gene therapy, but confined the work to the test tube. In 1992, Peschle became a full professor at Thomas Jefferson University's Cancer Institute in Philadelphia, Pennsylvania, while maintaining his lab at the Istituto in Rome.

As for Peschle's Italian collaborator, Velma Gabutti, she simply returned to her polyclinic in Torino and faded away. She had always been a clinician, not a researcher. Most of the fallout landed on Peschle. Gabutti simply went back to doing what she did best: taking care of thalassemia patients.

On the other side of the world, Cline's nemesis on the Human Subject Protection Committee, chairman Jeremy Thompson, also paid a price. After UCLA accepted its responsibility for the Cline affair, it decided to rewrite its General Assurance agreement with the federal government. It clarified questions about UCLA's responsibilities for its researchers, on campus or not. At the same time, UCLA mounted an educational campaign for its scientists so no one could make Cline's claim that he didn't know all the rules. At that point, Vice Chancellor Al Barber decided that if the new system was going to work, the human subjects committee would have to get along with the UCLA research community. He personally fired Jeremy Thompson from the chairmanship of the committee, in the fall of 1981. Thompson, a tenured professor, remained at UCLA for another decade, and then resigned to go into a private consulting practice.

．．．

After it was all over, Cline disappeared from clinical research. He stopped taking care of patients; he stopped teaching. Cline vanished into the world of the basic science laboratory, searching for vindication of his scientific ideas about gene therapy, looking to prove that treating the children with the beta globin gene was not so far-fetched. Cline always maintained that he managed to get some interesting results from the ill-fated experiment, though he never published them. Cline, along with Karen Mercola, Carol LeFevre, and Velma Gabutti, submitted a draft of their preliminary results to *The New England Journal of Medicine, The Journal of the American Medical Association, Lancet, Science,* and *Nature.* All flatly rejected it.

The draft of the paper does suggest that some genes did get into the bone marrow cells of both girls. They removed peripheral blood and bone marrow samples from both girls, and analyzed them for the inserted genes. Between one and two weeks after the treatment, small amounts of the TK gene could be detected in the blood cells from both patients. Between three and ten weeks, "multiple TK bands were observed in Southern blots of DNA from hematopoietic cells. . . . After ten weeks, no additional high-molecular-weight TK sequences were observed, and these sequences gradually disappeared from DNA of blood cells, and were undetectable after three months in one patient [Ora], and after nine months in the other [Maria]."

The scientists, however, could not tell if the TK gene had become part of the patient's cells, and had replicated, or were merely carried along passively in some blood cells, fading as those cells died off. Cline apparently neither looked for, nor found, evidence that the beta globin gene had entered any of the bone marrow cells, but Peschle did. The levels were very low, ridiculously low, and of no therapeutic value. But they were there. Tiny amounts of beta globin were oozing out of the genes transplanted into the defective cells.

Peschle never published the results because of the controversies about the human experiment, and even a decade later, he virtually denies they found the blood protein. "In a couple of a cases we got some results that looked interesting, but we never had solid evidence of expression of beta hemoglobin," Peschle has offered. "There was never a clear reproducible positive result in terms of beta globin expression."

Others in Peschle's laboratory remember it differently. "The sad part of the story was that they actually did the experiment in the cell cultures, and they actually did have good results," said Fulvio Mavilio, then a researcher in Peschle's group. "They showed some beta globin, but those data never got published because Peschle was scared of the reaction in the United States to Cline. They got synthesis in the human progenitor cell cultures. It was not a tremendous expression, but they could show that you could transfect that part of beta globin gene into cells, and have it produce beta globin."

• • •

Although the analysis was limited, Cline concluded that "at least some foreign gene sequences can be introduced into blood-forming cells for many months without causing detectable harm. This information provides modest optimism for the future of gene therapy." What Cline based his optimism on was less clear. After three years of follow-up, the scientists could see that the experimental treatment made no difference in the children's clinical picture. If it had worked, their bone marrow would have begun producing normal red blood cells instead of fragile, abnormal cells with thalassemic hemoglobin. There was no change in the children's need for on-going blood transfusions.

• • •

None of these preliminary results ever became public. Cline abandoned gene therapy. In the ten years, from 1980 until the first true gene therapy experiment was performed in September 1990, Cline published nearly 100 papers. Most were on basic research on cancer and oncogenes, though he wrote an occasional perspective piece on gene therapy. He became a part of history, but not the glorious triumph he predicted. There was no Nobel Prize for Martin Cline, though there might have been had he shown patience. Instead, there was only ignominy.

As for the two patients, Cline would never see them again. That probably was good, because Ora Morduch, the Israeli patient, became extremely angry when the news hit that the experiment lacked proper permission. She and her family felt that she had been treated like a guinea pig.

Morduch, a vivacious woman despite her disease, went on to become the moving force behind the thalassemia patients' society in Jerusalem. Intelligent and ambitious, she acted as the strong advocate for the group as a whole and for individual patients reluctant to demand the kind of care they needed. Whenever Ora Morduch perceived a medical transgression, she was quick to barge into the office of Eliezer Rachmilewitz, who was now responsible for all thalassemics in the Hadassah system, to lobby, cajole, plead, or even angrily demand whatever was needed. Although hobbled by the thalassemia-caused bone abnormalities, she traveled the world attending medical conferences, searching for the latest treatments. "She was a very tough customer," Rachmilewitz recalled fondly.

Ora Morduch did so well physically because iron chelation therapy finally came to Israel. She was among the first patients to get it, and the drug treatment removed the iron accumulated from the years of blood transfusions. In his *New England Journal* defense of starting the gene therapy experiment, Cline said: "Conventional chelation therapy holds little promise for reversing cardiac iron deposition or preventing early death." He was wrong. Ora Morduch responded dramatically. As the repeated drug treatments slowly drained the iron from her gnarled body, the erratic rhythm of her heart stabilized and returned to normal. "If she had not gotten Desferal, she would have died many years ago from her iron overload," Rachmilewitz said.

She herself would never be normal because of the permanent bone deformities caused by the thalassemia, but she would lead a long, productive life, that finally ended in 1992. With the fall of communism, Ora Murduch decided to travel to Eastern Europe in the summer of 1992. Because she kept kosher, she was eating a lot of cheese. In the unsanitary conditions of the crumbling communist culture, Ora contracted brucellosis, an infection caused by *Brucella abortus* that was probably in contaminated milk products she consumed. Most patients recover from the fever, severe headaches, diarrhea, and malaise in two to three weeks, but Ora's organs, already racked from thirty-three years of thalassemia, gave out. She could not fight off the infection, and she died.

. . .

Maria Addolorata had a much happier fate in Italy. After the treatment in Naples, she returned north with her family to Torino, where Vilma Gabutti continued to care for her. Like the Israelis, the Italians also began using iron chelation therapy shortly after the 1980 gene treatment. Like Ora Murduch, Maria Addolorata began to improve, slowly but steadily. With time, Maria's heart rhythms also steadied, and her frequent life-threatening, galloping heartbeats vanished.

At age sixteen, during the experiment, Maria looked like a twelve-year-old. The disease had dampened production of the hormones needed to go through puberty. With the needed hormones available in a synthetic form, Velma Gabutti decided to take her through puberty artificially, by injecting the hormones in the proper sequence. This treatment also worked. By her mid-twenties, Maria was married, and had moved to the southern part of Italy. By age twenty-eight, in 1992, she was trying to have a baby.

As for the gene therapy, it had no effect, Gabutti said. "We have not seen any change. We didn't see any increase in hemoglobin synthesis."

The miracle cure never happened. If Martin Cline had any hope of being rescued from his technical violations by triumphing over a fatal disease, hope slowly faded with the scant signs of genes in the girls' blood.

. . .

If the Cline affair had merely been a story about an individual scientist violating the rules of human experimentation, it would be only a footnote in the history of science. But the case was different. It was the first time anyone tried to put new genes into people. It was the first test of the regulations concerning human experimentation. Most important, it was the first to shake the leaders of biological science from their somnambulism about human gene therapy. It blared that the technology had arrived and the researchers were going to use it. Gene therapy was no longer a technology of the future, something that would have to be addressed some day. Gene therapy had arrived, and it needed to be confronted now.

If the scientific community did not work out guidelines under which it would proceed, then someone else would—probably Congress, acting as the agent of the public. It would be a repeat of the gene engineering wars. The Cline case became one of those turning points in history, where the actions of one man—good or bad—change the trajectory for many.

"In the early days of the recombinant DNA guidelines, gene therapy was not a central worry," said NIH's Talbot. "It was the Cline case that brought it to the fore."

In some ways, however, unspoken concern about changing genes in people was always in the background. When the Harvard Medical School team announced in 1969 that they isolated the first gene, they also warned about governments using these newfound abilities to change genes in people. When Paul Berg began SV40 experiments in the late 1960s and helped launch the recombinant DNA debates, the battles were always over putting genes into bacteria, and whether genetically altered bugs could crawl out of the laboratory. Many spoke cryptically about the underlying fear that one day, they would be putting genes in people. Would that be dangerous? Could that create some eugenic policy imposed by a revitalized Nazi movement in some country somewhere in the world—maybe even the United States?

Gene transfer was becoming practicable, making philosophers and bioethicists worry about its application to humans. Most scientists, however, argued that the application of gene transfer to humans was decades away, well beyond the turn of the twenty-first century. Cline proved them wrong. And he did it with astonishing timing.

On June 20, 1980, twelve days after Cline left Los Angeles for Europe and the Middle East, the general secretaries of America's three major religions crafted a prescient statement of concerns about the imminence of human genetic manipulations. Then they sent it to President Jimmy Carter. The religious leaders— Dr. Claire Randall, General Secretary of the National Council of Churches, Rabbi Bernard Mandelbaum, General Secretary of the Synagogue Council of America, and Bishop Thomas Kelly, General Secretary of the United States Catholic Conference—wrote, in part:

> We are rapidly moving into a new era of fundamental danger triggered by the rapid growth of genetic engineering. Albeit, there may be opportunity for doing good; the very term suggests danger. Who shall determine how human good is best served when new life forms are being engineered? Who shall control genetic experimentation and its results which could have untold implications for human survival . . . ?
>
> History has shown us that there will always be those who believe it appropriate to "correct" our mental and social structures by genetic means, so as to fit their vision of humanity. This becomes more dangerous when the basic tools to do so are final-

ly at hand. Those who would play God will be tempted as never before.

Worried about the "unforeseen ramifications," the religious leaders declared that "these issues must be explored, and they must be explored now." They called on President Carter to "provide a way for representation of a broad spectrum of our society to consider these matters, and advise the government on its necessary role."

Twenty days after the three religious leaders issued their warning about genetic engineering in people, Martin Cline secretly mixed the recombinant DNA of the human beta hemoglobin gene with Ora Morduch's bone marrow cells. The religious leaders' gift of prophecy had not failed them. But that would not be known for months.

Letters like the one the religious leaders sent to the White House often enough end up in the dead letter file. At that time in the Carter Administration, a presidential commission with an interminably long name surfaced to take on the challenge: the President's Commission for the Study of Ethical Problems in Medicine and Biomedical and Behavioral Research. It was a blue-ribbon committee, formed by congressional mandate at the recommendation of the then extinct National Commission for the Protection of Human Subjects of Biomedical and Behavioral Research. The presidential panel would follow up the work of its predecessor, concentrating on the delivery of health care to the public. The religious leaders' letter arrived right around the time that the presidential commission was refining its mission.

The skinny blond lawyer who had scared the scientists at Asilomar now served as the commission's executive director. Alexander Morgan Capron heard about the letter over at the White House. He brought it to his chairman's attention, wondering whether this might be something that the commission should take up. Biotechnology, especially human genetics, was not part of the commission's original mandate, but Congress allowed either the White House or the commission to add issues as it went along. At a regularly scheduled meeting in July, Capron proposed the issue to the entire commission. Member Arno G. Motulsky, a physician and a genetics researcher from the University of Washington in Seattle, urged the panel to take up the question. Frank Press, then presidential science adviser and chief of the White House Office of Science and Technology Policy, happily turned over the letter. The commission would look to see if this really was an issue they should take on.

The work of any presidential commission can be broken down into two basic activities: holding hearings and writing reports. The President's commission did both. Before the presidential commission went out of business in 1983, it produced eleven volumes—nine reports, the proceedings of a workshop on whistle-blowing in research, and a guidebook for the human subject protection committees. The commission also took on gene therapy as part of its overall project, "to examine the broader social and ethical issues in genetic engineering, and their significance for public policy." Mostly it was concerned about public anxiety over genetic engineering, what William Gaylin called "the Frankenstein factor" in a *New England Journal of Medicine* article in 1977. "This 'Frankenstein factor' conveys the public uneasiness about the notion that gene splicing might change the nature of human beings, compounded by the heightened anxiety people often feel about interventions involving high technology that rest in the hands of only a few," the commission concluded in its November 1982 report, *Splicing Life.* "Dr. Frankenstein was a creator of new life; gene splicing has raised questions about humanity assuming a role as creator." What's more, the story of Frankenstein focuses on "the uncontrollability and uncertainty of the consequences of human interferences with the natural order."

. . .

The objections to scientists "playing God" turned out to be more complicated than most had assumed. Religious leaders consulted by the commission said molecular biology raised issues of responsibility rather than matters to be prohibited. They feared genetic engineering would provide "powers that human beings should not possess." The forbidden fruit could be eaten. Religious leaders wanted the public to have a say in the decision if they were all going to get kicked out of Eden again. What's more, they concluded "that human beings have not merely the right, but the duty to employ their God-given powers to harness nature for human benefit. To turn away from gene splicing, which may provide a means of curing hereditary disease, would itself raise serious ethical problems." Even Pope John Paul II, who had been critical of genetic engineering, changed his mind, and told the Pontifical Academy of Sciences in 1982 that he supported the use of gene therapy to "ameliorate the conditions of those who are affected by chromosomic diseases. . . ."

In the end, the presidential commission concluded that gene therapy was not intrinsically immoral or unethical, provided it was properly regulated. "It seems more prudent to encourage its development and control under the sophisticated and responsive regulatory arrangements of this country, subject to the scrutiny of a free press and within the general framework of democratic institutions," than to ban it in America, and force scientists to work overseas. In the early days of genetic engineering, during the moratorium, several American researchers relocated their experiments to more hospitable regulatory climates in other countries. In some ways, Cline was merely following that precedent—except he was using humans for his experiments, not bacteria growing in a test tube.

The commission also recommended ways to regulate gene therapy. Some ideas on that already were floating around. Donald Fredrickson, who lost his job as NIH director when President Reagan won the White House, had suggested creating a "third generation" recombinant DNA advisory committee that would combine the activities of the current RAC with a dormant interagency committee responsible for coordinating the activities of all the federal agencies overseeing genetic engineering. Fredrickson's structure would take RAC out of NIH. He would make it a super RAC, answerable to a cabinet officer, probably the secretary of the Department of Health and Human Services. It would still get technical support from NIH, and it would have "a distinguished chairman, from the nongovernment sector."

The presidential commission expanded on Fredrickson's ideas. "Rather than creating additions to RAC, it might be preferable to redesign it," making it a "governmental body of greater breadth than the present RAC. . . ." The commission recommended "creation of a Genetic Engineering Commission (GEC) of 11 to 15 members from outside the government that would meet regularly to deal solely with this field." It would get scientific advice from technical panels on laboratory research, agricultural and environmental uses and dangers, manufacturing concerns, human uses, and international controls. It also would have close interactions with sixteen government agencies. The GEC, the commission concluded, would be "an agency for the protection of the future."

• • •

Rep. Albert Gore, Jr., chairman of the House Committee on Science and Technology's Subcommittee on Investigations and Oversight, held three days of hearings on human genetic engineering, in November 1982, the same month that *Splicing Life* came out.

Despite the congressional attention, nothing much happened with the commission's recommendations. There would be no super RAC, or genetic engineering commission, or any new agency to protect the future. These were the early days of the Reagan Administration, a conservative White House opposed to increased regulations of any kind. The plans for a grand governmentwide GEC were shelved, along with the presidential commission when its legislative mandate expired. Its recommended successor, a biomedical ethics board, made up of six senators and six representatives, never got off the ground.

That's not to say the presidential commission had no impact. Back at NIH, where the Recombinant DNA Advisory Committee still functioned, its leaders and the new NIH director, James B. Wyngaarden, began to move on the issue, though quite slowly. It would take four more years for the federal government—actually NIH operating nearly alone—to respond to the commission's recommendations for a new regulatory body. Even then, it would only be the addition of a subcommittee to the RAC—the Human Gene Therapy Subcommittee.

The RAC, and its subcommittee, would play a pivotal role in ensuring that when human experiments did begin, the public would have a ringside seat to the entire deliberative process. The exception would be the parallel reviews conducted behind closed doors at the Food and Drug Administration.

• • •

In the decade that followed, a parade of scientists and science commentators complained that Martin Cline's crude and premature attempt at human gene therapy had set back the field. That probably is not true. Two things needed to be worked out before the field was really ready for its first patient: The government needed to set up a review mechanism, and the gene transfer technology had to improve in efficiency. It would take the RAC's Human Gene Therapy Subcommittee several years to develop appropriate review procedures. Its yardstick, a document called the Points to Consider, laid out exactly what information a research team would need to provide to win approval to test gene transfer in people. The *Federal Register* published the first drafts of the Points in January 1985. Then, the subcommittee just cooled its heels for three years, waiting for a gene therapy proposal to hit their doorstep. The science was just not ready.

• • •

The Banbury Center at the Cold Spring Harbor Laboratory held an invitation-only, three-day conference on gene therapy in February 1982. Organized by NIH's French Anderson, Stanford's Paul Berg, and Theodore Friedmann, a gene therapy proponent at the University of California at San Diego, the conference assembled the leading geneticists to figure out where the field stood. Anderson insisted that they invite Cline, who came. "The purpose of this meeting," Berg said at the opening, "is . . . to make some assessment of whether the prospects of gene therapy are myth or reality."

The mood of the conference, however, was gloomy. Anderson admitted "that we are further away now than we thought we were a year ago. It seems that we are going in the wrong direction . . . maybe we are just starting to realize what the real problems are." After reviewing the three available techniques for inserting genes into cells, Anderson rhetorically asked whether "any of those techniques are adequate for actually putting genes into humans? My feeling at the present time is that none of them are. . . ." Yet, Anderson made it clear that "I am not against gene therapy. That is what I am devoted to."

Bob Williamson of St. Mary's in London was equally pessimistic, adding a slap at Cline: "I think that there is room in the long run for gene therapy, but jumping in with both feet, just in order to claim priority or to win grants in this field, would be a disaster. This haste must be avoided at all costs."

"The consensus of the conference," said Berg, "was that genetic approaches to treatment will probably be acceptable eventually; but there are many major technical and social problems that ideally should be solved before this occurs."

To the scientists, the biggest problem was not the ethical issues but the technical problems of getting genes into cells. The existing methods were just too inefficient. During all the discussion, the use of the most efficient gene delivery systems, viruses, was barely mentioned. Only Paul Berg pointed out that viruses offered a possible alternative. Clearly, however, the use of viruses as vectors for genes—an idea as old as the Phage Group—had not made much progress. That was about to change.

Chapter 8
Genetic
Messengers

"We live in a dancing matrix of viruses; they dart, rather like bees, from organisms to organisms, from plant to insect to mammal to me and back again . . . passing around heredity as though at a great party."

—*Lewis Thomas,* The Lives of a Cell

One afternoon in 1981, Edward M. Scolnick drifted into French Anderson's seventh-floor office in the NIH clinical center looking for advice. Scolnick, a physician-researcher like Anderson, worked for the National Cancer Institute, and was one of the leaders of the intense hunt during the 1970s for cancer-causing viruses.

Decades earlier, researchers had shown that retroviruses—the unusual viruses that convert RNA to DNA in order to successfully invade a host cell—could cause cancer in animals. Following that lead, Scolnick and many other scientists plunged into an intense search to find retroviruses that caused cancer in people. Researchers found many animal retroviruses, but only NCI's Robert C. Gallo found one that caused a rare white blood cell cancer in people—a virus called human T-cell leukemia virus 1, or HTLV-1. Years later, Gallo would become famous for helping isolate Human Immunodeficiency Virus, the organism that causes Acquired Immune Deficiency Syndrome.

By the end of the 1970s, it didn't look as if there was going to be a bevy of human cancer viruses, and political pressure was building to curtail the virus program because of its high costs. Scolnick, like others in the field, began looking for other things to study, including using retroviruses for more basic biological research. He dissected out proteins from many retroviruses, determining their role in the virus's life cycle.

Although the NCI tumor virus program failed to find an infectious cause for cancer, the work on retroviruses did lead to the accidental discovery that normal cellular genes—now called oncogenes—can be switched on in the wrong cell at the wrong time, turning the cell into a cancer. These genetic accidents can occur for many reasons, including random mutations or those caused by the chemicals in cigarette smoke, for example.

Scolnick helped figure out that retroviruses could pick up normal cellular genes from animal cells when they infected them. The abducted animal genes were then carried along in the retrovirus—just as bacteriophages and SV40 could pirate genes during their infection cycle. Sometimes, the retrovirus inserted a kidnapped oncogene into another cell and converted that cell into a cancer.

This observation led to a simple idea first stated independently by Scolnick, Howard Temin at the University of Wisconsin at Madison, and Robert Weinberg at the Massachusetts Institute of Technology: If researchers could control what gene the retrovirus pirated and carried to the next cell it infected, then they might be able to use retroviruses to transfer genes into mammalian cells.

Scolnick decided he would try to use his retroviruses to make a vector, from the Latin word meaning "to carry." His vector system would be designed to carry a gene of his choosing and to insert it into mammalian cells. Gene engineering techniques made it relatively easy to manipulate the retrovirus's genes, cutting out some and replacing them with whatever gene he wanted to transplant.

But that left Scolnick with a problem. Removing the genes would wreck the virus, making it unable to function normally and reproduce itself. In one sense, that was a good thing: Any virus-based gene-delivery system would have to be nonreplicating to make it safe for use in people. After all, a live virus that could multiply might cause an infectious disease. But destroying the virus's ability to grow made it impossible to mass-produce the billions of copies of the gene-carrying vector he would need to infect target cells.

Normally, when a virus attacks a cell, it unpacks its genetic material somewhere inside the cell—either the nucleus where the cell's chromosomes reside, or the gooey cytoplasm—and uses its genes to manufacture all the proteins needed to make new virus particles. But after Scolnick finished gutting the virus's genome to make room for new genes, the vector would no longer make all the proteins needed to produce a functional virus.

A virus infection is all about reproduction. It's like making thousands of duplicates of a document in a photocopying machine. The original document is the virus. It contains all the information needed to make copies of the complete document. But the document cannot copy itself. It needs a photocopying machine, and the photocopying machine needs to be loaded with the information-carrying pages and turned on to manufacture piles of the entire manuscript.

A virus cannot copy itself either; it needs the machinery of a living cell, which is the virus's biological photocopying machine. The virus document (the information-carrying genes) is fed in, the cellular photocopier runs, and piles of duplicate viruses come out the other side.

Scolnick wanted to make duplicates of his gene-engineered virus, but because he had pulled out so many of its genes, it was no longer a complete viral manuscript. It would be like putting one half of a book into the photocopier and hoping to get out a complete manuscript. It just wouldn't work. No matter how efficient the cellular photocopier might be, if only one half of the virus information was put in, only half a virus could come out—but more likely nothing would come out.

All of the research teams thinking about using retroviruses as vectors faced the same problem. The hint of a solution came from the discovery of defective retroviruses that could integrate their genes into a host cell but then could not reproduce themselves and make new viruses. Like the engineered viruses, the defective viruses lacked genetic information for some critical protein needed to make intact viruses. It caused a dead-end infection, where the viruses went into the cell but never came out again.

For the defective retroviruses to replicate, the infected cells had to be coinfected with a second, related retrovirus—a so-called helper virus. The second virus manufactured all the proteins needed by the first virus. This way, the disabled virus could borrow proteins from the competent virus and package its genes. The helper virus, in a sense, rescued the defective virus and allowed it to reproduce.

It would be like trying to photocopy a complete edition of a book by starting with only half the chapters from a damaged volume. To get a copy of the full manuscript, chapters would have to be borrowed and photocopied from another volume that either had all the chapters, or contained the chapters missing from the first copy. Either way, it would take this complementary system to generate a full edition of the book.

The biological systems, however, are not as neat as standing by a photocopying machine to mix and match book chapters. The photocopy machine could be set up to produce only one version of the book. The cell infected with two different types of related viruses—the defective and complete versions—would produce different kinds of viruses in a mixture. Some viruses would be complete editions of the virus; others, the genetically engineered versions, would be abridged. Some would contain the defective virus genome from the first virus, and some would contain the intact virus genome supplied by the second virus that could go on and cause an infection. Because the virus covers are identical, it would be impossible to tell which was complete and which was abbreviated.

Even with these limitations, Scolnick reasoned, it might be possible to get some genes transferred into target cells in the laboratory. He was going to give it a try, and he was going to use hemoglobin. He would infect cells growing in the lab with his engineered virus to insert the virus genome and its human gene cargo. Then, he would coinfect the cell with a completely normal helper virus that would produce all the proteins needed to make virus particles. The second helper virus would rescue his defective, lab-made virus designed to carry genes. If the system worked, some of the viral particles released by the cells would contain the globin gene.

Finally, he would infect target cells growing in the lab with this mixture, infecting some cells with the globin-carrying viruses and hopefully getting globin protein production. To show that gene transfer actually occurred in such a way that the protein was produced, Scolnick would need to be able to detect the globin protein in the infected cells.

That's why Scolnick came to Anderson. He wanted to learn the techniques for spotting globin protein should he manage to get the gene inside the target cell. Anderson had been working with hemoglobin for more than a decade. He knew as much as anyone on the NIH campus about globin chemistry and how to find the blood protein in the midst of all the other molecules within the cell.

The request surprised Anderson. Hemoglobin analysis is a long way from oncogenes. He asked Scolnick what he was up to. Scolnick described his plans to graft genes into the Friend leukemia virus, a retrovirus, and use it to transplant the gene into cells that normally don't make hemoglobin.

Anderson found Scolnick's ideas invigorating. Only the year before, he had basically abandoned the hunt for gene transfer technologies that could be used in patients. Over the past year, he had hardly done any serious work on the project, spending much of his time overseeing the workers in his lab and concentrating on tae kwon do and sports medicine. None of the available techniques were practical enough to allow the use of gene therapy clinically. Although scientists long had talked about using viruses to transfer genes to cells, no one had even gotten close to actually turning them into a functional system.

Now here was Ed Scolnick with ideas about a specific, unique approach. Anderson, still looking for a way to get into gene therapy for patients, wanted to know more about Scolnick's scheme, but he didn't know anything about retroviruses. So they set up a swap: Anderson would teach Scolnick about protein assays for hemoglobin if Scolnick taught Anderson a bit about retroviruses. "I started reading about retroviruses, and Scolnick set up and published the first paper using a retrovirus vector," Anderson recalled. "It didn't work well. He got rearrangements [in the inserted genes] and all, but he showed you could get a gene into a cell."

Scolnick's paper was published in 1982, showing for the first time that a gene could be loaded into a retrovirus and inserted into a cell. He, of course, used what everyone else was using at the time: the TK gene inserted into cells that were then grown in HAT medium. He had slipped the TK gene into the retrovirus, and the virus had put the gene into a small number of cells that could then grow in HAT medium, proving the gene transfer took place.

Before Anderson could start any collaboration with Scolnick to work on hemoglobin, however, Ed left NCI. He got caught up in a political bloodbath over the hundreds of millions of research dollars spent on the fruitless search for human cancer viruses. Scolnick left NIH and became president of the Merck Research Laboratories in West Point, Pennsylvania. The changes didn't allow Scolnick to continue pursuing gene transplants with retroviruses, and he mostly dropped out of the field.

Anderson was disappointed by Scolnick's departure. "The life cycle of the retrovirus makes it clear that this might work. The genes were in the middle, the controls [genetic on-and-off switches] were on the outside, and you ought to be able to engineer it so you can take viral genes out and put in another gene. But I didn't have any access. I didn't know how to work with retroviruses," Anderson lamented. "I had so little expertise that I could not pick up on it, and so nothing happened."

At least for 1982 and early 1983, there was little progress at National Institutes of Health in turning retroviruses into molecular delivery trucks for genes. Still, Anderson had been sensitized to retroviruses. He caught the scent of the technology to come. And he was excited.

So were other scientists around the country. Ideas often spread like virus infections. Once an idea gets started, it infects the next person and the next, until an idea has become widespread and its origins only dimly known without extensive epidemiological studies. The idea of using a virus to transfer genes had been around for decades, but the method for making it reality emerged only in the early 1980s, at the beginning of recombinant DNA's second decade. That's when a handful of highly competitive research groups on both coasts raced to work out the fundamental technologies needed to turn retroviruses into gene transporters. All the teams grappled with the same technical problems Scolnick faced. Each would add its own piece to the mosaic of knowledge about retroviruses, and how nature's strangest virus could be used to revolutionize medical therapies.

In the basic sciences, one researcher would dominate in the early going: Richard Mulligan. Others would contribute, and eventually, French Anderson would take the basic science to the clinic and his tiny patients, but in the beginning, Anderson had to work hard just to catch up.

• • •

At the Massachusetts Institute of Technology, the transformation of the retrovirus into a genetic truck was in full swing. After relocating from Paul Berg's lab at Stanford University in 1980, Richard Mulligan settled into the MIT cancer center where he got the freedom and resources he needed to pursue his own ideas. Using a "future research plans" paper he produced as part of his doctoral dissertation as a blueprint, Mulligan began to work on engineering retroviruses for gene transfer. Initially, he expected the studies to be a mere extension of the SV40 work he had done at Stanford, a basic scientific study. Instead, they revolutionized genetic transfer.

Retroviruses always have been a strange bug, even among viruses. They're "retro" because they carry their genes as a single strand of RNA, not as double strands of DNA like most other organisms. Once a retrovirus's RNA genes get inside a cell, a special enzyme called reverse transcriptase converts the viral RNA into double-stranded viral DNA, the same form as the cell's genes.

Even in the DNA form, however, the retroviruses' genome is special. The linear fiber of viral DNA carries several genes in the middle of it. On the ends are specialized genetic elements called long terminal repeats, or LTRs. These LTR structures allow retrovirus genes to be easily spliced into the infected cell's chromosomes. So after the retrovirus DNA migrates into the nucleus, the LTRs randomly insert the viral genome into one of the chromosomes when the cell undergoes division. This makes the viral genes a permanent part of the cell.

The retroviral LTRs also act like genetic on switches, forcing cellular enzymes to read the viral genes and produce viral messenger RNA. This mRNA makes proteins that get assembled into new virus capsules. And then the single strand of viral RNA, its mRNA, can be packaged into the viral capsules, producing mature, infectious retrovirus particles.

This cycle of viral gene integration and low-level viral protein production happens without the virus killing the cell. In evolutionary terms, the relationship between retrovirus and host cell has virtually evolved to the point of commensalism, a condition approaching symbiosis. Where most viruses kill the infected cell in the process of making new viruses, retroviruses are benign. The infected cell lives to reproduce itself, with its viral passenger safely and permanently on board within its chromosomes, quietly producing new virus particles at some slow rate.

The retroviruses' ability to persist in cells is remarkable. Through DNA analysis, researchers have found remnants of ancient retroviruses in the cells of many different animals—including humans. Presumably the cells were infected in some ancestral organism millions of years ago, and the retrovirus genes were passed down to each new generation, all the way to the present. The virus genes usually no longer function because mutations have disrupted them, so they don't cause disease. But the remaining DNA sequences among the host cell's chromosomes are like permanent viral footprints.

Some of these fossilized infections are so ancient that the same viral genetic sequence appears in what are now widely separate species, indicating that the retrovirus genes were carried along even as new species evolved. For example, domestic cats and baboons share one such evolutionary fossil that must have infected their common ancestor more than 10 million years ago. Cats also carry retroviral remains that were inherited from some common ancestor of rats, and rats share an ancient retrovirus with pigs.

This remarkable degree of cellular permanence made retroviruses look like nature's best offering to produce a stable gene transfer system. The challenge, however, was finding a way to exploit its rather complicated viral life-style. Richard Mulligan thought he had figured out a way through the formidable obstacles when he started his work at MIT.

First of all, the retrovirus carries its genes as RNA. Although the tools of genetic engineering had revolutionized the manipulation of double-stranded DNA, RNA chemistry is completely different—and more difficult. To get around the problem of trying to manipulate RNA genes, Mulligan and his growing list of collaborators, which included David Baltimore, used the reverse transcriptase enzyme Baltimore codiscovered to convert the viral RNA genome into complementary DNA, or cDNA. Scolnick had used a similar approach at about the same time.

Once in the cDNA form, the retroviral genes were easy to handle with the gene-engineering techniques. Mulligan quickly showed it was possible to cut out retrovirus genes, and splice in other genes, like globin. This produced naked bits of retroviral DNA that carried extra genes. To see if these extra genes functioned, Mulligan had to get this recombinant construct into a cell. Retroviral DNA, like any other bit of DNA, could be slipped into a cell by using the Axel-Wigler calcium precipitation technique. Once in the cell, the inserted retroviral vector produced the specific proteins for which it carried the code. The experiment mirrored Mulligan's SV40 expression vectors that he had built at Stanford.

Although this showed retroviruses could be used to express genes in cells and make protein, it was not much of an advance over the SV40 work. It still had all the inefficiencies of the Axel-Wigler technique, so it would not be possible to put the retroviral genes into very many cells this way. Mulligan wanted to make a gene transfer system with high efficiency, something as effective as a viral infection itself. He would have to custom-make his own virus particle, one packed full of genes of his choosing, not nature's.

Several ideas were floating around in the scientific community about how to make a gene transport system out of retroviruses. Mulligan took the same basic approach as Scolnick and others, who had started speculating about how to use retroviruses. But Mulligan added his own critical variation on the theme: He decided to genetically engineer a line of cells that provided all the viral proteins needed to make a retrovirus.

Scolnick and others had played around with coinfecting cells with normal viruses and engineered viruses. This produced an inseparable mixture of viruses. If Mulligan could find a way to make cells provide the same functions as a coinfection, then this messy step could be eliminated, and only one type of genetically engineered retroviruses would come out, not a mixture.

Mulligan's engineered cells would be called producer cells because they produced genetically engineered virus particles. These special retroviruses would carry genes of their designer's choosing, genes that could change the infected cell's characteristics—even, one day, heal human illnesses.

At least that was the idea. To make it work, Mulligan too turned to nature and looked at what happened to defective retroviruses. Most of the defective viruses got stuck because they could no longer manufacture one of the critical proteins needed to make a mature virus. Was it possible to make a viral genome that contained enough information to make all of the viral proteins but was missing a critical ingredient needed to make an infectious virus?

The MIT scientist began by deconstructing the process of making a new retrovirus. Once the virus inserts the DNA version of its genome into the cell's chromosomes, it has to get the viral genes back out in the form of RNA. To do that, it uses the cell's enzymes, which read the viral DNA in the chromosome, and transcribes a full-length copy of viral messenger RNA. The viral mRNA can be packaged into the virus capsule. Scientists were mystified as to why only viral RNA was jammed into the viral particle, but not messenger RNA from the cellular genes. Several lines of evidence suggested that there had to be a "packaging signal" in the viral RNA itself. Any piece of messenger RNA that lacked this packaging signal would not get packaged into the virus.

Mulligan set out to find the retrovirus signal, and discovered that one, indeed, existed. He used the Moloney murine leukemia virus, a retrovirus originally discovered at the National Cancer Institute by John B. Moloney, and shown to cause leukemia in mice. And then he applied a strategy tested by Scolnick in 1981 and by Howard Temin and colleagues in 1982.

The approach works like this: First, the Moloney RNA genome was converted into a cDNA copy. Then the DNA genome was inserted into a bacterial plasmid. Bacterial plasmids are rings of DNA that can be easily inserted and reproduced in bacteria growing in the test tube. This technique allowed Mulligan to make billions of DNA copies of the retrovirus genome to work with.

Next, he intentionally deleted small parts of the retrovirus genome at the left end of the molecule, between the left LTR and the beginning of the gene that made the virus's envelope protein. Envelope protein makes up the outer shell of the virus capsule. Mulligan was trying to eliminate whatever piece of genetic material contained the packaging signal for incorporating viral RNA into newly made virus particles. The rest of the virus's genetic material was needed for making all the essential proteins—reverse transcriptase, another enzyme, and the capsule protein—and had to remain intact.

After that bit of gene splicing, Mulligan had to put the viral genes into a cell to test whether the modified viral genome could make intact virus or not. Again, Mulligan used the Axel-Wigler method of transfecting a cell with calcium phosphate precipitation. Fortunately, because the viral genes were in the DNA form with their long terminal repeats in place, they acted like a normal retrovirus infection once they entered the cell and inserted their genes into the chromosome.

Once Mulligan got the defective retroviral DNA into the cell's chromosomes, it remained stable. What's more, after a few days, some of the infected cell cultures began to make virus particles that were normal in their ability to infect new cells, but abnormal since they carried no viral genes. The defective virus genome Mulligan engineered could not package its RNA into the newly made virus particles. Mulligan had found the packaging sequence—a stretch of DNA about 350 nucleotides long that he called the Psi sequence, represented as the Greek letter Ψ—in the Moloney retrovirus.

Now that the packaging sequence was found, Mulligan could make a line of producer cells that provided all requirements for producing infectious virus, except that the particles would carry no genetic material. Mulligan inserted this defective retrovirus genome into mouse fibroblasts, a type of skin cell, to create a line of producer cells he called Psi 2.

Now that he had created the packaging cell line, Mulligan only needed to infect the Psi 2 cells with a second retrovirus genome that contained the Psi packaging signal attached to some genes—like globin or TK—that he wanted inserted into the retrovirus particles. If everything worked, the result would be virus particles that carried the genes for globin or TK. He would, in effect, have created a photocopying system using chapters from two different books (two different viruses) for making one intact edition—a retrovirus vector that carried the gene intended for transfer.

To test his complementary system, Mulligan gutted a viral genome, pulling out the genes to make the envelope protein and the virus enzymes. All that remained of the original viral genome were the LTRs and the Psi packaging signal. He then spliced in a bacterial gene from *E. coli* that makes the enzyme xanthine guanine phosphoribosyl transferase, or XGPRT, often just called GPT. The gene is a dominant, selectable marker in mammalian cells, and works just like the herpes TK gene or the human HPRT gene. GPT gives cells the ability to survive in HAT medium.

Again using the Axel-Wigler transfection technique, Mulligan inserted the second retrovirus genome, now carrying the GPT gene, into the Psi 2 producer cells. A few of the Psi 2 cells picked up the Psi-plus genome that carried the GPT gene; they now carried two retrovirus genomes: The producer cells that absorbed GPT could now survive in HAT, making it easy for Mulligan to find the cells that now contained both the Psi-minus genes he started with and the Psi-plus genome that allowed the cell to produce GPT.

Once they were identified, he isolated the few altered cells and cloned them, growing them in normal medium. A few days later, the cells began producing retroviruses. To his delight, Mulligan discovered that the mature retrovirus particles contained the GPT gene but not the fully normal retrovirus genome. The engineered retroviruses were plucked from the culture fluid in which the Psi 2 cells grew, and could then be used to infect mammalian cells, and deposit the GPT gene.

Mulligan now had a complete system. The producer cells contained two retroviral genomes: a Psi-defective genome that contained all the genes needed to produce all the proteins used in making a virus, but unable to package its genes; and a Psi-competent vector that could package its genes in a viral particle, but only carried the GPT gene.

Because the GPT-carrying virus capsule was normal, it would enter a mammalian cell and insert its genes into the cell's chromosome. But it produced a dead-end infection because the engineered virus was defective and could not get out again. It lacked the viral genes needed to make envelope protein or the other viral enzymes. It did, however, incorporate the GPT gene into the infected cell's chromosomes, so it could now make the GPT protein that would make the cell resistant to HAT medium.

Mulligan had created a molecular Trojan horse. The virus particle looked and acted like a normal retrovirus, but once carried inside the cellular city gates, it unpacked a completely different genetic invader. What's more, the Psi 2 packaging cell line was a system for making retroviruses that could carry virtually any gene. All Mulligan had to do was hook a new gene to an eviscerated Moloney retroviral genome that still carried LTRs and the Psi packaging signal, and put the combination into the Psi 2 packaging cell line.

The published results, released in the May 1983 issue of *Cell* magazine, caused a sensation. It was apparent to everyone that this was going to be a powerful technique for efficiently moving genes into mammalian cells. Mulligan quickly set about demonstrating the power of the system and improving it.

For example, the virus particles produced by the first system Mulligan created could infect only mouse cells. Over the next year, Mulligan genetically engineered the proteins in the Moloney retrovirus capsule with genes from another virus so the hybrid could infect a much wider range of cells, including monkey and human cells. This would be a critical step in making the system useful for genetic therapy in people.

But Mulligan never talked about gene therapy for people. He was a basic scientist. He spoke of improving the system, getting the efficiency up so that one day it might be employable in patients. "Even at MIT, I was reluctant to push gene therapy. It represented a superficial discipline [in the early 1980s]. We were talking more about model systems for hematopoiesis [blood formation from the bone marrow] and learning about stem cell function." Still, Mulligan knew that all the work was aimed at using gene therapy to treat patients someday.

Although Mulligan had decided not to talk about the coming of gene therapy, others were starting to. French Anderson, for one, had caught the scent again. Theodore Friedmann at UCSD too, a longtime gene therapy proponent, began actively discussing progress in the field. And most observers quickly realized that with the development of the Psi 2 system, the stage was now technologically set to launch an attack on a human diseases with genes. Although there would be many technical advances over the Psi 2 system, it contained all of the essential elements needed to do gene therapy.

Psi 2 represented a major leap forward. Only a year earlier, in February 1982, the world's experts in gene therapy had gathered at Cold Spring Harbor to review the current status of gene therapy. The conclusion was grim: The technologies did not exist even to consider attempting to transfer genes into people.

Even Anderson, whom Ed Scolnick had stimulated to think about retroviruses in a new way, was deeply pessimistic at the Cold Spring Harbor meeting. But this *Cell* paper of Mulligan's was different. This was exciting. It was suggestive. The development of a packaging line that produced retroviruses capable of transplanting genes was a significant advance that might actually move the field closer to reality. What's more, as news began to spread about the successful use of retroviruses to transplant genes—even before the first papers were published—other research groups began to get on the retrovirus bandwagon. Most of the scientists moving into this new field had worked with retroviruses for years, either in oncogene studies for cancer or in the basic biology of viruses themselves.

"As that work was just starting to be reported, it became clear to me that this was the way to go," Anderson said, "but I didn't have any access. I didn't know how to work with retroviruses." It would take him years to catch up if he had to start from scratch.

That's when he ran into a bit of luck. On February 10, 1983, Anderson went to Princeton to give a lecture on gene therapy. Mostly he talked about the possibility of using microinjection to insert genes into fertilized eggs from humans to correct inherited illnesses. It was loosely based on the approach used by Yale's Frank Ruddle and others to make transgenic animals—but Anderson was talking about doing it in people.

The approach had many technical hurdles: First a genetically defective egg would have to be identified, removed from the woman, manipulated in the laboratory, microinjected with the right gene that had to get into a chromosome and function normally. Then the repaired egg had to be tucked back into the mother, who would carry the baby through a full pregnancy. The chances of success were small. What's more, this kind of genetic engineering affected not only the individual child but all future generations in that family. This approach, called germ-line gene therapy, continues to raise major ethical questions.

After Anderson's talk was over, a young assistant professor at Princeton, named Eli (pronounced *EL-ee*) Gilboa, came up and introduced himself to Anderson. Gilboa also was working on gene transfer, but he was more interested in somatic gene therapy—changing the cells in an individual's body, human or animal—than he was in germ-line therapy. What's more, he was doing it with retroviruses, just like Richard Mulligan. Anderson listened closely as Gilboa told his story.

Gilboa was born in Timisoara, Romania, and his family emigrated to Israel when he was eleven. After boarding school and Hebrew University, Gilboa earned a doctorate at the Weizmann Institute of Science in Rehovot, Israel, where he studied the SV40 virus. Because of historical linkages between scientists at Weizmann and MIT, Gilboa got invited to Cambridge and moved into David Baltimore's laboratory in 1977 for a stint as a postdoctoral researcher, a kind of internship for scientists.

Gilboa moved into the laboratory bench of Inder Verma, a bright India-born molecular biologist, who also passed through the Weizmann on his way to America and David Baltimore's circle of associates. Once at MIT, Gilboa quickly fell under Baltimore's charismatic spell and became fascinated by the retrovirus, and began studying the mechanism of the reverse transcriptase enzyme. Using Baltimore's enzyme, Gilboa was the first to convert the RNA genome of the Moloney murine leukemia virus into a cDNA, and clone it into a bacterial plasmid.

Just as Mulligan arrived at MIT in 1980, Gilboa was leaving for Princeton, which had accepted him as an assistant professor of biochemistry. Even with a research position, however, Gilboa still had to come up with his own government grants or other funding to support his research. After a couple of false starts, Gilboa decided to see if he could use the Moloney retrovirus to transfer genes into mammalian cells.

"There is no question that it was an independent thought," Gilboa said. His ideas, however, followed the general outline that even then was being discussed by retrovirus researchers like Temin and Baltimore: Gut the retrovirus genome, and insert the gene to be transplanted, infect target cells, and let nature take its course.

Since Gilboa already had his own clones of the Moloney DNA to work with, the genetic engineering to alter the virus's DNA was relatively straightforward. But this was before Mulligan had produced the first packaging cell lines for the Moloney virus, so getting the designer genes into the virus particles was difficult.

For his system to work at all, Gilboa would take the cloned virus DNA, insert the gene to be transferred, and then use the Axel-Wigler precipitation method to put the viral DNA into cells growing in culture—just as Mulligan would do. To coax the cells to produce retroviruses that carried the inserted genes, however, he had to infect them with normal retroviruses that produced infectious viral particles. During the infection cycle, some percentage of the newly made viruses contained the gene-carrying vector.

This approach produced the mixture of normal infectious virus and genetically engineered viruses. Just like the others before him, Gilboa could not tell which was which, so cells would be infected in the test tube with the entire batch, and then those that were transformed by the new genes would be identified in selection medium. The system was useful for basic studies, but it was too crude for gene therapy in humans.

Gilboa's early studies too showed that retroviruses could be used to carry genes no matter how crude the preparations. Eventually, after Mulligan, Baltimore, and Richard Mann worked out the details of the Psi 2 packaging system in 1982, they sent it to Gilboa. The spiderweb of connections between scientists who once worked or trained together often provides tools available nowhere else. Gilboa benefited from the social connection to the Baltimore group even though he was working in the same field and could be considered a competitor.

Because Gilboa was a basic scientist trying to understand how retroviruses functioned, he decided to work out the details of the Psi packaging sequence in the vector's DNA. Mulligan actually had done a quick and dirty experiment, lopping off pieces of the Moloney genome just to find what was needed for packaging. It was practical, direct. Gilboa wanted to know the particulars.

In 1982, he had one of his graduate students, a doctoral candidate named Donna Armentano, begin making systematic deletions in the virus's genes to identify the minimum bit of DNA that contained the packaging signal. She began by plucking the main genes out of the Moloney retrovirus with restriction enzymes, but left behind different-length stretches of DNA next to the left and right long terminal repeats (LTRs). Some of the DNA lengths actually included bits of the beginning of the gene that encoded viral capsid protein, a gene called *gag*. The packaging signal sat somewhere between the left LTR and the full *gag* gene.

In place of the virus's normal genes, Gilboa and Armentano spliced in a newly isolated bit of DNA called the neomycin resistance gene. It makes an enzyme that destroys the antibiotic neomycin. A chemical cousin of neomycin, a drug called G418, kills mammalian cells just as neomycin destroys bacteria. A culture broth laced with G418 prevents any mammalian cells from growing unless they carry the neomycin resistance gene, and it makes the protective enzyme. The neomycin gene–G418 drug combination produces a selection medium system that works just like the herpes TK gene and HAT.

Armentano cut out the retrovirus's normal genes for enzymes and envelope, and spliced in *neo* at different places in the virus's genome. "I did two series, so there were maybe seven in one and six in the other," she recalled. They were labeled N1, N2, N3, N4, etc.

Because it was an experiment, "some of the vectors had even more sequence than you should have logically in front of the *neo* gene, and they had retained some of the *gag* sequence," Gilboa explained, referring to one of the retrovirus genes. "They should not have worked, but we already made them so we put them in the series and tested them also. We expected to get no colonies [from the transformed cells that had the extra sequences when grown in G418-treated medium] because *neo* would not be expressed. The first three had some *gag* sequences, and we were sure they would give no colonies, but they gave more colonies than the others."

Somehow the beginning of the *gag* sequences was boosting the efficiency with which the *neo* gene made the protective enzyme. The boost was so great that Gilboa's vector made ten times more protein than Mulligan's.

"We realized that they could be used as vectors to give simply higher titers," the level of vector production, Gilboa said. "We chose N2 as the model vector because it was in the middle. We could have chosen N1, N2 or N3."

With the first high-efficiency vector in hand, Gilboa could have been a leader in the field of retrovirus gene transfer. But he wasn't. "I was sort of in a desert," Gilboa said. "Princeton was not a desert, but I made my own desert. I just did my own work." Although he had connections to the Baltimore dynasty, Gilboa was an outsider for purely social reasons. For one thing, he displayed flashes of a fiery temper, especially when he felt someone else was taking credit for his work.

At the same time, he was a shy, quiet Romanian Jew with bright blue eyes and light curly hair. Gilboa was a stranger in a strange

American culture where the language was familiar but not comfortable. He had trouble expressing himself—especially in the written form. Though he had done cutting-edge work in designing the vectors, he had difficulty getting it published because of his poor facility with the language.

But he had no trouble communicating with Anderson. The older scientist was very impressed by what he found in Gilboa's Princeton lab. During that first hour-long conversation, they formed a collaboration. Although the Armentano work would not come until later that year after the collaboration got started, Gilboa had much to offer Anderson in terms of retroviral technology. And Anderson's resurgent lab, now growing large again at the National Institutes of Health, had resources Gilboa couldn't equal. Anderson made available everything the younger man asked for. Initially, it would be supplies and reagents; soon it would be personnel and other collaborators.

"I was flattered that someone like French, a big investigator, was willing to even look at me," Gilboa remembers of his first reaction. What's more, Gilboa believes Anderson rescued him from his desert. "He brought me into the limelight," Gilboa said. "It is anyone's guess what would have happened to me otherwise."

Gilboa already was considering candidate genes to put into the vectors he was designing. An earlier collaboration with a European scientist in Philadelphia had started Gilboa thinking about using gene therapy for thalassemia. "We started working on putting the beta globin genes into vectors in a very naive fashion," Gilboa recalled. Nearly everyone, it seemed, had tried his hand at globin at one point or another.

Of course, Anderson had a lifelong interest in globin genes and gene therapy. He was enthusiastic about the combination of the two. Initially, the Anderson-Gilboa collaboration focused on globin gene therapy, and Gilboa spent the next year working on squeezing globin expression from his vectors. By then, Anderson had acquired a mouse model of thalassemia; it would provide the opportunity to test whether gene therapy could treat or cure, that deadly disease in animals.

Globin, however, would continue to be a dead end, even with the new, powerful retrovirus vectors. There were just too many problems with getting the right amount of globin expressed at the right time in the right cells. The amounts of alpha and beta globin would have to be carefully equalized, or the imbalance would merely be shifted in the other direction and the defective cells would not be cured. A simpler disease model was needed. One that involved a single gene. One that did not require complex genetic controls. One in which simply replacing the missing gene would cure the defective cells—and perhaps the patient.

Even as Gilboa and Anderson struggled one more time with the globin gene, another group of scientists, on the other side of the continent, were about to point gene therapy in a new direction.

• • •

Ever since his first crude efforts at genetic transformation in mammalian cells in the late 1960s, Theodore Friedmann believed in the dream of gene therapy, that it would be possible some day. During his time at the National Institutes of Health from 1965 to 1969, he and his colleagues tried mashing naked DNA into mouse cells that lacked the hypoxanthine phosphoribosyl transferase, or HPRT, gene. Friedmann hoped to find any transformed cells by selecting them in the then newly developed HAT medium. If it worked, it might open a way to treating mutations in the HPRT gene in people, a defect that causes a gruesome mental retardation called Lesch-Nyhan syndrome, whose victims eat their own fingers and lips.

The experiments, however, were a failure, but ever since then, Friedmann thought about finding ways to transfer genes into mammalian cells. He and French Anderson even had become friends back then at NIH because of their shared interest in gene therapy and considered working together to develop needed techniques. What's more, they were contemporaries. Friedmann, a physician, had even done his pediatric residency at Children's Hospital in Boston at the same time as Kathy Anderson, French's wife.

When Friedmann left NIH in 1969, he traveled to the Salk Institute to work on animal viruses with Renato Dulbecco, just like Paul Berg, who had preceded him only a year earlier. Friedmann decided to work with polyoma virus, Dulbecco's pet, not the SV40 virus.

Friedmann thought it might be possible to use polyoma as a vector to transfer genes. But these were the days before the genetic engineering revolution, so he took a fairly blunt-force approach: He isolated pure fractions of the polyoma capsules—just the loose protein, no viral genes—and then attempted to reassemble the proteins into a viral capsule around naked, purified cellular DNA. If that proved possible, then maybe he could use the reconstituted virus to infect target cells and transfer cellular genes. It never worked; he couldn't control the recombination of virus particles with cellular genes in general, and he certainly had no way to select out a specific cellular gene to treat a specific disease. Friedmann, like Anderson, was left to write speculative articles through the 1970s about the possible future science and its ethical considerations.

After his stint at the Salk, Friedmann's plan to return to NIH and work with Anderson on gene therapy fell through. He decided to stay in La Jolla where the University of California at San Diego was opening a new medical school, and it offered Friedmann a slot as a professor of pediatrics. He never much liked taking care of babies, so he spent his time in molecular biology laboratories doing research.

By 1976, Friedmann decided polyoma was never going to be an effective vector for gene therapy and abandoned it. As the genetic engineering revolution heated up around the same time, he decided to take a sabbatical in the Cambridge University laboratory of Frederick Sanger. Friedmann had worked in Sanger's lab in the early 1960s, before he went to NIH. Now, he would go back and learn how to sequence DNA with an enzymatic technique Sanger had just invented, for which he would receive his second Nobel Prize. (He won the first for finding a way to determine the order of amino acids in proteins. DNA sequencing determines the order of the nucleic acids in DNA, which determines the order of amino acids in protein.)

When the year was over, Friedmann returned to San Diego, but he still didn't know what he wanted to study. Having become a polyoma expert since his days in Dulbecco's lab, he chose to sequence the genome of that virus. When he completed his research in 1979, this was the third genome to be completely sequenced. The others were also viruses, SV40 and Phi X174.

After that success, Friedmann decided to go back to another system with which he was familiar: Lesch-Nyhan. He decided his group would use the new recombinant DNA techniques to isolate and clone the HPRT gene that caused that terrible mental malady. In the back of his mind, Friedmann still harbored the belief that if the gene could be isolated, then maybe some form of gene therapy could be used. The Axel-Wigler procedure was widely known by 1979, and Friedmann—like Martin Cline at UCLA—believed it offered a chance of transforming, even healing, defective cells.

Over the next two years or so, with luck and genetic engineering, Friedmann and his UCSD group managed to extract a fragment of the human HPRT gene, and clone it into bacteria. Then, using their cloned HPRT gene fragment as a probe, they searched through a library of human fibroblast genes that Paul Berg and others at Stanford University had created. From Berg's set of cloned genes, which he had shared with Friedmann, the San Diego scientists extracted an intact human HPRT gene. At that time, few human genes had been purified. The work took a while and wasn't published until August 1982, in the *Proceedings of the National Academy of Sciences*.

Once he had the gene in hand, Friedmann took the next step toward gene therapy and immediately used the Axel-Wigler calcium phosphate precipitation technique to insert the gene into HPRT-minus mouse cells, and selected for transformed cells in HAT medium. The inserted gene saved the mouse cell from HAT intoxication—and cured it of its genetic disease. Although the system worked, it was not very efficient, and certainly would not be useful for genetic therapy in human patients. But it did allow Friedmann to show that the HPRT gene could be inserted into defective cells and cure them of the biochemical abnormality.

"Around that time, we learned from the literature that people were beginning to put genes into different kinds of viruses," Friedmann recalled. "We already had given up on polyoma. It wasn't working."

Retroviruses seemed to be the new way to go. Neither Friedmann nor the senior postdoc in his lab, Douglas J. Jolly, knew enough about retroviruses to immediately turn them into gene transporters. Jolly, however, had been working on a project with Inder Verma at the Salk Institute to study the long-terminal repeats, LTRs, of retroviruses. This connection with Verma might be an opportunity for them to get into the gene therapy game with their newly cloned HPRT gene.

The Salk Institute is just down Torrey Pines Road from the University of California at San Diego. Inder Verma was a gregarious, well-known oncogene researcher, who somehow seemed connected in one way or another to most of the major players in retrovirology. Verma had preceded Eli Gilboa at the Weizmann and later in David Baltimore's MIT lab. Now he worked under the aegis of Renato Dulbecco and knew Paul Berg.

Even Ted Friedmann—who was neither an oncogene expert nor a retrovirologist—had known Verma for years. Their meeting was inevitable. First, Friedmann had a long-standing relationship with Dulbecco and many others connected to Salk. In addition, Robert Weinberg, an oncogene researcher at MIT and a close friend of both Friedmann and Verma, had introduced the two in 1972. And last,

Friedmann and Verma lived four houses apart on the same La Jolla street, Solana Beach, and often attended Sunday afternoon recitals at a neighbor's home. Both were researchers interested in genetics. "It was hard to miss each other," Verma laughed.

Inder Verma always had been one of the more colorful characters of molecular biology. Born in Sangrur, Punjab, India, in November 1947, Verma conveys the wise presence of molecular biology's Buddha. He grew up in a professional family: Both parents are economists. He entered university early, and quickly moved to graduate school at Delhi University. But he was too young at age nineteen to be allowed to graduate from Delhi's rigid system that demanded the maturity of a twenty-one-year-old. Verma fled south, to the university in Bangalore.

During a trip to the library, Verma had become enamored with a postage stamp he saw on mail from the Weizmann Institute of Science in Israel. He wanted his own stamp, so he sent in an application for a fellowship at the Weizmann. He got the stamp on the return mail—and he won the fellowship. India, however, lacked diplomatic relations with Israel, so no one could get a passport to enter the Jewish state. Inder wanted to go anyway, and his family helped make it a political issue that ultimately reached the Supreme Court of India. The court ruled that no Indian citizen could be denied a passport to anywhere. Verma got two passports: one good for the world and a separate one for Israel.

On November 2, 1967, Verma arrived in the Middle East with three pounds sterling in his pocket (Indian law wouldn't permit him to take currency out of the country) and a long black ponytail draped down the center of his back. A short time after his arrival in Israel, he and two Belgian friends decided to grow beards without mustaches, a traditional Hindu style.

At the Weizmann, Verma worked on SV40. In 1969, a year before he completed his doctorate, Robert Weinberg, the retrovirus expert from MIT, came into the lab for a sabbatical, and shared bench space with Verma, and the two became good friends. When Verma graduated in 1970, Weinberg introduced him to David Baltimore and got him an invitation to do his postdoctoral research on retroviruses at MIT.

That same year, 1970, Baltimore discovered reverse transcriptase, and later won the Nobel Prize. Because the lab was initially one of the few to have the enzyme, Verma began studying it. In June 1971, a

Russian researcher visited the lab to talk about his discovery that messenger RNA had long stretches of adenosine nucleotides on the end—a so-called poly A tail—but he didn't know on which end. Since DNA chemistry was easier than RNA chemistry, Verma figured he could use reverse transcriptase to make a DNA version of the globin gene mRNA to show where the polyadenosine sequence resided. This was the first time anyone had made a complementary or copy DNA out of RNA. It would become a standard technique in the molecular biology revolution and widely used for purifying genes.

Three years later, when Verma completed his postdoctoral training in Baltimore's lab, he had produced fourteen papers. "I was a very productive postdoc," he said. On the strength of his Baltimore connection and his productivity, Verma landed a job at the Salk Institute in 1973. He never left.

Verma's group continued to work with retroviruses, but by now oncogenes had been discovered in these strange intracellular parasites. Oncogenes are normal cellular genes turned on at the wrong time in the cell's life, or mutated in such a way that they make the cell cancerous. His group, conducting basic research, would become a leading team for isolating and identifying oncogenes, including being the first to detect the so-called *fos* oncogene. *Fos* stood for FBJ osteosarcoma virus, and it had picked up a human oncogen.

Along with the oncogene work, Verma's team sequenced *src*—a gene that causes sarcomas, or cancers of muscle cells—that had been picked up by a mouse retrovirus. With others in the field discussing the possibility of using retroviruses for gene transfer, Verma too began to consider the possibility. "If they [retroviruses] could contain *src*, why couldn't we remove *src* and put in the gene we want?" Verma wondered. "It was obvious that if oncogenes could be removed and substituted with a therapeutic gene, then we could make vectors."

In 1982, a new postdoctoral researcher named A. Dusty Miller came to join Verma's lab at the Salk. Initially, Dusty worked on the *fos* gene, describing, with others in the lab, how it turned a cell into a cancer. But soon he began spending most of his time thinking about genetic transplants and trying to answer Verma's question about the use of retroviruses as molecular trucks.

When Ted Friedmann and Doug Jolly approached Verma in late 1982 to propose a collaboration to put the HPRT gene into a retrovirus, Verma added Miller to the group. Miller would be the hands-on guy at the Salk; Jolly would do the same at UCSD. The two groups would become a powerful team, and produce one of the first successful systems for transferring a gene that had the potential for curing an inherited illness in humans.

Basically, Friedmann had a simple proposal: Let's find an efficient way to put the HPRT gene that he and Jolly had isolated into HPRT-negative cells. Perhaps they could use the retroviruses that Verma and Miller worked with. If they could make an HPRT-carrying vector out of a retrovirus, maybe they could get it into a high percentage of cells in such a way that it produced a large amount of protein. If they could do that, maybe they could cure the cells and patients.

Each team brought its own expertise. "It would be fair to say that he was introduced to retrovirology by us," Verma said of Friedmann. "We would not have been interested in HPRT. We had nothing to do with it. Doug and Ted got us into that."

Initially, however, in late 1982 and 1983, not even this sophisticated San Diego group had a good way to use retroviruses for carrying genes. Neither Mulligan's Psi 2 system, nor Gilboa's N2 vectors, were yet available. The San Diego team lacked a way to package the reconstructed retrovirus genome into a retrovirus particle so it could infect a cell.

As an alternative, they borrowed the approach Paul Berg and Richard Mulligan first took when they used SV40 to transplant the beta globin gene into monkey kidney cells. They would make an expression vector by extracting a retrovirus's genome, cutting out the retrovirus genes and inserting the HPRT gene. This produced a piece of DNA that contained enough information to get the genes into the cell in such a way that they would function. But to get this naked piece of DNA into cells, they would have to use the calcium precipitation technique to smuggle the reconstructed vector into HPRT-minus cells. If the retrovirus vector worked, it would make the HPRT protein and cure the defective cells.

Again, it would not be an advance over the SV40 work Mulligan had done to insert the hemoglobin gene, and certainly it would not be efficient enough for treating people. But it would be a first step. It might, for example, show that retroviruses had advantages over SV40 expression vectors because of their ability to stably insert genes into the target cell's chromosomes. It also might prove something that had only been a theory: Inserting a normal copy of a mutated, ineffective gene into a defective cell would cure it. The Mulligan experiment with hemoglobin had shown only that the beta globin gene could be inserted into a cell so it produced the protein. The cell normally did not make hemoglobin, so there was no cellular defect to fix.

By mid-1983, the San Diego team had produced an expression vector made from a retrovirus plasmid, and had successfully transfected it into a handful of HPRT-minus cells with the calcium phosphate precipitation technique. The corrected cells had enzyme activity restored to 4 to 23 percent normal levels, and accumulation of the toxic byproducts caused by the mutation were "partially to nearly completely corrected," they wrote in the August 1983 *Proceedings of the National Academy of Sciences*.

But this approach was too crude to be useful for anything other than lab studies. They knew that from the start. The San Diego team, like everyone else, wanted to develop a highly efficient transfer technique that spread genes as quickly as a natural infection. Even as they did the transfection experiment, Miller and Verma began working on ways to package their recombinant retrovirus genome into infectious virus particles.

To start, they used the same approach that Scolnick and others had proposed a year or so earlier: Insert the recombinant virus genes into a cell by the calcium precipitation method, and then infect the same cells with a normal retrovirus. As the wild-type virus reproduced, it would make whole virus particles that could package the vector virus's genes. The cell would then produce a mixture of viruses: Some of the virus particles would contain the recombinant viruses; some would be the natural, infectious version.

The mixture would then be used to infect a cell culture, where some of the cells would absorb the vector carrying virus. Since the vector carried a selectable gene that could protect the cell from some toxin—HPRT gene in this case, which allowed the cells to grow in HAT medium—the cells that had received the vector could be readily identified, and the cells infected with the normal virus destroyed.

This worked well enough to produce some recombinant viruses, and to show that a retrovirus could be used to transplant the HPRT gene. But the technique, contaminated as it was with live retrovirus, could never be used to treat patients—Friedmann's fondest hope.

But the way to go was clear. Only three months before the San Diego team published its PNAS paper, Mulligan reported his work on the construction of Psi 2 producer cell lines that could manufacture genetically engineered retroviruses. What's more, Miller learned about the Princeton advances with the N2 vector, and called Gilboa to ask him for sample vectors. Gilboa, who had not yet published a word about the work, reluctantly agreed to provide the vectors, but only on the condition that Miller not report anything about them until Gilboa published. Miller consented, got the N2 from Gilboa, and set out to make his own line of producer cells. Later, he would make his own line of vectors based on the N2 concept as well.

• • •

These were the critical years—1981 to 1984—in which the foundations for modern genetic therapy would be laid. All the needed components fell into place, including the basic retrovirus genetic backbone (used to transport a gene of interest into the cell and integrate it into the cell's chromosomes) and a line of producer cells that could mass-produce the engineered virus particles. There were many technical problems, including the accidental production of helper virus—native viruses that contained all the normal genes needed to cause an infection—from the producer cell lines. But the basic concepts existed by 1984.

What's more, the HPRT experiment pointed the researchers in a new direction. All of the groups previously had struggled with the beta globin gene—trying to get it into the retrovirus and trying to get it to produce enough of the protein in the cell to begin really thinking about gene therapy for thalassemia. No one was able to make it work. Not Mulligan. Not Anderson and Gilboa.

Clearly, if retroviruses were going to be the weapon to attack genetic diseases, then they had to be pointed at a different disease. The California study strongly suggested that a simpler disease would be easier. Maybe one in which a single gene was damaged. Merely restoring the function of that gene should be enough to reverse the illness.

There were plenty of candidate diseases: cystic fibrosis, Duchenne's muscular dystrophy, hemophilia. All are caused by mutations that knock out the function of a single gene. Restoring that function by adding in the gene should be a cure. But in the early 1980s, the gene had yet to be cloned for any of these. In fact, there were only a handful of isolated genes to pick from.

HPRT was one of the first purified genes associated with a human disease. But HPRT probably was not going to be a good candidate. Clearly, the gene worked in cells grown in the test tube. With Miller's new vectors in hand, even contaminated with helper viruses, Friedmann and the rest of the team reported in the June 1984 issue of the *Journal of Biological Chemistry* that it was possible to heal white blood cells taken from a Lesch-Nyhan patient and grown in culture. The cured lymphocytes produced between 3 and 23 percent of the normal amount of enzyme, about the same as in the transfection experiments.

"Researchers believe that this amount of activity in the appropriate cells of the body may be sufficient to correct the behavioral symptoms of Lesch-Nyhan patients," said a UCSD press release about the finding. The press release went on to say that the team would be attempting to improve the viral vectors, insert the gene into mouse bone marrow cells, and prove the system could be used safely.

"Bone marrow cells are likely to be the first targets of clinical gene therapy, since they can be removed from the patient, infected with new genes and reimplanted," the San Diego release stated. "However, it remains unknown whether HPRT produced by such cells could correct the brain effects seen in Lesch-Nyhan syndrome, and there is currently no animal model in which this can be tested."

Friedmann, indeed, had worried that this approach would not help patients. Lesch-Nyhan afflicts the central nervous system. Despite the promise of retroviral vectors, they were not going to work in nerve cells because the cell receiving the gene transplant had to be dividing for the virus to integrate its genes. Mature nerve cells never divide, so retroviruses could never get in them.

Friedmann speculated that merely replacing the gene in blood cells—say, bone marrow cells—would produce enough HPRT enzyme to remove the toxic substance from the body, and reverse the disease process. That's why the team began trying to put the HPRT gene into mouse bone marrow.

That strategy, however, proved ineffective. Although there was no way to test genetic repair of Lesch-Nyhan in animals, Robertson Parkman, head of pediatric bone marrow transplantation at Children's Hospital in Los Angeles, found a way to test the concept in people. He performed a bone marrow transplant a few years later, in 1985, on a patient with Lesch-Nyhan syndrome in an attempt to reverse the disease. Since bone marrow cells from a healthy donor carry the normal HPRT gene, the transplanted cells would function normally in the patient and reduce the levels of toxin.

The transplant worked perfectly and reversed the accumulation of poisons in the patient's blood, but the benefits from the bone marrow transplant did not extend to the central nervous system. The patient remained profoundly retarded and self-destructive. Putting the normal HPRT gene into a patient's own bone marrow could be expected to do no better. There would have to be a way to get the genes into the brain earlier in life to prevent the brain damage that appeared to be irreversible. HPRT would not be the way to prove the power of gene therapy.

For Friedmann, the failure was profoundly disappointing.

• • •

Back in Bethesda, Anderson and Gilboa had spent an equally frustrating time attempting to get beta globin expression out of the N2 vector. Philip Kantoff, a physician-researcher, had joined Anderson's NIH group in July 1983 to work with the mouse model of thalassemia that Anderson had acquired. The model might have been an adequate way to demonstrate the ability of N2 to carry the beta globin gene into the mouse bone marrow cells and generate enough globin protein to reverse the disease. But it just never worked. Although the Bethesda scientists could show that N2 entered the cells, Anderson's group couldn't get much protein production. They didn't understand the control mechanisms of globin any better than the rest of the field. It would take several more years for Thomas Maniatis at Harvard to figure out that the on-off switches for globin expression were—in genetic terms—miles away from the stretches of DNA that carried the globin code.

Finally, as 1983 came to a close, Anderson and Gilboa decided to give up on globin. In early 1984, Anderson started looking for a better disease. He surveyed the available genes—those that had been isolated—and hit on one of the few truly workable possibilities: adenosine deaminase deficiency, one of the mutations that causes severe combined immune deficiency, or SCID. Babies born with SCID fail to produce a functioning immune system, and are vulnerable to every passing germ. In 1984, most SCID babies died before age two from some overwhelming infection. The only treatment was a bone marrow transplant which, unlike Lesch-Nyhan syndrome, cured the disorder if a matched donor was available.

SCID is a rare disease with a famous case: David, the boy in the bubble. David lived for twelve years in a spotless plastic enclosure at the Baylor College of Medicine in Houston, Texas. He would die later that year, in 1984, after a bone marrow transplant, intended to cure him, lethally infected the boy with Epstein-Barr virus.

The gene for David's disease wasn't known. The defect was on the X chromosome, and it wouldn't be isolated until 1993. The ADA gene, however, located on chromosome 20, was independently isolated in the early 1980s by three separate research teams at roughly the same time.

Stuart H. Orkin, a well-respected scientist from Harvard University who works at Children's Hospital in Boston, led one of the groups that first isolated the ADA gene. Richard Mulligan, who knew Orkin as a friend, immediately recognized that this was yet another of the few cloned genes that could be placed into one of his vectors. Mulligan and Orkin would later write in a published paper that "adenosine deaminase appears to be an ideal candidate for somatic genetic therapy." The two men decided to collaborate, so Mulligan set about putting the ADA gene Orkin had isolated into his own vector system.

The second ADA gene was cloned in the Netherlands, by a young scientist named Dinko Valerio, a graduate student who worked in the Radiobiological Institute laboratory of Alexander J. van der Eb, who in 1972 had discovered that mammalian cells could absorb genes if the DNA was precipitated out of solution with calcium phosphate. Van der Eb's discovery led to the development of the Wigler-Axel technique.

Valerio's ADA gene—which at first he guarded jealously—would follow a tortuous trail through America's gene therapy labs. In the early 1980s, Valerio managed to land a postdoctoral fellowship at Genentech, Inc., working in the lab of David W. Martin, Jr., vice president for research. Martin had decided that gene therapy was going to be a coming technology with a big potential market for the already successful biotech company.

Initially, Martin focused on the calcium precipitation technique as the way to insert genes into defective cells. It was only a way to get started, though everyone recognized its inefficiency. As retroviruses started to come on the scene, however, Martin decided to speed up the work by hooking up with Inder Verma at the Salk. The San Diego team had become a leading laboratory for genetic transfer once it completed the HPRT work and showed that it was possible to "cure" genetically defective cells in the test tube. In that collaboration between Genentech and the Salk, Dinko Valerio, on fellowship from the Netherlands, worked with Dusty Miller.

At this point, Miller was building retroviral vectors to carry just about any useful gene on which he could get his hands. Valerio offered the ADA gene, which Miller promptly put into a retrovirus vector. Initially, however, the San Diego team was not interested in pursuing gene therapy for ADA deficiency. And like many collaborations, the relationship between the Salk and Genentech dissolved before the work could progress terribly far.

Valerio's fellowship at Genentech ended, and he went back to Europe. Miller himself was nearing the end of his postdoctoral time at the Salk, and was promptly wooed by the Fred Hutchinson Cancer Center in Seattle, Washington, a job he ultimately took. And Martin too got out of the gene therapy business, leaving it for others to pursue. And even Ted Friedmann, disappointed by the failure to find a clear way to use HPRT in gene therapy for humans, became less active in the field for a time. The relationship between the Salk workers and the UCSD team sank to a low ebb.

The California collaboration was one that Anderson had watched with some anxiety. He considered the Salk-UCSD team to be the most powerful and impressive of the groups pursuing gene therapy. He knew of Friedmann's desire to make it work. He knew Verma's skills and reputation. When the collaboration essentially ended, Anderson was relieved. "If that group had not broken up, there never would have been any room for me," he said.

The same would later happen of the Orkin-Mulligan collaboration. Mulligan successfully put Orkin's ADA gene into the pZIP vector and the Psi 2 cell line, but they never had much luck getting good ADA expression. After several years of work, it turned out that the gene Orkin cloned had a mutation that prevented it from making a normal version of the protein. Faced with those difficulties and not really ever intending to do gene therapy with the vector, Orkin and Mulligan went their separate ways. Orkin continued to work on the basic biology of genes, how they are regulated and control the differentiation of cells. Mulligan continued designing better and better vectors.

The third ADA gene was cloned by John Hutton, dean of the University of Cincinnati School of Medicine. Hutton's team actively studied how normal genes worked, what turned them on and off. The isolation of the ADA gene came out of those ongoing studies.

By 1984, Anderson had convinced himself that ADA was the best disease for testing gene therapy. Others too believed that it was a good candidate for many reasons. Now that a number of groups had cloned it, all Anderson had to do was get his hands on the gene some-how and stick it in the vectors Gilboa and the NIH team had worked on.

But Anderson was not in the ADA field. He wanted to do gene therapy. Even if he was right—that ADA was the perfect gene to test gene transfer—it would take him years to redirect his lab and do the hard work of cloning the gene. The only shortcut was to get it from someone. Science is supposed to be collegial, but it also is highly competitive. Although scientists are expected to share their basic ma-terial once they have published on it so other scientists can confirm and extend their work, Anderson didn't believe Stu Orkin would share the ADA gene. It also seemed clear at the time that Dinko Valerio would not give away his ADA gene.

John Hutton, however, might be a different story. Hutton and An-derson both went to Harvard College and Harvard Medical School around the same time, though they did not know each other particu-larly well. Still, there was enough of a connection for Anderson to feel that he could approach Hutton to talk about ADA. Besides, An-derson was hard at work writing a review article on the future prospects of gene therapy going into clinical trials, and he was con-centrating on ADA as the leading candidate. He had spent much of the past few months reviewing the literature and interviewing the leaders, like those in Boston and San Diego. Most had been candid about where they were going, discussing work yet to be published. On September 28, 1984, Anderson called John Hutton.

The Ohio researcher described the studies going on in his labora-tory and how they had isolated the ADA gene—a major effort that in-volved the work of many people in the lab and took years. Cloning the ADA gene was a major feat for a relatively small lab at one of the country's less famous schools. During the phone call, Anderson de-cided to go for broke. He wanted the gene; Hutton had it. No one else was going to send it to Bethesda.

"John," Anderson asked, "what are you going to do with your ADA gene? Are you going to build vectors with it?"

No, Hutton replied. He recently had held a lab meeting to discuss the question of pursuing gene transfer with their ADA gene, but the group decided it was not their strong suit, that they would not be able to get into the gene therapy field easily because they just didn't have the resources. They would just use their ADA gene to conduct basic studies, not aim it at a therapy.

Anderson couldn't believe his ears. He took the next leap, and asked: "Will you send us the gene so we can try putting it into a vector that we can use for gene therapy?"

Sure, Hutton said, when do you want it?

Anderson drew a deep breath. He couldn't believe Hutton was going to just turn over the gene. It would make it possible for Anderson to enter the ADA field. A few days later, a Styrofoam cooler packed with ice and test tubes delivered the ADA gene to French Anderson's doorstep.

What happened next was classic Anderson. Since he had been interviewing all the field leaders for his *Science* article, Anderson worried that he would offend some of them if he jumped into the field after they had taken the time to explain how to do it. Once he got the ADA gene, Anderson was most worried about Stu Orkin, and that he would be upset because Anderson was taking a shortcut into the field Orkin had helped create with the gene's discovery. Clearly Hutton knew of Anderson's plans, and Anderson considered Valerio to be sufficiently junior not to worry about.

The versions of what happened next differ. Anderson says he called Orkin to tell him that he was going into the ADA field. "I said, 'Stu, I gotta tell you what happened. Hutton sent me the plasmid. That means that we will go into ADA,' and I said, 'I wanted you to know, and I will never ask again what you are doing,' " Anderson recalled, adding, "Stu was furious. He spent two years cloning the gene, and I got it in two days by making a phone call."

Orkin, while confirming the conversation, remembers it differently. "He did tell me that it [ADA research aimed at gene therapy] was not going to be done in competition," Orkin recalled. "He said we have it [the ADA gene]; we are not trying to compete." But Orkin said he was not furious; he was indifferent. "It was in character for French to say I was furious. It is obvious that if one lab is working on it, and so is another, they are in competition. There is no other way to interpret it."

Orkin, however, did not consider Anderson much of a threat. In fact, none of the retrovirologists working to develop genetic transfer techniques—especially Mulligan—held Anderson in high esteem. "There is considerable controversy out in the real world about the quality of the science that is backing up the work that he [Anderson] is doing," Orkin said. The reason for their disdain was fairly simple: "He hasn't done experiments that are highly informative or clear-cut or important. That is the judgment of the scientific community," Orkin said.

It certainly was not Anderson's view of himself. And now that he had all the pieces—the ADA gene and the N2 vector—he was going to put it all together and make a final drive to be the first to perform a successful gene therapy experiment in human patients.

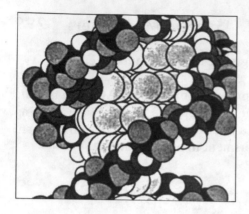

Chapter 9
The Politics
of Genes

*"What did the first TILs out of the intravenous cannula say?
'That was one small step for a gene, but a giant step for genetics.' "
—Sign on a Laboratory of Molecular Hematology wall,
Hall 7D, NIH Clinical Center, May 1989*

In October 1984, French Anderson published "Prospects for Human Gene Therapy" in *Science* magazine, the review article for which he had interviewed all the field leaders to complete. The article was, as much as anything else, a stake in the ground, a scientific manifesto for what had to be done to make gene therapy a reality. And it was Anderson's game plan. He described and discarded the various ways genes could be put into mammalian cells, explaining why retroviruses would be the most effective approach. What's more, thanks to the retroviral packaging cell lines created by MIT's Richard Mulligan and the Salk's Dusty Miller, the retrovirus vectors were safe, since they couldn't cause an infection unless some normal virus was accidentally produced.

He described several genetic disorders for which the genes were cloned, settling on ADA deficiency as the most likely disease for several reasons: The gene was active in blood cells, and experiments showed that correcting the bone marrow, which gives rise to all blood cells, could reverse the disease. "This observation offers hope that defective bone marrow can be removed from a patient, the normal ADA gene inserted into a number of cells through gene therapy, and the treated marrow reimplanted into the patient where it may have a selective growth advantage," he wrote.

This concept of selective growth advantage would prove controversial but also important in getting final approval to eventually try this in humans. Since white blood cells that lacked the ADA gene accumulated toxins, they were sickly and short-lived. White cells with the normal gene inserted would not be sickly and should be able to outgrow the deficient cells, outcompeting them and eventually displacing them in the bloodstream. It made Darwinian sense, but it had never been tested, and many scientists didn't believe it would work.

Anderson also argued that ADA was a good candidate because expression of the transplanted gene, the production of the ADA protein, did not have to be tightly controlled. It only had to be turned on and left on. What's more, experiments suggested that there was a huge effective range: 5 percent of normal ADA levels would help, and as much as 5,000 percent of normal would not cause harm.

"It now appears that effective delivery-expression systems are becoming available that will allow reasonable attempts at human gene therapy," Anderson wrote. "These systems are based on treatment of bone marrow cells with retroviral vectors carrying a normal gene."

And once the details were worked out, Anderson argued, "I believe it would be unethical to delay human trials . . . [because] patients with serious genetic diseases have little other hope at present for alleviation of their medical problems."

But for the patients who hoped to benefit and the scientists who struggled to make this all happen, Anderson raised the key question: "What criteria should be used in evaluating gene therapy protocols?" he asked, and then answered with "It should be shown in animal studies that (i) the new gene can be put into the correct target cells and will remain there long enough to be effective; (ii) the new gene will be expressed in the cell at an appropriate level; and (iii) the new gene will not harm the cell, or by extension, the animal."

Once he drew this line in the sand, Anderson would have to meet the standards he set and that other scientists agreed were essential. Even though he now had Eli Gilboa's N2 vector through their collaboration, and the ADA through John Hutton's generosity, Anderson was a long way from proving that ADA could be treated by replacing the gene. First, he would have to put the ADA gene into one of Gilboa's vectors. Next, he would have to show that the ADA vector could get into mammalian cells and produce the ADA protein. Then he would have to show it could correct ADA-deficient cells in culture—preferably human bone marrow cells—and then in animal experiments, mice then monkeys, progressively rising up through the evolutionary tree until he reached humans. Like evolution itself, it would be a long, hard climb.

Besides the scientific hurdles, there would be a legal one as well. On April 25, 1984, the *Federal Register* carried a notice from National Institutes of Health that any federally funded gene therapy experiment involving recombinant DNA would have to be approved by the Recombinant DNA Advisory Committee (RAC) and the NIH director. RAC was established in 1975 in response to public fears about the cloning of genes in bacteria. NIH expanded RAC's mandate as part of the government's response to the presidential commission's report *Splicing Life,* which called for special oversight of any gene therapy experiments in humans. In 1983, RAC set up the Human Gene Therapy Subcommittee to establish criteria for reviewing human experimentation with genes and then to conduct the initial evaluations. The RAC, and its subcommittee, would ensure that the public had a window through which it could watch scientists deliberate about the technology—and the wisdom—of changing genes in people.

For most scientists, this added bureaucratic gauntlet was intimidating. Most were content to work in their labs, answering questions judged to be important by their peers and wanting to be left alone. Anderson, however, was not. He realized early on that if he was to be the first to do genetic experiments in humans, he would have to confront this regulatory barrier. Initially, he joined the RAC working group looking at gene therapy and later the subcommittee. Critics said he did that only to co-opt the review process, to develop good relationships with the committee members, and to grease the skids for his own future approval. Several scientists pointed out that it could be seen as a conflict of interest for Anderson to serve on a committee that would have to review the first gene therapy experiment he publicly proclaimed he wanted to do it. Anderson agreed and resigned, but continued to offer technical advice and written comments.

• • •

Long before he would have to worry about justifying his work before the RAC, Anderson and his colleagues had many technical details to resolve. Philip Kantoff, a physician-researcher, had joined Anderson's lab in July 1983 and initially worked on the mouse model for thalassemia with the intention of using it to test the transfer of the globin gene. In January 1984, Anderson asked Kantoff to join the gene therapy project and start working with Eli Gilboa to learn about making vectors, the molecules that can carry genetic information into new cells. Kantoff agreed, and in the spring went to Princeton for a couple of weeks to work in Gilboa's lab and learn the fundamentals of retrovirus vectors.

Gilboa and Kantoff hit it off right away; Kantoff was impressed with Gilboa and the vector work, and became very excited by its potential power. Initially, Gilboa and Kantoff continued trying to build a vector that carried the beta globin gene, but the work was labor-intensive and frustratingly slow, consuming much of 1984. None of the vectors worked very well, though; several were put into mice, but none of them produced very much protein—certainly not enough to consider a clinical study, or even much work with the mouse thalassemia model. "It was obvious to everyone that a disease for the first human trials would need to be easier than globin," Kantoff said.

When the ADA gene arrived in a plasmid (pBR322) from Cincinnati in September 1984, Anderson gave the gene to Kantoff and Gilboa to stick it into the N2 vector. In a few weeks of intense work, the team had popped the ADA gene out of Hutton's plasmid and into a derivative of N2 that they called SAX. It had the basic N2 backbone, with the long terminal repeats (LTR) on each end, and a neomycin-resistance gene to make transformed cells selectable in medium laced with G418. Between the LTR and the *neo* gene, Gilboa and Kantoff inserted the ADA with its own genetic on switch, a promoter from the SV40 virus that switched on protein production in mammalian cells.

The completed SAX vector was inserted into the different producer cell lines manufactured by Richard Mulligan and Dusty Miller. The cell lines produced high quantities of genetically engineered Moloney retroviruses that carried the SAX vector, and it did transplant the gene into target cells. By the end of 1984, the basic vector was completed, and ready for testing in animal cells growing in culture and then in whole animals. The work was proceeding nicely, and it made Anderson all the more confident that he and the team were on the right track and that human trials were imminent.

• • •

With molecular genetics going well, Anderson had another problem to confront. He had spent most of his career studying globin diseases and knew relatively little about immunologic disorders like ADA deficiency. He needed an expert who knew the biology of the white blood cells and would also be able to care for the first gene therapy patients, who were most likely to be children. During a dinner conversation with his wife, Dr. Kathryn Anderson, deputy chief of surgery at the National Children's Hospital in Washington, D.C., Anderson asked if she knew any pediatric immunologists. She pointed out that there was a good one just down the hall from Anderson's own office in the clinical center, NIH's massive hospital: R. Michael Blaese of the National Cancer Institute.

Blaese was an expert in Wiskott-Aldrich syndrome, a rare, inherited X-linked immunodeficiency found only in boys. They usually died by age three from bleeding in the brain, chronic lung infections, or, in those who live long enough, blood cancers like lymphoma and leukemia. After years of research, Blaese deciphered enough about the basic disease process to realize that the boys suffered constant bleeding because their spleens were destroying their already abnormal platelets, a blood element involved in clotting.

If the spleens were removed, Blaese reasoned, then the bleeding problems should decrease. NIH, however, didn't have a pediatric surgery program; this forced him to search around for a pediatric surgeon who would take the risk of the experimental surgery, especially since medical textbooks said removing the spleen in these boys was a bad idea. The name he kept hearing was Kathy Anderson, so he approached her and asked if she was interested. After explaining the basic biology, Blaese suggested a collaboration in which he provided the patients and she took out their spleens. The approach made sense to her, so she agreed in the late 1970s to try the experimental surgery in a clinical trial. It would either prove or disprove Blaese's hypothesis—and it might help children in the process.

Surgery wasn't a cure, but the experiment proved successful; the bleeding problems in these children subsided once the spleen was removed. Blaese added chronic preventive treatment with antibiotics and intravenous immunoglobulins—therapies designed to prevent opportunistic infections. The combination of treatments added a decade or two to the children's lives. It also cemented a relationship between the bear-sized Blaese and the diminutive surgeon.

. . .

Robert Michael Blaese always had been an immunologist. Born in Minneapolis on February 16, 1939, he graduated magna cum laude from Gustavus Adolphus College in St. Peter, Minnesota, and got his

M.D. from the University of Minnesota School of Medicine in 1964. In medical school, Blaese was hired as a technician for Robert A. Goode, a physician and research immunologist who performed, in 1968, the first successful bone marrow transplant, which cured an inherited immunodeficiency disease. Goode got Blaese excited about the possibilities of medical research. During Blaese's first year in Goode's lab, the function of the thymus gland was discovered there.

Goode himself was a towering figure in immunology, though he would later be tainted by a fraudulent experiment performed by a protégé, William T. Summerlin. Goode left Minnesota to be director of the Sloan-Kettering Institute at the Memorial Sloan-Kettering Cancer Center in New York City, and had brought Summerlin along to New York in 1974. Summerlin continued work begun in Minneapolis designed to understand the biology of transplantation. By grafting black-pigmented skin onto a white mouse, he hoped to show that the mice could be made immunologically tolerant to unrelated tissue, but the experiments had problems. At one point, while at Memorial Sloan-Kettering, Summerlin used a felt-tipped pen to blacken an area purported to show the location of successfully transplanted black skin. The ensuing scientific fraud became a celebrated case, and it cost Goode his credibility and his position, even though he was never accused of wrongdoing and actually had launched the investigation.

After graduation from medical school, an internship, and residency in pediatrics, Mike Blaese came to the National Institutes of Health in 1966 as a clinical associate in Thomas Waldmann's metabolism branch of the National Cancer Institute. Blaese was part of that wave of new recruits who came to NIH during the Vietnam doctor draft to fulfill his military obligation; he never left.

In 1984, Blaese was getting burned out. He had worked on Wiskott-Aldrich syndrome for a decade and a half, and was struggling with the cellular immunology of chronic fatigue syndrome and Epstein-Barr. "I was getting tired of it," Blaese said. "They are miserable problems."

That's when Anderson called to ask if Blaese was interested in a new problem: gene therapy for children with SCID caused by ADA deficiency. "I was very excited about the whole idea of working on this kind of approach," Blaese recalled.

After the collaboration began in 1984, Blaese started searching around the country for children with ADA deficiency. He wanted to get defective white blood cells from the children to see if he could find a way to grow them in the laboratory, so the SAX vector, which Anderson's team was just completing, could be tested in cells actually missing the ADA enzyme.

Early in 1985, Blaese interviewed a physician-researcher named Donald Kohn from the University of Wisconsin at Madison who was looking for a place to do postdoctoral work. Kohn got referred to Blaese by a former Minnesota colleague who had moved to Madison and also happened to have two children with ADA deficiency. In February 1985, Blaese jumped on a plane to Madison to gather blood and bone marrow cells from the two patients so he could get cultures of their cells growing in the laboratory. Blaese and Anderson would need the ADA-deficient cells to test the restorative powers of the SAX vector. As a first step toward human trials, the team would need to prove that it could transfer the ADA gene into defective cells, that the gene would make the protein, and that the protein would reverse the effects of the disease in those cells.

ADA children, however, make few white blood cells because of the enzyme deficiency, making it difficult to isolate any white blood cells to begin with and virtually impossible to grow them in culture. To solve that problem, Blaese collected bone marrow from the kids, and infected the marrow cells with HTLV-1, the human T-lymphocyte virus isolated by NCI colleague Robert Gallo and shown to cause leukemia in humans. The virus has the ability to convert mature blood cells, which normally have a limited life span, into immortal cancer cells that can grow forever in culture. When the HTLV-1 virus infected the bone marrow cells, it converted a few of the sick, ADA-deficient lymphocytes into immortalized cancer cells. These cells were not normal, however, since they grew well even though they still had the genetic disease, and they lacked a functioning ADA gene. Still, they would give the NIH scientists cells that lacked the ADA enzyme for testing. Blaese also made B cell lines (they're the lymphocytes that make antibodies) from the ADA-deficient children by immortalizing them with Epstein-Barr virus.

"By the early summer of 1985, we had several T cell lines growing from ADA deficient kids," Blaese said. "That was an incredibly important thing because it gave us cells to work with, cells that we could try to put the gene into, and look to see if we could cure the biochemical abnormality."

But would the SAX vector get into the T cells and cure them? The answer awaited Don Kohn's arrival in July 1985. When Kohn walked into Blaese's lab, there were two reagents waiting for him that had never existed before: the SAX vector that carried the ADA gene and the ADA-deficient cell lines in which to test the vector. All Kohn had to do was mix the two together to see if SAX would fix the T cells.

Since the cells already grew, despite the defect, because they had been turned into an immortal cancer, Blaese had to devise a unique test to prove the addition of the ADA gene made a physiological difference. Compared to cells that made the normal amount of the ADA enzyme, the immortalized T cells remained sensitive to the chemical adenosine, and were easily poisoned when it was added to the growth culture. Restoring the ADA enzyme through a gene transplant, Blaese reasoned, should give the immortalized cells normal resistance to the addition of adenosine.

Kohn took the producer cells that Gilboa and Kantoff had modified to make SAX, and placed a layer of transformed lymphocytes on top of them in a cocultivation culture where two kinds of cells are grown together. As the producer cells spewed out SAX-containing retroviruses, the viruses floated through the culture fluid to infect the immortalized T cells and insert the ADA gene. The efficiency of gene transfer wasn't great—about 20 percent of the cells took up the ADA gene. But those that did produced near-normal levels of the ADA protein, and demonstrated near-normal resistance to the addition of adenosine to the growth medium.

"And of course, it went beautifully," Blaese recalled. "We had data within about five days, which showed that the gene was getting in really high levels and that we were reversing the metabolic abnormality with gene transfer. The experiments all went perfectly that summer; we cranked out an enormous amount of data. By the end of the summer of 1985, we knew that gene transfer would cure these kids."

The results were published a year later in the *Proceedings of the National Academy of Sciences,* showing for the first time that it was possible to reverse ADA deficiency—at least in the test tube—by transplanting a normal copy of the mutated gene. It was an exciting time in the laboratory, and the NIH scientists seemed to be advancing faster than the other groups. The team of Stewart Orkin at Harvard and Richard Mulligan at MIT attempted to use the ADA gene Orkin had isolated to produce protein inside an ADA-deficient cell after the gene was inserted into one of Mulligan's vectors, but they were never able to get protein production. They later showed there was a mutation in the ADA gene that Orkin had isolated. In the quest to do gene therapy for ADA-deficient patients, Anderson's team edged ahead.

Now, however, the NIH team faced their most formidable task: proving they could get ADA protein made in the bone marrow cells of living animals. It was one thing to transfer the gene into cells growing in a laboratory dish, but making it work in an intact animal would be entirely different. Anderson, Blaese, and the others needed to show they could use SAX to transplant the ADA gene into bone marrow cells that lived forever: the stem cells. Only by correcting the stem cells would the researchers have a hope of permanently curing SCID children—and that, after all, was the goal.

Stem cells had, in fact, become the holy grail of gene therapy. It was not enough to be able to transfer a gene; the gene had to be inserted into the right cell or else it would not help the patient. And the only cell that made any difference to diseases of the blood was the stem cell.

Changing these cells with new genes, however, would not be easy. There are few of them in the bone marrow, probably less than one in a thousand bone marrow cells. And, in the mid 1980s, there was no way to identify any of them. The only reasonable approach seemed to be taking a large sample of bone marrow cells and infecting all of it with the gene-carrying retrovirus vector in the hopes that enough stem cells would be altered to make a difference. But with infection efficiencies of only 10 to 20 percent, transplanting genes into every cell seemed unlikely.

Anderson, however, believed he had a solution. In diseases like ADA-deficiency, he argued, correcting just a few stem cells might be enough because their progeny would have a growth advantage over the uncorrected cells. The still-sick cells would continue to be poisoned by the toxic buildup that the ADA enzyme prevented. Therefore, even a few primordial stem cells could eventually give rise to millions of stem cells that could make all the white blood cells in the patient's body. At least, that was the theory.

In the field of getting genes into bone marrow cells, however, Richard Mulligan had the edge. He was the first to report in 1984 that it was possible to use retroviruses to put genes into the stem cells of mice, but the gene-transfer technique was so inefficient, and the expression of the transplanted genes so temporary, that he considered it too ineffective to begin thinking about using it in patients.

Despite Mulligan's lead in working with bone marrow, the NIH team moved to catch up, and rapidly ran into the same problems of getting efficient gene transfer and protein production. They could get the retrovirus vectors into mouse bone marrow cells, but they got either little or no expression of the protein for which the gene coded.

The exception was the neomycin resistance gene, which seemed to work fine in most cells, proving that a transplanted gene could function if it operated under the right conditions. But what were those conditions? Perhaps, the NIH scientists wondered, the SV40 promoters, the genetic on switch that proved so successful in getting ADA protein production in human T lymphocytes, was somehow turned off in mouse bone marrow cells. No one knew, and as the number of failed experiments piled up, the NIH team began to worry that they might never find a way to get gene expression in mouse bone marrow.

In his *Science* paper in 1984, Anderson had laid out a progression of experiments that had to be done before anyone could move on to his goal of doing human trials. His list started with cells growing in culture, then mouse studies to test for effectiveness and safety, and then primate trials that would exactly replicate a gene therapy protocol planned for people. His team seemed to be in danger of getting stuck at the second step, well short of human trials.

It was at this point that Anderson made a fateful decision: He chose to skip the mouse studies and go straight for monkeys. "If you look down the road and see you need to take steps 1, 2, 3, 4, and 5, and you can see that step 5 is important, but steps 3 and 4 can be skipped, then you can speed things up by skipping to what's important," Anderson said.

Back in 1984, when Anderson and Blaese decided to make a full-court press on human ADA trials, Blaese called up a former colleague from the University of Minnesota, a bone marrow transplanting center, to see if he was interested. Blaese's friend, Richard O'Reilly, had moved to Memorial Sloan-Kettering along with the general influx of Goode protégés. O'Reilly was a physician-researcher who ran Sloan-Kettering's bone marrow transplantation program, and since inserting the gene into bone marrow was the goal, O'Reilly had a lot to offer. In addition to running an established facility at a major research center, O'Reilly had monkeys right in midtown Manhattan.

The connection to Sloan-Kettering had another impact on the team. Once the collaboration began between the NIH scientists and

O'Reilly, Eli Gilboa left his junior position at Princeton for a research slot at Memorial Sloan-Kettering.

In August 1985, Anderson began shuttling to New York City to begin inserting SAX into monkey marrow to produce human ADA. The protocol itself was fairly simple: A bone marrow sample was removed from the animal and mixed with the retrovirus in a culture dish. Meanwhile, the animal received a lethal dose of radiation that wiped out all the cells in its bone marrow to make room for the genetically treated cells when they were transplanted back into the animal's body. To do that, the treated bone marrow cells were merely injected back into a vein, and the cells found their way back to the bone marrow. The animals received intensive medical care, like any bone marrow recipient, to fight off infections until their marrow recovered and restored their immune system.

Using various approaches, the NIH–Memorial Sloan-Kettering team showed they could get the human ADA gene into monkey bone marrow cells at a very low efficiency—up to 0.5 percent of normal monkey ADA levels—but even this low level did not persist. After four or five months, ADA expression completely disappeared from the animal's blood and bone marrow. They could, however, show the presence of the active *neo* gene in T lymphocytes for up to 256 days after the transplant.

Since there was some expression of the transplanted ADA gene, the problem was not the vector; it was delivering the gene to the bone marrow cells. It was possible, however, that the cells were turning off expression of the foreign gene with time, but the more likely explanation was that the gene was not getting into stem cells, but rather only peripheral blood cells or short-lived bone marrow cells that already had been set on the developmental pathway to become some kind of mature white blood cell. Once these cells fully differentiated and died out, the expression of the ADA gene was lost.

Despite the negative results, Anderson found a ray of hope. First, Anderson believed, even 0.5 percent efficiency might be enough to treat an ADA patient if that half percent hit some stem cells. Because of the selective growth advantage that Anderson believed would come with the normal ADA gene, even a few stem cells might outcompete the defective cells in the patient's body and eventually repopulate the patient's bone marrow. What's more, all of the monkey studies—some twenty-five animals in all—never showed any ill effects from the genetic treatment. As far as Anderson was concerned, this proved that the gene therapy experiments in animals were a success, both in producing some protein and in being safe.

By 1986, Anderson believed that the time had come to launch an attempt to insert the ADA gene in to the bone marrow of human ADA patients. Almost no one else in the now-expanding field of gene therapy agreed with him—including his colleagues who participated in the monkey studies. Nearly all of them, from O'Reilly to Gilboa to Blaese, considered the monkey studies a failure. And the more Anderson talked publicly about the coming gene therapy revolution, the more disapproving, even contemptuous, became his scientific colleagues.

Harvard's Stuart Orkin, for example, didn't believe the technology was mature enough to begin thinking about human trials. "We didn't think we had the experimental evidence in our own hands that [trying to treat a patient] would be worth doing," Orkin said a few years later.

Anderson, however, saw it quite differently. He felt there was nothing else to prove in the animal experiments. True, they provided disappointing evidence of gene transfer and expression, but he believed that the animal models were irrelevant. Only in humans with ADA disease could the technology be truly tested. The time had arrived to push for people, and he was excited.

Anderson was not alone in this belief. A cadre of venture capitalists decided that it was time for gene therapy to come of age. And they were willing to put up the money to make it happen. Wallace Steinberg, chairman of Healthcare Ventures of Edison, New Jersey, wanted Anderson to be in charge of a new company he intended to start. Anderson declined, but agreed that he would help get the company, to be run by James Barrett, off the ground.

A new law, the federal Technology Transfer Act of 1986, had just been passed to help move technology developed in government labs at taxpayers' expense into the marketplace, where it could help the American economy. The NIH was a government lab. Following the law's provisions, Anderson formed a Cooperative Research and Development Agreement, a CRADA, with the nascent company. It was the first, and one of the biggest CRADAs NIH signed. On the strength of that agreement, Genetic Therapy Inc. (GTI) of Gaithersburg, Maryland, opened its doors in July 1987, with $2.5 million in venture capital and 30 employees.

In the next several years, GTI would become an extension of Anderson's NIH lab. Things that couldn't be done easily through the government could easily be done in the private sector. GTI would make significant contributions to Anderson's efforts, including producing many of the key reagents needed in the experiments to follow.

What's more, the national media were getting excited, too. In 1984, the country's leading newspapers, including *The New York Times, The Washington Post,* and *The Wall Street Journal,* and even the news section of *Science,* began to carry stories about the progress of several gene therapy research teams, including those led by Anderson, Mulligan, and Friedmann, and how they seemed to be moving toward launching experiments in humans.

Anderson, of course, was fueling the speculation, sometimes shooting himself in the foot by making predictions that exceeded the science. With the Martin Cline story only four years old and a presidential commission issuing *Splicing Life* in 1982, reporters frequently asked the scientists when human tests would begin. Most scientists declined to make a prediction; Stanford's Paul Berg would emphatically deny that it would be anytime soon. But Anderson was more confident: By the end of next year, he told *The Washington Post* in 1984.

Not only was he wrong but he failed to appreciate the amount of enmity such a prediction would engender among other scientists. This led many of the others in the field to dismiss Anderson and the NIH team, branding him a lightweight researcher who didn't know what he was talking about, someone who clearly didn't appreciate the complexity and the limitations of the science.

"Anderson's hype gave people a funny sense of the progress in the field," said MIT's Mulligan. "You would hear about it, and draw the conclusion that we were very close [to human trials], when I knew that we were not."

Although Mulligan may have been right that the science was not close to being ready to put genes into people in 1984, more and more people were getting excited by the prospect of it. RAC organized its Human Gene Therapy Subcommittee, and asked it to lay out the criteria—which it pleasantly named the Points to Consider—by which the subcommittee would judge a human gene therapy protocol. Even the Institute of Medicine got caught up in the raised expectations of gene therapy. In 1986, the institute devoted its sixteenth annual meeting to gene therapy, even though major questions remained unresolved, and there was no consensus in the field that the time to proceed had arrived.

David W. Martin, Jr., vice president for research at Genentech Inc., who, like Anderson, considered ADA deficiency to be one of those diseases that might make an excellent test case to launch the field of gene therapy, set a hopeful tone early in the Institute of Medicine meeting. He walked through the key questions that anyone wanting to do gene therapy with ADA would face, starting with the questions about how much protein would have to be produced from the transplanted gene.

"A few percent, maybe 5 percent ADA activity, would be sufficient to reverse the disease," Martin said. And as for an upper limit, studies suggested that 5,000 percent of normal would do no harm. "That is a pretty good-sized window," Martin said. "We have a good chance of hitting that window."

It is, however, a therapeutic dose window that would have to be hit every day, for the rest of the patient's life. But the gene transporters were having problems with long-term expression. They would put the genes in with some vector and initially get good protein production. But then, after a few weeks, the cells somehow seemed to shut off the transplanted genes, and protein production disappeared.

"This has been the biggest surprise for those of us in gene therapy," Martin told the IOM meeting. Somehow, the scientists would need to find better genetic on-off switches, called promoters, that would stay stuck in the on position.

And there was the question of which cells in the body should receive the transplanted genes. Ideally, Martin said, they should go into stem cells, but he laid out the difficulties with that approach.

In addition, some of the questions might be related. Was it possible that scientists would find a way to keep the genes turned on in some cells, but not in the right cells, like the bone marrow stem cells? "Is there some unique property of stem cells that precludes expression from an internal promoter of a retroviral vector which is capable of efficient replication?" Martin wondered. "We must learn more about hematopoiesis, and more about retroviral replication and gene expression of genes carried by retroviruses."

Martin told the IOM that tissue-specific expression probably was not necessary in ADA deficiency, since transfusions of red blood cells—which for unknown reasons are rich in the ADA enzyme—could detoxify the patient and at least keep the immune system functioning at some level. But the question remained: "How to stably introduce the gene in self-renewable cells, and have it be safe."

Safety was a question. Although the researchers now knew that the retrovirus could be engineered so that it would not reproduce and cause an infection, it still had to integrate the genes it carried into the host cell's chromosomes. Work with oncogenes and retroviruses showed that this integration process could—randomly—disrupt the genetic control mechanisms for genes that controlled the growth of cells. Some cancers, it turned out, could be caused by the random insertion of an animal retrovirus into the wrong part of a cell's chromosome.

Even if the insertion didn't cause cancer, it might knock out the function of some other important gene in that particular cell. That might or might not be important. If gene insertion simply caused a cell to die, then it probably didn't matter. But if it gave rise to a population of now deranged cells, it might. No one knew.

In the end, Martin concluded, "We have a long way to go. It's incremental . . . slow. What we need is more communication between the active groups."

But that was unlikely. Animosity was growing between Mulligan's and Anderson's groups. By the 1986 IOM meeting, it had reached a crescendo.

French Anderson came next on the Institute of Medicine program, and he immediately claimed his territory: "ADA deficiency is almost certainly going to be the first candidate for gene therapy." Anderson pointed out that it met most of the criteria and answered most of the questions raised by David Martin.

But even Anderson had to admit the early experiments were not going well. The monkey trials at Memorial Sloan-Kettering were completed, and the results disappointing. They were able to insert the human ADA gene into bone marrow cells, and get a low level of protein production. But "once it hits its peak, it totally dies out again," Anderson lamented. "They are all negative now," he said of the monkeys.

In fact, in only four monkeys did Anderson's group show it could get ADA production out of the treated monkey marrow, and that level of expression—0.1 to 0.5 percent human ADA expression compared to natural monkey ADA levels—was lower than the 5 percent level found in some healthy humans. What's more, the ADA protein could be found in the monkey blood only for short periods of time, peaking between 60 and 80 days, and disappearing completely by 170 days.

Although Anderson expected an ADA trial to open up the field of gene therapy, it was not clear how success with ADA would be extrapolated to other diseases. In 1986, everyone expected gene therapy to be aimed at inherited illnesses like cystic fibrosis and muscular dystrophy. But "there is a very big gap between ADA deficiency and nearly all the other genetic diseases," Anderson admitted. "We are a long way from treating other genetic disorders." Even if a genetic treatment worked for ADA, it might not work for any of the others.

Even with the problems, by the 1986 IOM meeting, Anderson was convinced the time had arrived to try gene therapy in patients. "We may be able to help people long before we really understand what we are doing," Anderson argued. "The question before the RAC is going to be, 'How much do we have to understand what we are doing before we try to help a patient?' "

Anderson, a physician, saw it as a perfectly reasonable question. Many medical treatments are attempted empirically in medicine long before underlying biology is understood. A mature body of philosophical discourse concluded that it is not only *ethical* to try a desperate treatment, it is *required*—if there is something that offers the slightest chance of saving a life. Anderson believed gene transfer had reached that point.

To the Ph.D.s in the audience, the ones like Mulligan with hard-won doctorates in biochemistry and molecular biology, understanding is everything. Experiments exist for the pure pleasure of knowing nature, of deciphering its ways, and then, with knowledge, intervening. For the basic scientists, Anderson's rush to the bedside was ridiculous and unwarranted.

After the morning break, Mulligan rose to speak. First he laid out his criteria for gene therapy candidates and how researchers would know success was achieved: A high number of cells had to be infected with the vectors, and they had to be stem cells. Do the inserted genes function properly in the proper cells? Will they be stable? Will they cure the disease? Success would be a numbers game: A million cells could be used to reconstitute a mouse, but "in humans or monkey, you need orders of magnitude more cells," Mulligan said.

And then he turned to Anderson's work. The bone marrow transplants that included the new ADA genes worked only in about half of the animals, he said of the monkey studies. And in those few animals, in which the ADA gene was successfully transplanted, the gene made only the ADA proteins for a short period of time.

Mulligan dismissed Anderson's results as unimpressive; Anderson simply failed to prove that he had transferred the gene to monkey stem cells—he couldn't even detect the ADA gene within the animals—and what little protein he could see was not enough to cure a patient. Before any human trial should go forward, Mulligan argued, significant expression would have to be shown in the right cells for many months. He didn't believe Anderson, or anyone else, would be meeting those criteria for years to come.

"If I were an M.D., I could see why you'd want to try anything," Mulligan told the *Los Angeles Times*. "But let's just define what we are saying and doing. There is absolutely no shred of evidence that it [an ADA experiment] will work. Maybe French just has faith in miracles."

Others too were skeptical. Philip Leder, the scientist whose bench Anderson took over when he joined Marshall Nirenberg's NIH lab in 1967, had become chairman of genetics at Harvard University and a gene therapy skeptic. "It is something that is feasible," Leder said as the summation speaker at the IOM meeting. But "this field is very young. The initial recombinant DNA experiments are only just behind us . . . a decade ago. . . . It has not yet fully achieved its final goal."

Leder wanted to see some of the basic questions resolved, like finding a way to improve the efficiency of gene transfer with the retroviruses, especially into stem cells, and he wanted to see the transplanted gene come under better regulatory control instead of just being turned on all the time. "There are troubles, but we are beginning to understand the basis of these problems," he said. "One day the problems will be overcome."

But, he added at the end, "there are concerns [about whether] the ADA protocol is the appropriate one with which to go forward."

Leder was looking beyond the rare, relatively simple recessive disease of the immune system to consider how gene therapy might be used for more common and more complicated diseases such as thalassemia, cystic fibrosis, and muscular dystrophy, and those caused by several genes working together.

What's more, he wanted to "be able to use gene therapy to replace the defective gene in the site of the normal gene [as had proved possible in yeast but would] be hard to do with mammalian cells." This so-called homologous recombination was terribly difficult and inefficient, even in yeast. It might, however, allow an inserted gene like beta globin to function normally and produce just the right amount of protein.

Besides the bevy of unanswered scientific questions, Anderson's critics worried that an unwarranted experiment, propelled by ego and ambition, might be more than just a spectacular failure—like Cline's. It might hurt a patient, even kill one. That might turn the public—which was barely paying attention to the intense debate among the scientists—against gene therapy, blocking any of the research physicians from proceeding to a clinical trial even if enough progress had been made. Everyone in the field remembered the recombinant DNA debates of the 1970s and the City of Cambridge moratorium on genetic engineering at Harvard and MIT, and the Cline affair, which all believed set back the field by a decade.

"Nobody wants to provide a target for the people to say, 'Gee, it doesn't work, and what are we doing to the environment, and what else is going on here?' " Dusty Miller, who by now had left Salk for the Fred Hutchinson Cancer Research Center in Seattle, told the *Los Angeles Times*.

Harvard's Stuart Orkin was even more emphatic: "The only urgency is competition of labs. There is not a crying need elsewhere. This is not like AIDS. These are rare diseases. Gene therapy is a stunt. Face it. Our position is to try to understand the science. If we learn enough, it will apply. Intellectually, it is more interesting to understand. We are just plodding along right now. We don't know why or what works." If the ADA experiment Anderson was pushing so hard failed, then "lots of explanation will be needed. It will give ammunition to lots of critics. We would be dead for ten years."

• • •

In his heart, Anderson didn't believe it would fail, though he couldn't prove with his animal data that it would work. Increasingly, Anderson became convinced that the only way to test gene therapy for ADA deficiency was going to be in people. No animal models of this disease existed, so there was no way to prove convincingly that this would or would not work until it was tried in patients. Sure, healing ADA kids with gene therapy was not going to make much difference to many people since there were only about forty patients in the world—far fewer patients than there were scientists working on the problem—but it would be a start. It would open the door to a new era of molecular healing, and Anderson intended to be the first to do it.

In the face of all the resistance from the other scientists at the Institute of Medicine meeting, Anderson decided to force the issue on the RAC anyway, to make them decide about whether or not enough was known to go ahead with patients. He had followed the sequence of experiments he predicted would be needed, but came up with ambiguous answers. The genes went into bone marrow cells, but they didn't do it efficiently or produce much ADA protein. He had gone from cell cultures to monkeys, but didn't have definitive answers. The NIH team had reached the end of the road with animal experiments. The only thing left was to try it in humans.

• • •

Six months after the October 1986 IOM meeting, Anderson submitted a 1¼-inch-thick, 500-plus-page document to the NIH Recombinant DNA Advisory Committee, RAC, to force the board to make a decision about his desire to put genes into people. Fearing a flat-out rejection, Anderson called it a "Preclinical Data Document" that "is not a clinical proposal, nor do we request any form of approval for a clinical protocol at this time," according to the introduction. "Furthermore, this document is not a final submission, but only an initial communication." It was, however, a complete review of where the field stood, a listing of the key unanswered questions, and an overview of how a human experiment would be conducted if one was to be submitted—which this wasn't.

In chess, this move would be called a gambit. In Washington, D.C., such releases—usually in the form of leaks to the news media—are called trial balloons. And Anderson had every intention of having this idea balloon-tested as harshly as possible by drawing as much attention and criticism as he could from the RAC and his competitors.

"One approach is to ignore critics," said Art Nienhuis, a longtime Anderson friend and collaborator. "French draws them out. In getting this to move forward, if you get people to commit to their criticism, then you eventually are going to neutralize them."

That was Anderson's goal. Anderson played his hand, laying his cards on the table to force his opponents' hand, get them to commit so their scientific complaints could be answered. Once the objections had been answered, Anderson reasoned, the RAC would have no choice but to let him go forward with the first human trial.

But this was a dangerous game, played by argument and logic, supported by data and extrapolation. The winner built a consensus in the minds of the RAC members who judged the experiment. The game could be won or lost on the turn of a single, crucial idea. Anderson began, disarmingly, by pointing out the three weaknesses of his own protocol:

1. [Only a small number of monkey bone marrow cells produced the ADA protein—and really only one monkey showed any significant level of ADA production, and that was only half a percent.] Will the postulated in vivo growth advantage of ADA (+) cells in ADA-deficient patients be sufficient for a clinical improvement in patients receiving ADA gene therapy? If not, how much activity (in how many monkeys) would be necessary to determine that an adequate level of ADA expression could be expected in an ADA-deficient patient?

2. Will the vector-delivered human ADA gene . . . be expressed at a sufficiently high level in stem cells . . . so that the postulated in vivo growth advantage would be maintained throughout the life cycle of the T cell?

3. [Since the NIH vectors were contaminated with a small amount, about 0.1 percent, of helper viruses—retroviruses that had the normal virus genome and were able to reproduce, and possibly cause, disease—] must the retroviral vector particle preparation used to treat the patient's bone marrow cells be totally helper virus free in order to make the risk/benefit ratio satisfactory?

He then spent the next 100 pages of the document addressing those very questions. The remaining 400 or so pages were devoted to supporting material from research published by his lab and others. Then, as he handed in the document, Anderson asked RAC's longtime executive secretary, Dr. William J. Gartland, Jr., to send the document to his harshest critics and competitors, eleven in all, including Richard Mulligan and Stuart Orkin, in Boston, Dusty Miller and Theodore Friedmann, from San Diego, Howard Temin from the University of Wisconsin, and Dr. Michael Hershfield from Duke University Medical Center, who had just developed a potentially revolutionary drug treatment for ADA deficiency. Gartland was happy to go through this "dry run" to give the RAC, and its consultants, a sense of what a real gene therapy proposal, and its review, might look like. What's more, since Anderson was pushing hardest to be the first to do gene therapy, it was likely that this was the protocol on which they would ultimately have to decide. This was a chance for

RAC to review it without actually having to make any risky decisions.

The reviews were slow—it took almost half a year before all the critiques came back, and they were scathing. Only Robert M. Cook-Deegan, a physician with molecular biology training and a policy analyst for the Congressional Office of Technology Assessment, concluded that "if the preproposal were a real proposal for research on adults, I would likely give it the go-ahead," but he would not have approved it for children. The rest of the reviews were borderline neutral to strongly negative—so much so that several were anonymously released into the RAC's public record.

There were comments like "It is disappointing to see so little characterization of the monkey bone marrow system *in vitro*" and "First, I object to the form of the document . . . [and] Second, no amount of paper can rectify two major deficiencies. . . ." And, again from the same person, "In the absence of adequate tests of the safety of such contaminating helper virus, there is little likelihood that we would approve a human trial with a stock of this sort."

The longest, at ten pages, and arguably the most thoughtful review, came from Richard Mulligan, which he signed. Mulligan treated it as though it were a formal review, and came away with somewhat predictable conclusions: "Since the data obtained so far provides no indication of efficacy, there is virtually no chance that useful information would be obtained from such a clinical study, other than demonstrating that, indeed, the system is not yet ready to be tested."

Mulligan addressed the deficiencies that Anderson knew all too well:

On the efficiency of gene transfer, Mulligan said: "In the case of the mouse, the answer to the question is that depending on the viral titers, the delivery system can be very efficient (close to 100 percent). In the case of the monkey or dog, the system is extremely inefficient. In the case of human cells, the efficiency cannot currently be assessed."

On what cells should get the genes: "It is reasonable to conclude that the target for gene transfer should in fact be the reconstituting stem cell. . . . These studies have not clearly shown that the ADA gene can be appropriately expressed after long term reconstitution."

On safety: "Since packaging cells do exist that don't give rise to wild-type helper virus, the virus producing cell lines used for clinical studies should be derived from them. . . ."

Mulligan's conclusion: "To proceed with clinical studies at this time would undermine the necessary pre-clinical studies that will be critical to success in the clinical setting. . . . Based on our current understanding of the gene transfer process and the results obtained *in vivo* in the mouse, monkey and dog, the proposed studies are not warranted."

Seattle's Dusty Miller agreed that there were existing cell lines that would make the retrovirus vectors without producing infectious helper viruses. Therefore, there was no reason to use any contaminated vector preparation. In the end, despite disagreeing with the NIH team on different issues, both Mulligan and Miller offered their improved packaging cell lines to give the experiment the highest chance of success and safety.

But perhaps the biggest problem had much less to do with the NIH team's data than with a revolution in the treatment of ADA disease itself. Suddenly, this uniformly lethal disease could be controlled with weekly ADA enzyme replacement injections. In the fall of 1987, Michael Hershfield from Duke University had published a preliminary report of the treatment of two ADA children with the injectible cow enzyme that had been treated with polyethylene glycol (PEG-ADA) so it could be used in humans—and apparently with great success. Hershfield's data showed that the enzyme injections reversed the immune deficiency.

"It seems to me that a major factor for original consideration of ADA deficiency for initial trial of gene therapy was that a distinct subgroup of patients . . . were ideal candidates from an ethical standpoint—ideal because there was no effective therapy for them. . . . This situation may no longer exist if ongoing, expanded trials with PEG-ADA continue to show it to be safe and effective in restoring immune function . . . ," Hershfield wrote in his review of Anderson's document. "Simply stated, my view is that *there must now be an increase in the stringency of the requirement for convincing preclinical evidence of efficacy and safety of gene replacement before undertaking experiments in humans* [his emphasis]."

Where it might have been ethically acceptable to try a totally experimental treatment on dying children because nothing else could save them, there was suddenly a plausible alternative that was proving effective and less risky. The development of PEG-ADA dramatically altered the ethical balance away from gene therapy, especially since Anderson couldn't prove in animals that it worked in animals.

The consensus among the reviewers was put most simply by Dusty

Miller: "In brief, the data here would not be sufficient to approve a human clinical trial."

For most people, such withering attacks would be unnerving, but they were just what Anderson wanted. Black had taken the pawn; Anderson moved to take the center of the board. Now that his severest critics had laid out their complaints, Anderson could attempt to answer them. His response came on November 20, 1987, in the form of a ten-page letter.

"Overall," Anderson began, "we agree with essentially all of the concerns raised by the reviewers." He then systematically refuted each concern.

On expression: "The low level of infection . . . is clearly a problem. . . . And there was no convincing evidence that the totipotent stem cells had been infected," Anderson admitted. "The very low infection rate would be therapeutic for a patient only if the corrected cells had a strong selective advantage in the patient's body. . . . There is, of course, a substantial amount of evidence that ADA+ T cells have a strong growth advantage in ADA-SCID patients." That evidence comes from matched bone marrow transplants for ADA patients in which all the blood cells in the patient's body remain his own, except for the T cells which come from the donor.

On PEG-ADA: Because only two children had received PEG-ADA for at least a year, and only one had developed objective signs that the immune system was partially restored, "no treatment has proven to be fully satisfactory as yet."

On safety: The NIH group tested the safety of its retrovirus vectors by intentionally injecting huge amounts of infectious helper virus into monkeys, up to 20 percent of the animal's blood volume, without getting an infection—even when the animal was immune-suppressed with drugs.

Anderson's conclusion: "In the specific clinical situation where all available therapies (including repeated bone marrow transplants and PEG-ADA) have failed for an ADA-SCID patient and there is nothing else to offer, would it be ethical to deny the potentially lifesaving gene therapy protocol in the face of certain death? We would certainly be far more comfortable if we understood the gene transfer procedure better, had helper-free virus supernatant of high titer, had 10 percent or more human ADA activity in several monkeys over a several-year period, etc. However, a clinical decision in this setting, as in other complex medical situations, should be based on an assessment of the risk/benefit ratio for the patient, in this case relative to the threat of impending death. Because we are aware of at least three infants who may soon fall into the above 'last hope' category, and because our safety studies thus far indicate that there should not be a risk to individuals who come into contact with the treated patient, we feel that a limited clinical protocol should be submitted in the near future to initiate a definitive review," Anderson wrote. "Our protocol now meets the minimum requirements for efficacy and safety for a 'last hope' procedure, i.e., for compassionate use only."

The cases were now made on both sides of the issue. Anderson had gotten his opponents to declare their objections, and he had answered them as best he could—though even he admitted he wished he had better answers. Still, he rationally maneuvered the debate to the point where he was able to justify that now was the time to start—at least, in a "last hope" patient. All that was required was convincing the Human Gene Therapy Subcommittee during its public meeting in December 1987 and getting them to agree that a formal proposal could be submitted. Depending on what the RAC members said, Anderson planned "to begin preparation of a limited 'last hope' clinical protocol after the December 7 meeting."

• • •

The Human Gene Therapy Subcommittee, as well as the Recombinant DNA Advisory Committee of the National Institutes of Health usually meet in Conference Room 9, on the sixth floor of Building 31, C wing. The room is dominated by an enormous, dark, rectangular table that holds the committee's piles of paperwork. The rest of the room is ringed by chairs, three and four rows deep; genetics aficionados, NIH staff, reporters, and the public, which seldom comes, can watch the debate take place in the open.

For its part, the RAC and its subcommittee has been a remarkable institution, conducting its public business with little day-to-day impact on public, or governmental, perceptions about genetics. Yet, in many ways, it is the most democratic of bodies, where one-person-one-vote rules, as seen by split votes over key issues, but where consensus is common. Scientists like Anderson, who want to push the envelope of acceptability and who often suffer rejection in the social side of science, get a fair hearing to state their case, and sway the committee to their viewpoints.

On December 7, 1987, Anderson would get to make his case, seeking permission to launch the field of gene therapy with a "last hope" protocol. His jury was the fifteen members of the gene therapy subcommittee, by design a mixed group including three basic scientists, three clinicians, three ethicists, three law and public policy experts. The subcommittee had asked Nobel laureate Howard Temin to be an expert consultant to help it evaluate Anderson's preclinical document and impending protocol, but Temin could not attend the December 7 meeting. So the subcommittee hired an "ad hoc consultant": Richard Mulligan. Anderson would get his wish about testing the protocol under the harshest conditions.

Anderson began by saying his group was not ready to submit a formal protocol but would discuss the issues. The debate initially went over the same ground of efficiency and safety. Charles Epstein, a thoughtful researcher from the University of California at San Francisco who led the review for the committee, concluded that "it would be preferable to avoid the use of helper viruses." In addition, he concluded that "the treatment method that employs polyethylene glycol–ADA (PEG-ADA) will be examined more closely before other treatments are attempted."

Even as the other scientists and physicians chimed in to dissect the technical details, James F. Childress, an ethicist from the University of Virginia's department of religious studies, raised the bombshell Anderson had tossed out at the end of his answer to the reviewers: Even if the safety and efficacy issues are not sufficiently resolved to justify a trial at this time, Anderson was talking about a "last hope" procedure rather than a standard clinical trial. Whereas it's "unclear what level of evidence was needed for a compassionate use protocol," Childress said, "a trial could be [ethically] justified in a last hope case."

Childress's position stirred the committee. "Is there a boomlet for last chance here?" asked Alexander Capron, an attorney from the University of Southern California and formerly the staff director of the presidential commission that wrote *Splicing Life*. Later in the day, Capron would argue that there was a "danger of exploitation association with 'last hope cases.' " First, premature submission could occur, that is, when the likelihood of benefit is very small, and, second, the potential for harm remains great, even if the benefits to others— but not the first patients—should appear.

The discussion stunned Mulligan. It started to look like Anderson would slip through on an ethical technicality with what Mulligan considered to be bad science. Besides, he thought all candidates for gene therapy were so desperately ill with a fatal disease that it was always "last hope."

"Excuse me," Mulligan said. "As I understand it, French never was going to do more than two patients, so what's the difference? I don't see how the issue has changed." Maurice J. Mahoney from Yale University agreed, but the debate about "last hope" continued to crop up throughout the day.

When Mulligan got his chance, he led Anderson through a review of the scientific issues, peppering him with questions for which Anderson didn't always have a good answer.

"I'm curious, do you have data that shows infection in stem cells?" Mulligan began.

That is a crucial point, Anderson said. "We have no evidence that we have entered totipotent cells." And the debate went on again about whether the gene-treated bone marrow cells would have a selective growth advantage.

But "0.5 percent expression is very low," Mulligan said. What level of efficiency would you need to get selective advantage?

"That's also a critical question," Anderson was reported as saying in the *Los Angeles Times* account. "There are data and studies, but they still don't answer the questions. We don't really know."

The debate went back and forth over the possible dangers from helper viruses, and whether a mouse model could be made for ADA deficiency to test gene therapy before going on to people.

Finally, Mulligan wheeled on Anderson and asked point-blank: "I am curious, French. Do you think it would work if you did it?"

"That's the question," Anderson admitted. "No one really knows. I don't think it is likely."

Then what would you learn? Mulligan wanted to know.

We'll learn whether a selective growth advantage exists for ADA corrected cells, Anderson replied.

And if nothing happens?

That's a good question.

By 3 P.M., the meeting wound down with both combatants feeling they could claim victory. The reality was closer to stalemate. Mulligan believed he had shown decisively that there was no evidence of sufficient efficiency or safety to support a human clinical trial. Anderson believed the debate was now clearly defined, and that everyone understood the questions that had to be resolved and ruled on.

"After listening to this today," Anderson announced at the end, "my group will get together and draft a limited last-chance protocol. Then the committee will have to make tough decisions."

Around the time of the subcommittee meeting, Anderson polled his staff during a regular gathering over pizza in the lab's conference room. He asked them by a show of hands how many would participate in the ADA gene therapy experiment if it was their own child. One third said they would, one third said they would not, and the remaining third was unsure. If even his own people weren't completely sold on the ADA protocol, Anderson concluded, then it probably should not go forward, no matter how badly he wanted to test gene therapy in patients. The NIH team had to find a way to improve the likelihood of success. A "last chance" protocol would never be submitted to the Recombinant DNA Advisory Committee.

. . .

In the summer of 1987, months before Anderson and Mulligan squared off before the RAC, Michael Blaese reached a state of frustration he had not felt before. He sat in his office with all the ADA monkey data spread across the constantly buried desk in his cramped, crowded office, and glared at it. For two years, since 1985, Anderson, along with O'Reilly and the rest, had been hitting their heads against a biological brick wall: the totipotent stem cell. All the labs had been running into the same problem. They could get genes into cells in the bone marrow, but it was either terribly inefficient or the genetically engineered cells died out after a few months, suggesting that they just were not getting into the stem cells—or both.

Yet, the laboratory experiments with his transformed T cell lines taunted him. In those cells, it was dead easy to stably insert just about any gene they tried. Why? Blaese wanted to know. One possible explanation had to do with the retrovirus itself: It can only integrate the genes it carries if the cell is dividing—a time at which it duplicates its DNA and then splits in half, with a full set of the duplicated DNA going into each of the daughter cells. Stem cells, apparently, just sit there in the bone marrow, seldom going into a reproductive cycle when its DNA would be susceptible to retroviruses. "So if we take lymphocytes in culture that are actively cycling, it is a piece of cake to put the gene in," Blaese said of his sudden epiphany. "But if the stem cell is just sitting there, it is not going to be susceptible."

Since Blaese knew that bone marrow transplants cured SCID kids by providing them with a normal source of T lymphocytes, then maybe all they had to do was genetically correct what few lymphocytes the children had in their circulation. Even though these peripheral T cells were thought to be short-lived, they might have enough of a growth advantage over the diseased T cells to do some good, he thought.

Blaese decided to do some studies to see if it was possible to put the ADA gene into normal T cells. The initial work on gene transfers Don Kohn had done used the immortalized T cell lines Blaese had made by infecting bone marrow from SCID kids with the HTLV-1 virus. These cells were great for doing basic lab studies, but they were abnormal. Before he could propose attempting to put genes into the mature T cells of patients, he had to show it was possible in the laboratory.

But the summer of 1987 was a time of transition for the Blaese lab. Don Kohn had completed his postdoctoral training and had taken a position at Children's Hospital in Los Angeles. His replacement, Dr. Kenneth Culver, would come in July. Initially, Ken Culver had never heard of Mike Blaese or gene therapy. Born, raised, and educated in Iowa, Culver had moved to the University of California at San Francisco to become a pediatric bone marrow transplant expert. Culver met Kohn on the West Coast during a scientific meeting, and heard about the cutting-edge work planned at NIH—and was intrigued. Kohn offered Culver a place to stay in Bethesda if he was ever passing through, and offered to introduce him to Blaese. Almost on a whim, Culver took up Kohn's offer, met Blaese, decided to apply for Kohn's job, got it, bought Kohn's house and his car. It was almost a complete takeover.

When Culver first arrived in the lab in July 1987, Blaese started him out drawing his own blood, and trying to produce lines of his own normal peripheral T cells, into which they would try putting the ADA gene. Culver, young and impetuous, was unimpressed. Since the lab already had transformed T cell lines from children with ADA deficiency, what was the point of trying to put the gene into normal T cells that already made human ADA protein? He chafed to get started on something important. Blaese put him on a project to see if he could remove T cells from mice, insert the human ADA gene, and then put the genetically engineered cells back into the mouse and detect human ADA production. This was going to require time, and Culver couldn't see where it was taking them.

All the focus of gene therapy, in all the research groups, had been bone marrow stem cells. All of the scientists were swinging for the bleachers, they wanted a home run with the first gene therapy experiment. Only by getting the gene into the stem cells would the patient be permanently helped, even cured. But Blaese was a physician who knew that most medical treatments don't cure the disease, only relieve symptoms, and often that's enough. He knew they could get ADA-deficient T cells out of the children, especially now that PEG-ADA was available to lower the toxins that killed off the T cells. If they could get the ADA into the defective T cells, and get the gene to function, then maybe they could provide a treatment that was better than PEG-ADA because the protein would be made within the cells where it was needed—and it could get the field of gene therapy off the ground.

There were, however, two problems: T cells did not live forever, though Blaese argued that some T cells, so-called memory cells, lived a long time, years to decades. For example, he often said, when a tetanus vaccine is given, some T cells are taught to recognize the tetanus antigen. The reason the vaccine works is that these "trained" cells persist in the blood for years. That's why, a decade later, giving a skin test against tetanus can still raise a nasty red bump on the skin. If they could get the ADA gene into those long-lived memory cells, then the treatment might actually do some long-term good.

The second problem was French Anderson. In the fall of 1987, as Culver started getting his good results, and Blaese was formulating his ideas about a new direction for gene therapy, Anderson was obsessed with ADA and bone marrow. The regulatory chess game was in full play, with the critical reviews of the preclinical document coming in and Anderson formulating his reply, all gearing up for the final showdown at the December 7 RAC subcommittee meeting. Blaese kept telling Anderson about Culver's data, and that maybe they should switch from bone marrow to peripheral T cells, since they seemed to be working much better. For three months, Anderson refused to listen. He wasn't interested in gene therapy for peripheral T cells, and told Blaese so repeatedly.

But after the stalemate at the subcommittee meeting in December, the ambivalent vote over pizza by his lab members, and growing good news about PEG-ADA which might make it more difficult to argue for a highly experimental treatment in children who no longer had a uniformly fatal disease, Anderson had become pessimistic about the possibility of getting approval to do ADA gene therapy in bone marrow. As a result, Anderson became more open to hearing about alternatives, and Blaese finally got through to him. Suddenly, it all made sense to Anderson, and he seized on it. Culver's mouse work did show that the scientists could take T cells, grow them in the laboratory, insert the ADA gene so that it made high levels of ADA protein, put the T cells back in mice where they would continue to make the ADA protein. They had discovered a reliable system that delivered the gene to a useful cell that persisted in the body for some time (though not as long as stem cells) and made the protein. And it was a logical step toward getting the gene into humans.

What's more, Blaese and Culver continued the mouse experiments for four months to make sure gene expression remained stable. It did. Next, with Anderson fully on board, they moved to monkeys: removing their T cells from the circulating blood, growing them in the lab, inserting the ADA gene, and putting the T cells back in the monkey's body. From single treatments, the NIH team eventually got continuous expression of the ADA gene in the monkeys for more than two years from a single treatment.

And finally, in the fall of 1987, Blaese called around and had the doctors with ADA patients—including Ashanthi DeSilva and Cynthia Cutshall, who would become the first children treated by the NIH team—send samples of peripheral blood from which Culver extracted defective T cells. These cells, which were not transformed with HTLV-1 virus, grew poorly in the laboratory, but with the increasing list of growth factors for human lymphocytes, Culver was able to keep them cooking along in the lab long enough to show that he could insert the ADA gene, and get it to produce protein for a reasonable amount of time. Everything that could be done scientifically to show this was a valid approach was now completed or underway.

There even was an effort that winter to deescalate the growing hostility between the NIH group and Mulligan's MIT team. In January 1988, one of the UCLA conferences, an annual series of meetings on different research topics, was held in Tiburon, California, on the gene transfer techniques. Anderson, Blaese, and company came to present their work, as did Mulligan. One evening, Mulligan invited the NIH team over to his lab's suite for beer and talk about genes. French sat on the floor; Mulligan swilled his beer. Everyone made their best, if awkward, effort to be friendly, and talk about the problems in the field.

Anderson thought the social effort went reasonably well, and often referred to it as an example of how he and Mulligan really were not personal enemies, that they just disagreed about the science. To Mulligan, the meeting convinced him that Anderson was a lightweight, but that some of the people on his team actually knew some science, even if they were not the best and the brightest. In the end, that effort in rapprochement and détente did little good.

But it didn't matter to the NIH team. They were about to form their own strategic alliance that would jump-start their faltering efforts to launch human gene therapy trials. Besides pushing Anderson toward T cells, Blaese believed they ought to involve Steven A. Rosenberg, the chief of surgery at the National Cancer Institute and the developer of a revolutionary new approach to fighting cancer.

In the 1970s, Dr. Robert Gallo, later of AIDS fame, discovered a lymphokine, or growth hormone for the white blood cells, called interleukin-2, or IL-2, that allowed him to grow normal white blood cells in the laboratory. Rosenberg took IL-2, and began using it to boost the immune system in cancer patients to fight their disease. Direct injection of IL-2 showed some signs of activity, but the doses were severely restricted because IL-2 is toxic to humans, causing fluids to accumulate dangerously in the lungs and doing other damage.

In the next round of experiments, Rosenberg removed white blood cells from the patient's circulation, and stimulated them to grow in the laboratory with IL-2. His idea was that the immune system was able to identify the cancer cells and kill them, but there were just not enough of the white blood cells, or lymphocytes, to destroy all the cancer. So he used IL-2 stimulation to produce billions of lymphocytes—he called them lymphokine activated killer cells, or LAK—and then returned them to the body of patients with melanoma, a deadly form of skin cancer. In a handful of cases, the results were dramatic: The patients were cured. But most of the time, they died either because the disease didn't respond or because the patient relapsed after a partial response.

To boost the effectiveness, Rosenberg began cutting out pieces of the melanoma tumor to isolate the white blood cells found just in the cancerous mass. These tumor-infiltrating lymphocytes, or TIL, were the ones most likely to be able to attack and kill the cancer. He stimulated the TIL with IL-2 and produced billions of TIL that could be given back to the patients. TIL was better than LAK: About 10 percent of the patients had a complete response; another 30 to 40 percent had a partial response, but the rest didn't respond at all.

Rosenberg had no idea why. Clearly, not all TIL were equal. He couldn't tell where they went in the body, or how long they persisted. His group labeled some TIL with a radioactive chemical so they could see where the cells went in the body with a gamma camera, and those studies showed that most of the cells concentrated in the tumors. But the radiation experiments could not tell how long the TIL persisted, or how active they might be, because the radiation damaged the TIL and could harm the patient, so only a short-lived isotope—indium[111] with a half life of 2.8 days—was used.

The inadequacy of the TIL treatments already had started Rosenberg himself thinking about gene therapy. Would it be possible, he wondered, to supercharge the TIL cells with a gene that made the growth hormone IL-2? Or with some other biologically active protein, like tumor necrosis factor, or TNF, that caused cancer cells to die in culture and in animal models? In 1986, Rosenberg had tried a brief collaboration with Werner Green, an IL-2 expert then at NIH, in which they tried to insert the IL-2 receptor gene into T cells by calcium phosphate precipitation. Receptors are like satellite dish antennas on the rooftops of the cells that allow them to receive chemical signals that tell them to do something, like grow or attack a cancer cell. Rosenberg hoped that this would allow the T cells to produce more IL-2 receptors than normal, making them more able to respond to IL-2 signals and fight cancer. The experiment never worked because of the inefficiency of this transfer method, which never got the gene into TIL. Then, Werner Green left NIH for Duke University.

Blaese's reason for wanting an alliance with Rosenberg was simple: Rosenberg knew more about growing large quantities of human T cells than anyone else in the world. His lab, three floors below Blaese's, was fully equipped and staffed for the work.

Once Anderson had signed onto the idea, Blaese walked down to Rosenberg's wing for a quick twenty-minute conversation in a hallway. He suggested that all the main players meet to formalize a collaboration and lay out what experiments were needed to get regulatory approval. Rosenberg, noncommittal at first, agreed to meet and hear Blaese's plan.

On March 17, 1988, Anderson, Blaese, and Culver tramped down to the surgery branch's crowded conference room to make their pitch to Rosenberg. Culver began by describing his results in mice, which showed they could put genes into T cells, and that they persisted and expressed for months. Blaese took over, explaining that, with one eye on the politics, they would start slow, initially only proposing to put the neomycin resistance gene in the TIL cells to "mark" the cells so Rosenberg could take biopsies from time to time and figure out where the TIL cells went in the body and how long they persisted. The *neo* marker gene would give Rosenberg the power to differentiate the TIL cells from all other white blood cells in the body. He finally might be able to figure out why TIL worked sometimes and failed at other times.

Blaese's proposal, however, was a dramatic departure from what everyone had expected to be the first human gene therapy. In fact, it wasn't even therapy. The *neo* marker gene would not boost the benefit of TIL. The experiment would do nothing for the patient, and that might present an ethical problem when it came to getting permission.

It also wasn't what Rosenberg wanted. As much as Anderson was driven to do the first gene therapy experiments, Rosenberg was obsessed with finding the cure for cancer, and he believed that his TIL approach, or some variant, was the key. If he was going to join up with the gene therapy boys from other parts of the clinical center, then he wanted to put in genes that would boost the benefits of TIL. Blaese thought that was possible. Once they had taken the first steps by putting *neo* in the TIL, and showing that gene therapy could work in humans, the team could take the next step, and insert genes for IL-2, interferon, and tumor necrosis factor in the hopes of making TIL a more effective cancer killer.

Rosenberg decided to sign on. Anderson walked him through their gene-transplanting technology and immediately started planning a series of experiments. "At that first meeting," Anderson recalled, "we laid out the entire protocol." And they launched an intense three months of work, where they'd meet every Monday afternoon at 3 P.M. to review the previous week's progress with mouse and monkey experiments, all aimed at proving the ideas were working.

• • •

By June 1988, the NIH team was ready to seek formal permission from the NIH Recombinant DNA Advisory Committee and the Food and Drug Administration to try gene transfer into human TIL cells that would be used to treat cancer patients. And winning approval was going to take political dexterity.

With Rosenberg in their camp, Anderson and Blaese had acquired a powerful ally. Intense in his quest and politically connected to the power structures within the cancer field and at the NIH, Rosenberg usually got what he wanted. What's more, the switch from proposing cutting-edge gene treatments for vulnerable children to dying cancer patients shifted the perceived ethical balance of the experiment.

Instead of worrying about genetic engineering in children who might have lived decades with the consequences of a failed experiment, the RAC would now have to make decisions about people with a fatal disease that had failed all previous treatments. Medicine, historically, has been more lenient toward doctors who wanted to try a desperate treatment for desperate patients. But this experiment would walk an ethically fine line because it wouldn't help the individual patient: "Although the procedure would be of no immediate benefit to the patient in which it is used, the information obtained should be of value in helping future patients and may, in fact, be beneficial in later therapy of the same patient," the NIH scientists wrote in the initial proposal.

What's more, the NIH team acknowledged, "This technology has never before been used in man. . . . [It] must undergo a series of extensive reviews. . . ." But what came next was not expected.

On July 29, 1988, the RAC's Human Gene Therapy Subcommittee met to consider, for the first time, an experiment in which genes would be put into people. Richard Mulligan turned up as one of the subcommittee's three ad hoc consultants, the argument being that he was best qualified to evaluate the vectors used to transfer the genes into the TIL cells.

No one had expected a TIL marking experiment in cancer patients to be the initial attempt at transferring genes into a patient, so the first subcommittee meeting began on an odd note. The first question was whether this was even gene therapy and whether it should be evaluated by the RAC subcommittee. After wrestling with the question, the subcommittee decided "the protocol was 'very similar to human gene therapy' and that the subcommittee should review it."

French Anderson took the podium and began to explain the rationale for using the neomycin resistance gene to mark TIL cells, and how it would be an ethically reasonable way to open the field of gene therapy. A Rosenberg colleague, Michael Lotze, explained the work in the NCI surgery branch to turn the patient's own immune system into a treatment for cancer. And finally, Michael Blaese described the procedure for marking the TIL cells, which were just specialized white blood cells, and how marking them with a gene would be superior to labeling them with some other technique, such as a radioactive compound.

The subcommittee listened politely until the team was finished, and then began its interrogation. The questions reflected the same issues Anderson faced the previous December when he made his "last chance" effort to treat ADA deficient children with gene therapy. They even echoed the challenges faced by Martin Cline nearly a decade earlier.

In what animals had this been tested? Monkeys? Mice? Since the team had organized only three months earlier, they were only able to show the gene transfer worked in human TIL cells in culture. For reasons no one ever figured out, they were never able to insert the *neo* gene into mouse TIL cells, but it clearly went into human cells.

Did *neo* make the neomycin protein in TIL cells, and how much did it manufacture? Yes, the gene did produce the neomycin protein effectively, but that didn't really matter because the gene did not have to produce protein to generate therapeutic effect; the gene only had to get into the cells to mark them because the gene could be detected with a relatively new laboratory test called polymerase chain reaction, or PCR.

And what about the safety issue that had truly plagued Anderson the year before, the production of helper viruses by the cell lines designed to produce the vectors? The NIH team already had switched to a new line of producer cells based on the work of Dusty Miller at the Hutchinson cancer center in Seattle. The cell line was believed to eliminate the possibility of helper virus production. But Richard Mulligan rose to challenge whether the safety tests used by Anderson and the others was actually sensitive enough to detect an occasional helper virus.

The argument went back and forth over safety, dwelling on minute details of the technology, until the emotion between Anderson and Mulligan reached a point where neither was hearing what the other was saying.

With no resolution in sight, the subcommittee finally voted to defer approving the study until the NIH team met several conditions: Animal testing was done in the mouse to show they could get the *neo* gene into mouse TIL cells, that they could detect the gene once it was inside the mouse TIL, and that there were no undesirable side effects, such as the production of infectious virus in the mouse or the formation of cancer. The committee also asked for additional information. They wanted to know whether the marked TIL cells were "representative of the cell populations," and they wanted a "dry run" of human TIL to show that there were no infectious viruses.

The vote was a bitter and emotional disappointment for Anderson and the others. Although Anderson knew that there was little chance that the first experiment would be approved very easily, he and the others were anxious to begin. The deferral by the RAC subcommittee in July could cause the experiment to be delayed for half a year because of the rules of government procedures under which the RAC operated. There would be no subcommittee meeting between July and October 1988, when the RAC itself would next meet. If the subcommittee didn't make a recommendation either way, the scientists worried that the full RAC would not even consider the experiment. The subcommittee would not next meet until December 1988, with the following RAC meeting in January 1989.

Anderson pressed for a compromise: If the NIH team got the subcommittee the information it requested, could it have a telephone conference call and make a recommendation before the full RAC met on October 3? The subcommittee agreed to do that if the NIH scientists provided the requested information. At least that offered some hope, but not much.

"There was an overwhelming feeling on the part of the molecular biologists [on the committee, mostly Mulligan] that it wasn't ready, and that it was too premature," Anderson said. "The motion was that the protocol be deferred until the following were done, but the obligations they set were impossible to meet, and are still impossible. We had a subcommittee that was not going to approve us for who knows how long," Anderson said, "but we had a firm belief that this protocol was ready."

The NIH team set out to get the information the subcommittee requested, though it would never be able to show they could put the *neo* gene into the mouse TIL cells, so part of the requested information could never be acquired. Still, they answered the other questions about the human TIL cells and the presence of infectious virus. Then Anderson and the others made a fateful decision: They chose not to send all of the information they had to the Human Gene Therapy Subcommittee.

"Our feeling was that it [the new information] would be set aside, and we would still be required to meet impossible standards," Anderson said. "We felt our only real hope was to go directly to the full RAC, the parent committee, even though the primary opponents on the Human Gene Therapy Subcommittee were on the RAC."

Anderson, Blaese, and Rosenberg did not flout the subcommittee; they actually sent some information for it to consider, but they did not attempt to answer all of the questions. Certainly they could not. The mouse TIL remained stubbornly resistant to gene transfer.

At 4 P.M., September 29, 1988, the Human Gene Therapy Subcommittee held a telephone conference to consider Anderson's answers. They found them wanting. According to the minutes of that meeting, "Dr. Mulligan stated that issues raised previously by the subcommittee concerning an appropriate animal model system had not been resolved. . . . Dr. Mulligan [wanted] to see the PCR data in order to assess their sensitivity. . . . Dr. Mulligan expressed some general reservations about data . . . showing that marked TIL cells do not differ from unmarked cell populations. . . . Finally, Dr. Mulligan questioned the sensitivity of the assay for helper virus as no data had been provided."

Several other scientists also had concerns that mostly reflected the issues raised by Mulligan. When LeRoy Walters, the subcommittee's chairman, called for a final sense of the committee an hour later, "there was . . . strong support for the view that this first clinical trial of gene transfer into humans must be conducted under the most stringent conditions. The first hurdle is the development of appropriate animal model systems. Most participants remained convinced that such efforts had not been exhausted. . . . In addition, there was consensus that many of the questions raised in July had not been resolved, such as the need for more quantitative data; more sensitive assays for viral replication and reconstruction, and tumorigenesis; and better assurance that marked TIL cells are representative of the relevant cell populations."

In the end, the subcommittee again voted to defer the experiment. Anderson was not surprised when he learned the results of the conference call. He was looking forward to October 3. Strategically, it was the next best chance to get approval to put genes into people and launch the field of gene therapy. It would be the showdown, with data, between Mulligan and Anderson in front of the assembled public, principally represented by the media. It would be a historic meeting, and the place was packed: Nearly a hundred observers crowded the ring of seats around the nearly thirty committee members and consultants at the gigantic central table.

Anderson began the team's presentation with an explanation of why they had not submitted the requested information to the subcommittee: The researchers, he said, had called the editors of the country's two most prestigious science journals, *Science* and *The New England Journal of Medicine,* to find out whether releasing their data at a public meeting like the RAC, or its subcommittee, would jeopardize their chances of publishing in these important publications. They feared the media would gain access to their unpublished research through the RAC, and release it. Both magazines tended not to publish any research that already was widely known, or published somewhere else, even newspapers.

"The data was not handed out because of the problems of jeopardizing publication," Anderson said. "The compromise is we will show you the data" in the form of slides, but not handouts. This is a problem that the NIH will have to resolve with the major journals for future reviews, he said.

This explanation caused a major furor, leading NIH director Dr. James Wyngaarden to call the journal editors, Daniel Koshland of *Science* and Arnold Relman of *The New England Journal of Medicine,* to complain. Both swore it was a misunderstanding and that releasing scientific data to a duly constituted government review committee as part of its oversight would never compromise a researcher's chances to publish in their journals.

Anderson ran through his arguments: These are consenting adults dying of advanced cancer, an incurable disease. They would be treated with a procedure that had minimal risks with enormous potential to gain information. There was a reasonable chance of success, based on animal data. At least, it wouldn't be dangerous. There would be little risk to the patient and no risk to the public or health care workers.

Anderson then summoned Rosenberg to make his case why this experiment would improve cancer treatment and urgently needed approval. Rosenberg rose to the front of the room, taking the podium and staring at the committee through round, metal-frame glasses.

"Four hundred and eighty-five thousand Americans died of cancer last year, and one out of every six Americans now alive will die of cancer if no new treatment modalities are developed," Rosenberg began. "That's more Americans dying [from cancer] than in World War II and Vietnam together. Obviously, we are in desperate need of finding new modalities to treat cancer." TIL, the organized attack of the immune system on the tumor, was that new modality. But it wasn't working as well as it could—two thirds of patients did not respond—

and they needed new ways to make the therapy more effective. "This is a treatment approach very much in its infancy," Rosenberg said. "One of the crucial pieces of information we need is, What is the long-term survival of these human TIL? The best way is to tag the cells with a marker gene."

But even in his presentation for the marking experiment, Rosenberg was looking down the road, giving the committee a glimpse of his plans. "These TIL represent packets of material that we can deliver to cancer within the patient. If we can add genes, we can add TNF or alpha interferon, we have the opportunity to dramatically improve these treatments. Very quickly, I hope, we will come back to you for approval to put these kinds of genes in cells."

Mike Blaese went next, describing the rationale for the basic gene tagging experiment: How long do TIL persist in the body, where do they go, and does longevity or location correlate to clinical effect? By labeling the TIL with a gene that can be tracked in the body, they should be able to answer these questions. In Blaese's mind, this was a terribly satisfying experiment, partly because it was his idea, partly because it showed the power of gene transfer technology, and partly because it was a surprise. Everyone had expected the first gene therapy to be some inherited illness, not cancer.

Then Blaese ran through some of the details of how the experiment would be done, and how they had put the *neo* gene into the TIL cells from fifteen patients to test their technology, and followed the protocol in six patients to complete a dry run of the experiment. "We have done everything except give it back to the patient," Blaese said. "We have answered all the questions."

And last, Anderson reviewed the newly acquired safety data from the new packaging cell lines.

Now it was the committee's turn to ask the questions again. Richard Mulligan, who by now was actually on the RAC, and no longer just acting as an ad hoc consultant, went back to the lack of animal data showing that the team could not get the gene into mouse TIL cells at any significant level. And then there was the question of safety. Mulligan and Anderson rehashed the sensitivity of the tests used to look for retrovirus vectors that might actually cause an infection.

In addition, Anderson went through the animal safety studies one more time, including a study where he intentionally infected five monkeys with infectious versions of the retrovirus vector. "We put enormous amounts of virus into three [of these five] monkeys, and nothing happened," Anderson said. Even after seventeen months, there were no illnesses in the monkeys, including one where nearly a quarter of its blood was replaced by infectious virus. That monkey got swollen lymph nodes, but suffered no long-term effect. "We can't get these animals infected," Anderson said. "This is a safe procedure."

Some RAC members were beginning to tire of the scientific minutiae. Harvard Medical School bacteriologist Bernard D. Davis, a respected physician and researcher, started to speak for the first time: "I am disturbed by the nit-picking [of the experiment and the NIH team]. That's what I see. It seems to me that the subcommittee is acting like a study section asking for more data. . . . The purpose for animal experiments is to get data we can't get from humans, or is too dangerous for humans. Animal experiments are in no sense a substitute."

Then Davis added the critical argument: "There has always been a principle in medicine that the sicker the patient, the higher the risk you are entitled to take in experimenting on the patient." What's more, he added, "we are not even close to any kind of risk that would justify the RAC holding this up."

During the coffee break, before the debate resumed, someone asked Anderson, "French, what's the rush?" Anderson responded to the RAC. "It would be useful for this committee to visit the oncology service and talk to patients who are dying. Ask the question of a patient whose life expectancy is two months: 'What's the rush?' "

"I am not comfortable with the emotional appeal that Dr. Anderson made," said Harvard's Davis, "but I feel we cannot disregard it. With patients at the end of their lives, we can move along. Future human gene therapy cases will be evaluated on their own merit."

Donald C. Carner, a lay member of RAC and president of Carner Inc. of Tiburon, California, made a motion that the proposed experiment be approved, but be limited to ten cancer patients whose life expectancy was ninety days or less, and who could understand they were taking risks with little chance of benefit. When the votes were counted, sixteen RAC members voted in favor, but five, mostly the molecular biologists, including Richard Mulligan, opposed it.

"I think it is a historic decision," said the new RAC chairman, Gerard J. McGarrity, president of the Coriell Institute for Medical Research in Camden, New Jersey. "It set a precedent."

A jubilant NIH team hung around the meeting room to talk to reporters and well wishers. "It was the only reasonable decision," Steve Rosenberg beamed. "I am hopeful that we will be doing this before the first of the year. Ten patients is not enough to answer the questions, but it is enough to get started. We are ready to do it today. I see no reason the approval can't be completed in the next month."

Rosenberg, however, was wrong. Although the RAC had recommended the project, its conclusions were only advisory to the NIH director, James B. Wyngaarden, who was out of the country when the decision was made. He was very unhappy with it when he returned. The way the review process had been conducted by his advisory committee was all wrong. First, the RAC had ignored the advice of its expert subcommittee specifically set up to review the first human gene therapy protocol. The subcommittee itself may have violated federal meeting rules by holding the telephone conference call. And then the final vote on the RAC was split: 16 to 5 with the real experts disagreeing with the decision. Wyngaarden overturned the RAC decision and told them to go back and do it again—the right way this time, with the subcommittee being satisfied enough to recommend approval, and then the RAC. In addition, the Food and Drug Administration also had to review the study and give its assent.

Anderson and the others scrambled for spin control. "He has not sent it back to the RAC; he would not do that," Anderson said of Wyngaarden with some disbelief in his voice. "He wants to review all the data himself, which is very appropriate. I think it is good that he takes the time and the interest to look at it [the data] himself."

Anderson too was dead wrong. Storm Whaley, the NIH's venerable public spokesman, confirmed that Wyngaarden had indeed sent the decision back to the RAC and its subcommittee. They had to do it all over again—and follow the rule book this time. It was the very worst that Anderson, Blaese, and Rosenberg could have feared. They would have to go through the entire review process again, and that would delay them for months, maybe a year. The next subcommittee meeting was in December 1988, their next chance to make their case. Then the RAC would meet in January 1989 to review the subcommittee decision. Wyngaarden not so much objected to the RAC's approval decision—he too thought it was reasonable to proceed—but he didn't want the process to be short-circuited, especially for the first human gene transfer experiment.

When the subcommittee reconvened in December, the momentum

was clearly toward approval. Anderson and the others worked throughout the fall to resolve as many of the outstanding questions as they could, though it never would be possible to put genes into mouse TIL cells. There were few technical issues left, though the subcommittee spent the day reviewing it all again.

"We apologize to the subcommittee for the rough water we have been through," Anderson began when he reexplained what they planned to do. The discussion went much more smoothly than past encounters; Richard Mulligan was not at the meeting. But Anderson admitted that the review process, no matter how rough-and-tumble, had an unexpected benefit: Where it was intended to protect patients and the public from unnecessary risks, it also improved the experiment itself. As the assembled experts on the RAC, including Mulligan, commented on the protocol, Anderson and the others used their ideas to modify and improve the experiment. In the end, a consensus was reached, and everyone was happy with the final experiment— even Mulligan. "Certainly, from a scientific standpoint, I think that we're very clear on the issues, and I think we're happy at this point in time," he said during the RAC meeting a month later.

At long last, in December 1988, the subcommittee voted 12–0 to approve the TIL marking experiment. Wyngaarden polled the RAC by mail to produce a document for the record, but didn't ask them to fully reconsider the experiment since they had already voted their approval in October 1988. In early January 1989, the RAC gave its final approval and so did Wyngaarden. The FDA would soon follow, but the scientists faced one more unexpected hurdle: Jeremy Rifkin, president of the Foundation on Economic Trends and an opponent of most types of genetic research, filed a suit in federal court accusing NIH of violating federal meeting rules for allowing the subcommittee to have an unannounced telephone conference that excluded the public from its deliberation on the TIL experiment. The suit was settled out of court, and the NIH team was finally ready to insert the first *neo* gene into a human patient.

Eager to begin, Anderson and the others expected to treat a patient immediately. Instead, they ran into myriad technical hurdles. Cultures of TIL cells became contaminated; candidate patients were too sick to undergo TIL treatment, or the researchers couldn't insert the *neo* gene in enough cells. Finally, in one patient, they seemed to have everything in place: The cells, which initially grew poorly, began reproducing rapidly. One third of them had been treated with the gene-carrying vectors, and seemed to be producing adequate amounts of *neo* protein. The patient, whose name was not released, came back to the NIH's hospital on Sunday, May 21, 1989, five months after the scientists received approval to proceed. A final battery of tests showed that everything was ready to go. At 10:47 A.M., Monday, May 22, 1989, the gene-carrying TIL cells started flowing into the body of a fifty-two-year-old truck driver from Indiana who had malignant melanoma, a particularly deadly form of skin cancer.

The first genetic treatment proceeded without a problem. Nine more patients would follow the truck driver over the next year or so, but the experiment was anticlimactic. With all the buildup of the first gene transfer into a person, virtual miracles were expected, but none were produced. A year later, on Easter Saturday 1990, the truck driver was dead, killed by the progression of his unchecked melanoma. Most of the others too died of their disease. Adding the *neo* gene did nothing to boost the effectiveness of TIL, and in the end, the historic experiment proved very little about why some patients respond better to TIL treatment than others. The underlying idea, that the *neo* gene could be used to mark a cell and demonstrate its continued presence in the body, proved correct. The TIL trials also proved that retroviral gene transfer into humans was completely safe. There was never a side effect in the first ten patients from the vectors or the transplanted gene.

As 1989 came to a close, French Anderson decided it was time to go back to his original plan of treating ADA deficiency with gene transfer. The experiments with TIL went well enough to convince Anderson and Blaese that they could insert the ADA gene into human lymphocytes in patients, and that it was safe. They decided to dive into the review process again to seek RAC and FDA permission to try putting the ADA gene into circulating white blood cells, but not into bone marrow or stem cells. This fundamental shift in cell targets meant that they were not trying to cure the disease so much as treat the children and restore their immune system, at least temporarily. As their ideas evolved, they envisioned treating the children several times a year with the ADA gene, putting it in fresh white blood cells every few months as the old ones died off.

Meanwhile, the collaboration between the team of Anderson and Blaese and Steve Rosenberg fell on hard times. Anderson struggled to maintain contact between the different teams of researchers, but in the end, relations cooled between Blaese and Rosenberg. What's more, Rosenberg had intended all along that the marking experiment would only be a first step toward supercharging TIL cells with some gene, like the one for tumor necrosis factor or interleukin-2, to make them a more effective cancer killer. Rosenberg expanded his collaborations with a biotech company called Cetus Inc., which attempted to build him a vector that made TNF and IL-2, and Rosenberg decided to enter his own gene therapy protocol.

The collaboration between the two teams became a competition, a race to see who would be first. Although the gene-marking experiment was the inaugural treatment of humans with genes, it really wasn't gene therapy, only a gene transfer experiment. The proposed treatments for ADA deficiency and the treatment of melanoma with TNF-enabled TIL would be real therapy. Both Anderson and Rosenberg wanted to be first. Both would have to run the regulatory gauntlet again. The first available subcommittee meeting was March 30, 1990.

Anderson and Blaese went first to present the ADA proposal, and ran into a fire storm of criticism. "Why do we need this treatment?" Richard Mulligan wanted to know. "And why is it different than PEG-ADA?" ADA deficient children were now receiving the new drug and appeared to be doing relatively well, even though they did not have completely normal immune systems. And then there were all the same questions about safety of the vector system used to insert the ADA gene, and a rehash of the worries about the accidental production of infectious viruses, especially since these were going into young, stable if not healthy children, who could be expected to live many years, not cancer patients with ninety-day longevity. And the discussion also turned to animal models of the disease, of which none existed for ADA deficiency. The subcommittee deferred the ADA experiment, and gave Anderson and Blaese a list of questions to answer about safety and efficacy.

Then came Rosenberg's turn. Where the TIL marking experiment and even the ADA gene therapy treatment used benign genes, his plans to put the TNF gene into TIL was likely to cause significant side effects because TNF is so toxic to humans. There were technical questions about the level of TNF expression, and what unwanted effects TNF might have in the body, especially if the engineered cells began to grow in the body and overproduce TNF. The subcommittee also deferred Rosenberg's experiment. Because of the toxicity questions, Rosenberg ultimately ran into more questions than Anderson and Blaese. In May, the NIH Institutional Biosafety Committee flatly rejected the TNF-TIL study.

Both teams of scientists struggled through the spring and early summer of 1990 to come up with whatever data they needed to satisfy their critics on the respective review boards. The RAC and its gene therapy subcommittee elected to hold a joint meeting on July 30, 1990, to listen once again to the reasons the investigators should be allowed to proceed. Anderson and Blaese went through their data again, but this time got help from an unexpected source: Italy. Dr. Claudio Bordignon, a physician-researcher who had been at Memorial Sloan-Kettering in the mid-1980s, and had helped with the monkey studies Anderson and Blaese had done with Richard O'Reilly, had moved back to Milan, to the Department Laboratory of Medicine at H. S. Raffaele, a hospital with research facilities. Bordignon too was interested in gene therapy for ADA, but was working on a model system that used so-called SCID mice that had no immune system of their own but could be manipulated to carry human immunity. Bordignon showed that it was possible to give SCID mice an ADA-deficient immune system and then correct it with gene transfer into the lymphocytes. He came back to the States just to give his testimony before the RAC; Bordignon's work was widely credited with turning the tide for Anderson because his animal studies showed there was a hope that correcting circulating white blood cells with the ADA gene might actually do some good.

The RAC was convinced and voted 12–1 to approve the ADA experiment. The one holdout was Richard Mulligan, who registered a protest vote because he believed "that whole scientific standards [for the experiment] were so low, so superficial, that it didn't meet a minimum standard that a graduate student should have. It would not have bothered me if French Anderson had gotten up and said, 'We have no data'—which he didn't—'We have no evidence to believe that this will work'—which he didn't—'but it sounds good. Let's just do it.' It wouldn't bother me in the same fashion."

The RAC also approved Rosenberg's TNF-TIL experiment at that same July meeting. Both teams crossed that finish line together, and both sought FDA clearance simultaneously, but Rosenberg hit a roadblock at the FDA. FDA staffers had many technical questions, and wanted the protocol significantly changed in ways Rosenberg resisted. Where Anderson and Blaese were able to proceed with their first treatment of Ashanthi DeSilva in early September 1990, Rosenberg was delayed nearly six months from RAC approval, until January 1991, before he could test the first TNF-TIL treatment in a patient. By then, Anderson and Blaese were ready to treat their second patient, Cynthia Cutshall.

Anderson and Blaese had won the race. They were the first to successfully perform human gene therapy. True, Martin Cline had attempted it earlier, but his was neither sanctioned nor successful. The success of the Anderson-Blaese experiment would not be known for some time, but the experiment proved symbolic just the same.

It opened up the field. Anderson repeatedly said that the first treatment was more of a cultural breakthrough than a scientific one. The techniques used in the trial had existed since 1984, and in many ways, the experiment could have been done much sooner. And when challenged on the worth of bothering with such a high-tech and expensive treatment for an extremely rare disease, Michael Blaese argues that rare diseases have often pointed the way to the development of treatments that become widely useful. Indeed, ADA deficiency was the first for which a bone marrow transplant was performed, proving the power of that technique. It was the first for which an enzyme replacement drug therapy was developed, and the first to be treated with a genetic transplant.

With the power of gene transfer demonstrated in children, lionized in headlines, and the imagination stimulated, dozens of scientific labs rushed into the field. Within the next three years, more than 100 patients worldwide had genes put into their bodies as part of more than fifty different gene therapy studies across the planet. Cancer remained the leading, and most unexpected, category of treated diseases. Inherited illnesses also were attacked genetically, including cystic fibrosis, hemophilia, and an inherited form of extremely high cholesterol, called familial hypercholesterolemia.

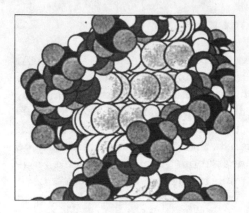

Epilogue

In May 1993, Cynthia Cutshall came to the National Institutes of Health, as she had so many times before, for yet another genetic treatment. This one, however, would be different. French Anderson and Michael Blaese had teamed up yet again with another NIH team, this one led by Arthur Nienhuis, a longtime Anderson colleague. Nienhuis's team had been searching for ways to identify the bone marrow's stem cells, and believed they now had a technique—based on the discoveries of other labs and a small biotech company in the Pacific Northwest—by which they could isolate stem cells out of the bloodstream.

In the two years since she first received new genes into her white blood cells in January 1991, Cynthia Cutshall had steadily improved clinically, though not as much as Ashanthi DeSilva, according to the laboratory studies. Both children started making antibodies to various skin tests and had grown tonsils. The girls had remained healthy despite an intentional, nearly year-long break in their genetic treatment. This apparently confirmed Blaese's hypothesis that there were long-lived white blood cells that could persist for years: They had gotten the normal ADA gene into some of them. Ashi DeSilva, for example, made the ADA enzyme in about 25 percent of her white blood cells, and that seemed to be holding steady.

There were, however, problems with the experimental treatment.

The children continued to receive injections of the drug PEG-ADA. Blaese was still not comfortable with the idea of discontinuing the drug, and that complicated the interpretation of the experimental results. What's more, the treatment of the peripheral white blood cells needed to be repeated from time to time. The children were not cured.

Now, however, the NIH scientists thought they might actually be able to cure her. The earliest ideas about genetic treatments envisioned putting the gene in the immortal, self-renewing stem cells of the bone marrow. If these cells could be healed, then all of the blood cells that they give rise to would also be normal—and for the rest of the patient's life. The individual would truly be cured.

The treatment, developed by Nienhuis and Cynthia Dunbar, a research physician, first stimulated Cutshall's bone marrow with a drug that mobilized stem cells in her bone marrow and moved them into the bloodstream. After several days of drug treatment, her white blood cell counts went up dramatically. The physicians then used a technique called plasmaphoresis to remove just white blood cells from her circulation. Then they ran the white blood cells through a glass column packed with white material, a kind of biological filter, that was able to capture just the cells that carried a unique marker on their surface called a CD34 receptor. Scientists at Johns Hopkins University and elsewhere had shown that the population of cells carrying the CD34 marker included a large percentage of stem cells. By isolating CD34 cells, the NIH scientists also collected a large number of stem cells.

With this enriched fraction of Cutshall's stem cells in hand, the NIH scientists treated them with the same ADA gene-carrying vector they used to insert ADA into her mature white blood cells. If the treatment works as expected, the gene-corrected stem cells that were injected into her body will settle in her bone marrow, and produce normal white blood cells for the rest of her life, eliminating the need for both the weekly injections of PEG-ADA and the repeated gene treatments of her mature white blood cells. Cynthia Cutshall will be cured. By the summer of 1993, the NIH team repeated the stem cell treatment in Ashanthi DeSilva.

What's more, two teams of California researchers, one at the University of California at San Francisco and one at Children's Hospital in Los Angeles, treated newborns who inherited ADA deficiency with a variant of stem cells gene therapy. Dr. Donald Kohn, the physician who once worked in Mike Blaese's lab on the ADA gene, led the Children's team.

. . .

As for the NIH scientists, all collaborations break up eventually. Steven Rosenberg went his own way, focusing exclusively on using gene transfer techniques to boost his immunologic attack on cancer. In the summer of 1992, after more than a quarter century, French Anderson left the National Institutes of Health for the sunshine of Southern California. He left because his wife, surgeon Kathryn Anderson, became chief of surgery at Children's Hospital in Los Angeles. Anderson joined the University of Southern California, and set up a gene therapy institute in the university's cancer center.

Kenneth Culver left NIH in that same summer to join Genetic Therapy, Inc., a small biotech company Anderson had helped start through a cooperative research and development agreement between the NIH and the company. In June 1993, Culver returned to his native Iowa to set up his own gene therapy institute and continue the development of a novel and promising genetic treatment for uniformly lethal brain tumors. Culver discovered that it was possible to inject the mouse cells that made a retrovirus carrying the herpes thymidine kinase (TK) gene directly into a patient's brain. The retrovirus vectors then inserted the TK gene into the tumor cells, making the cancer susceptible to the anti-herpes drug gancyclovir. Animal studies showed convincingly that it was possible to kill brain tumors with this treatment. Human tests of this method began in December 1992 as a collaboration with NIH neuroscientists, and by mid-1993, studies showed five of the first eight patients were helped by the treatment. While their tumors clearly shrank, the researchers would not know for some time if the patients were cured.

Only Michael Blaese remained at NIH, spending much of his time traveling the world to describe the progress of gene therapy in the United States.

Notes

Chapter 1

16 Michael S. Hershfield, et al., "Treatment of Adenosine Deaminase Deficiency with Polyethylene Glycol–modified Adenosine Deaminase," *The New England Journal of Medicine* 316 (1987): 589–96.

17 Interview with Paul Van Nevel, National Cancer Institute.

18 Interview with W. French Anderson, National Heart, Lung and Blood Institute.

18 Conversations between officials at the Food and Drug Administration and the National Institutes of Health were reconstructed from phone logs.

19 Interview with Kenneth Culver, National Cancer Institute.

19 Interview with R. Michael Blaese, National Cancer Institute.

22 Interview with Van DeSilva of North Olmstead, Ohio.

25 Interview with Raj DeSilva of North Olmstead, Ohio.

29 Interview with Susan Cutshall of Canton, Ohio.

29 Interview with Ricardo U. Sorensen, Louisiana State University School of Medicine, New Orleans, La.

32 Interview with Raj DeSilva.

35 Interview with R. Michael Blaese.

42 Interview with William T. Shearer of Baylor College of Medicine and Texas Children's Hospital, Houston, Tex.

43 Interview with Jay J. Greenblatt, National Cancer Institute.

43–48 Description of the first treatment based on videotape made by National Institutes of Health's audiovisual department.

Chapter 2

53 *The New York Times* (November 23, 1969). Front-page article describes the isolation of the first gene.

54 Jonathan R. Beckwith, "Gene Expression in Bacteria and Some Concerns About the Misuse of Science," *Bacteriological Reviews* 34 (1970): 222–27.

55 *The New York Times* (November 30, 1969). Follow-up story to the isolation of the first human gene in the Sunday review section.

55 Marshall W. Nirenberg, "Will Society Be Ready?" *Science* (August 1967).

56 Louis Harris and Associates, March of Dimes survey of attitudes of American adults toward using gene therapy. Poll conducted in the spring of 1992; published in September 1992.

57 Charles Darwin, *On the Origin of Species by Means of Natural Selection, or the Preservation of Favoured Races in the Struggle for Life* (New York: Macmillan Publishing Co., 1962).

58–62 Daniel J. Kevles, *In the Name of Eugenics: Genetics and the Uses of Human Heredity* (Berkeley: University of California Press, 1985).

62 L. C. Dunn, *A Short History of Genetics: The Development of Some of the Main Lines of Thought: 1864–1939* (Ames, Iowa: Iowa State University Press, 1991). Originally published by McGraw-Hill, Inc., 1965.

63 Franklin H. Portugal and Jack S. Cohen, *A Century of DNA: A History of the Discovery of the Structure and Function of the Genetic Substance* (Cambridge: MIT Press, 1977).

63 E. B. Wilson, *The Cell in Development and Inheritance* (New York: Macmillan, 1900).

63 F. Griffith, *Journal of Hygiene* 27 (1928): 113.

63 Oswald T. Avery, Colin MacLeod, and Maclyn McCarty, "Studies on the Chemical Nature of the Substance Inducing Transformation of Pneumococcal Types. I. Induction of Transformation by a Deoxyribonucleic Acid Fraction Isolated from Pneumococcus Type III," *Journal of Experimental Medicine* 79 (1944): 137–58.

64 A. D. Hershey and M. Chase, "Independent Function of Viral Protein and Nucleic Acid in Growth of Bacteriophage," *Journal of General Physiology* 36 (1952): 39–56.

64 J. D. Watson and F. H. C. Crick, "Molecular Structure of Nucleic Acids: A Structure for Deoxyribonucleic Acid," *Nature* 171 (1953): 737–38.

64 F. H. C. Crick, "Central Dogma of Molecular Biology," *Nature* 227 (1970): 1209–11.

64 M. Meselson and F. W. Stahl, "The Replication of DNA in *Escherichia coli*," *Proceedings of the National Academy of Sciences* 44 (1973): 671–82.

64 M. W. Nirenberg and H. J. Matthaei, "The Dependence of Cell-free Protein Synthesis in *E. coli* on Naturally Occurring or Synthetic Polyribonucleotides," *Proceedings of the National Academy of Sciences* 47 (1961): 1589.

67 Rollin D. Hotchkiss, "Portents for a Genetic Engineering," *Journal of Heredity* 56 (1965): 5. Paper was originally given as a talk on August 17, 1965, at the annual meeting of the American Institute of Biological Sciences, University of Illinois, Urbana.

68 Elizabeth Hunter Szybalski and Waclaw Szybalski, "Genetics of Human Cell Lines, IV. DNA-Mediated Heritable Transformation of a Biochemical Trait," *Proceedings of the National Academy of Sciences* 48 (1962): 2026.

68 Interview with Theodore Friedmann, University of California at San Diego School of Medicine.

68 Interview with J. Edwin Seegmiller, University of California at San Diego School of Medicine.

69 M. L. Morse, E. M. Lederberg, and J. Lederberg, "Transductional Heterogenotes in *Escherichia coli*," *Genetics* 41 (1956): 758

69 T. Friedmann, J. H. Subak-Sharpe, W. Fujimoto, et al., paper presented at Society of Human Genetics meeting, San Francisco (October 1969).

70 C. R. Merril, M. Geier, and J. Petricciani, "Induction of Galactose Operon in Human Cells Using Lambda Phage," *Nature* 233 (1971): 398.

71 Stanfield Rogers, "Induction of Arginase in Rabbit Epithelium by the Shope Papilloma Virus," *Nature* 183 (1959): 1815.

71 Stanfield Rogers, "Shope Papilloma Virus: A Passenger in Man and Its Significance to the Potential Control of the Host Genome," *Nature* 212 (1966): 1220

72 Interview with Joshua Lederberg.

72 H. G. Terheggen, A. Schwenk, M. Van Sande, et al., "Argininemia with Arginase Deficiency," *Lancet* 2 (1969): 748.

72 Stanfield Rogers, "Reflections on Issues Posed by Recombinant DNA Molecule Technology," *Annals of the New York Academy of Sciences* 265 (1976).

73 Yvonne Baskin, *The Gene Doctors: Medical Genetics at the Frontier* (William Morrow and Co., 1984).

74 Theodore Friedmann and Richard Roblin, letters to the editor, *Science* (1972).

74 Theodore Friedmann and Richard Roblin, "Gene Therapy for Human Genetic Disease?" *Science* 175 (1972): 949–55.

75 Ernst Freese, report from the conference on gene therapy, published by the Fogarty International Center, NIH, Bethesda, Md. (1972).

Chapter 3

76–80 Interviews with W. French Anderson.

80 W. French Anderson, "Arithmetical Computations in Roman Numerals," *Classical Philology* 51 (1956): 145–50. Paper describes how to perform mathematical calculations with Roman numerals.

80 *Time* (March 19, 1956). Describes French Anderson as a prodigy.

80 *Harvard Alumni Bulletin* (March 3, 1956). The description of French Anderson's first year at Harvard.

81 Ernie Roberts, "Dawn-to-Dark Runner, Mathematical Miler Pacing Harvard," *Harvard Crimson* (March 22, 1956).

81 *Boston Traveler* (June 11, 1958), covered Harvard track meets, including references to French Anderson.

82 Robert Kanigel, *Apprentice to Genius: The Making of a Scientific Dynasty* (New York: Macmillan, 1986).

82 Harriet Zuckerman, *Scientific Elite: Nobel Laureates in the United States* (New York: The Free Press, 1977).

83 Kistiakowsky story from Anderson interview.

83 Richard Rhodes, *The Making of the Atomic Bomb* (New York: Simon & Schuster, 1986).

85 W. F. Anderson, J. A. Bell, J. M. Diamond, et al., "Rate of Thermal Isomerization of cis-butene-2," *Journal of the American Chemical Society* 80 (1958): 2384–86.

85 Interview with Paul Doty, Harvard University.

90 Interview with Lauren Chang.

90 Larry Thompson, "NIH at 100 Where Big Government Meets Big Science," *The Washington Post* (January 13, 1987), Health section.

91 NIH Almanac 1986, U.S. Department of Health and Human Services, NIH Publication No. 86–5 (September 1986).

91 Interview with Marshall Nirenberg, National Heart, Lung and Blood Institute.

92–93 Horace Freeland Judson, *The Eighth Day of Creation: The Makers of the Revolution in Biology* (New York: Simon and Schuster, 1979).

93 Robert G. Martin, "A Revisionist's View of the Breaking of the Genetic Code," *NIH: An Account of Research in Its Laboratories and Clinics*, DeWitt Stetten, Jr., ed., and W. T. Carrigan, ass. ed. (San Diego: Academic Press, 1984).

94–97 W. French Anderson, "From the Genetic Code to Beta Thalassemia," *NIH: An Account of Research in Its Laboratories and Clinics*, DeWitt Stetten, Jr., ed., and W. T. Carrigan, ass. ed. (San Diego: Academic Press, 1984).

98 *The Eighth Day of Creation*.

98 Interview with W. French Anderson.

99 Interview with Donald S. Fredrickson, former director of the National Institutes of Health.

99 Interview with Marshall Nirenberg, National Heart, Lung and Blood Institute.

100 *Pediatric News* 2:6 (June 1968): 1. News story quotes a French Anderson talk in Chicago describing his vision of gene therapy.

100 Letter from Franz J. Ingelfinger, M.D., ed., *The New England Journal of Medicine,* to W. French Anderson (August 19, 1968). It rejected a review article Anderson offered on the possibility of gene therapy.

104 P. M. Prichard, J. M. Gilberg, D.A. Shafritz, et al., "Factors for the Initiation of Hemoglobin Synthesis by Rabbit Reticulocyte Ribosomes," *Nature* 266 (1970): 511–14.

104 J. M. Gilbert and W. F. Anderson, "Cell-free Hemoglobin Synthesis. II. Characteristics of the Transfer Ribonucleic Acid–dependent Assay System," *Journal of Biological Chemistry* 245 (1970): 2342–49.

108 Arthur W. Nienhuis, clinical record, narrative summary, on Nicholas Lambis (January 9, 1971).

109 Joseph L. Goldstein, M.D., clinical record, narrative summary, on Julia Lambis (February 10, 1969).

110 Barbara J. Culliton, "Cooley's Anemia: Special Treatment for Another Ethnic Disease," *Science* 178 (1972): 590–93.

110 Natalie Davis Springarn, *Heartbeat: The Politics of Health Research* (Manchester, N.H.: Robert B. Luce, Inc., 1976).

Chapter 4

116 E. P. Fischer and C. Lipson, *Thinking About Science: Max Delbruck and the Origins of Molecular Biology* (New York: W. W. Norton & Co., 1988).

116 Max Delbruck, "Experiments in Bacterial Viruses (Bacteriophages)," *Harvey Lectures* 41 (1946): 161.

117 Renato Dulbecco, "Reactivation of Ultraviolet-inactivated Bacteriophage by Visible Light," *Nature* 163 (1949): 949–50.

119 Interview with Renato Dulbecco.

119 Paul Berg, *Le Prix Nobel* (1980). Berg's Nobel lecture and biography.

120 *The Eighth Day of Creation.*

121 Jane S. Smith, *Patenting the Sun: Polio and the Salk Vaccine, the Dramatic Story Behind One of the Greatest Achievements of Modern Science* (New York: William Morrow and Co., 1990).

122 David Baltimore, *Le Prix Nobel,* reprint of his Nobel Prize address and autobiography, 1975.

122 Tim Beardsley, *Scientific American* 266 (1992): 35–36. Biographic notes on David Baltimore.

124 *Nature* 228 (1970): 609.

125 Interview with Inder Verma, Salk Institute, La Jolla, Calif.

126–54 Michael Rogers, *Biohazard* (New York: Alfred A. Knopf, 1977).

126–54 Sheldon Krimsky, *Genetic Alchemy: The Social History of the Recombinant DNA Controversy* (Cambridge: The MIT Press, 1982).

126–54 Michael Rogers, "The Pandora's Box Congress," *Rolling Stone* (June 19, 1975).

126–54 Nicholas Wade, *The Ultimate Experiment: Man-Made Evolution* (New York: Walker and Co., 1977).

126–54 Nicholas Wade, "Microbiology: Hazardous Profession Faces New Uncertainties," *Science* 182 (1973): 566.

126–54 Donald Fredrickson, "Asilomar and Recombinant DNA," *Bio-Medical Politics*, The Institute of Medicine, Committee to Study Decision Making Staff, Kathy E. Hanna, ed. (Washington, D.C.: National Academy Press, 1991).

131 J. Shapiro, L. Machattia, L. Eron, et al. (including Jon Beckwith), "Isolation of Pure Lac Operon DNA," *Nature* 224 (1969): 768–74.

131 *Genetic Alchemy*.

132 Daniel Nathans, *Le Prix Nobel*, 1978.

133 Judith P. Swazey, James R. Sorenson, and Cynthia B. Wong, "Risks and Benefits, Rights and Responsibilities: A History of the Recombinant DNA Research Controversy," 51 *Southern California Law Review* 1019 (1978): 1021–29; reprinted in *Law, Science and Medicine* (1984).

133 Nicholas Wade in "Microbiology: Hazardous Profession Faces New Uncertainties," *Science* 182 (1973): 566.

135 Quote is from Andrew Lewis, transcript of an interview by Rae Goodell (July 30, 1975), "Recombinant DNA Controversy," Oral History Collection, IASC, MIT, 7.

135 A. M. Lewis, Jr., M. H. Levin, W. H. Wiese, et al., "A Non-defective (Competent) Adenovirus-SV40 Hybrid Isolated from the AD2-SV40 Hybrid Population," *Proceedings of the National Academy of Sciences* 63 (1969): 1128.

136 *Rolling Stone*, ibid.

138 S. E. Luria and S. L. Human, *Journal of Bacteriology* 64 (1952): 557–69.

138 W. Arber and D. Dussiox, *Annual Review of Microbiology* 19 (1962): 365–78.

139 Janet E. Mertz and Ronald W. Davis, "Cleavage of DNA by RI Restriction Endonuclease Generates Cohesive Ends," *Proceedings of the National Academy of Sciences* 69 (1972): 3370–74.

140 John Lear, *Recombinant DNA* (New York: Crown Publishing Group, 1978).

142 J. P. Swazey, J. R. Sorenson, and C. B. Wong, "Risks and Benefits, Rights and Responsibilities: A History of the Recombinant DNA Research Controversy," *Southern California Law Review* 1019 (1978): 1021–29.

143 Michael Crichton, *The Andromeda Strain* (New York: Alfred A. Knopf, Inc., 1969).

147 Nicholas Wade, "Recombinant DNA: NIH Group Stirs Storm by Drafting Laxer Rules," *Science* 190 (1975): 767.

147 Victor McElheny, *The New York Times* (December 9, 1975).

148 Erwin Chargaff, "On the Dangers of Genetic Meddling," *Science* 192 (1976): 938.

148 Robert Sinsheimer, "An Evolutionary Perspective for Genetic Engineering," *New Scientist* (January 20, 1977).

150 Stephen S. Hall, *Invisible Frontiers: The Race to Synthesize a Human Gene* (New York: The Atlantic Monthly Press, 1987).

151 Charles Gottlieb and Ross Jerome, "Biohazards at Harvard," *The Boston Phoenix* (June 8, 1976).

154 Burke K. Zimmerman, "The Gene-Splicing Wars," *Science and Politics: DNA Comes to Washington,* eds. Raymond A. Zilinskas and Burke K. Zimmerman (Washington, D.C.: American Association for the Advancement of Science, 1986).

Chapter 5

160 Interview with W. French Anderson, National Heart, Lung and Blood Institute.

162 Interview with Joshua Lederberg, Rockefeller University.

163 J. B. Gurdon, "Adult Frogs Derived from the Nuclei of Single Somatic Cells," *Developmental Biology* 4 (1962): 256–73.

163 J. W. Gordon, G. A. Scangos, D. J. Plotkin, et al., "Genetic Transformation of Mouse Embryos by Microinjection of Purified DNA," *Proceedings of the National Academy of Sciences* 77 (1980): 7380–84.

163 Yvonne Baskin, *The Gene Doctors: Medical Genetics at the Frontier. How the Breakthroughs in Gene Therapy Will Change Your Future* (New York: William Morrow and Co., 1984).

165 Interview with W. French Anderson, National Heart, Lung and Blood Institute.

167 Interview with Thomas P. Maniatis, Harvard University.

169 S. Kit, D. Dubbs, L. Piekarski, et al., *Experimental Cell Research* 21 (1963): 297–312.

169 W. French Anderson and Elaine G. Diacumakos, "Genetic Engineering in Mammalian Cells," *Scientific American* 245 (1981): 106–21.

171 Interview with Thomas P. Maniatis, Harvard University.

171 Richard Pearson, Obit: "Nicholas Lambis, Patient in NIH Anemia Research," *The Washington Post* (November 1979).

172 Harold M. Schmeck, Jr., "Injection of a Gene Cures Flaw in Cell," *The*

New York Times (Oct. 10, 1979): A14.

173 Interview with Mario Capecchi, University of Utah, Salt Lake City, Utah, and the Howard Hughes Medical Institute.

173 James D. Watson, Jan Witkowski, Michael Gilman, et al., *Recombinant DNA,* 2d ed. (New York: W. H. Freeman and Co., 1992), 256.

174 J. W. Gordon, G. A. Scangos, D. J. Plotkin, et al., "Genetic Transformation of Mouse Embryos by Microinjection of Purified DNA." *Proceedings of the National Academy of Sciences* 77 (1980): 7380–84.

174 Elizabeth Antebi and David Fishlock, *Biotechnology: Strategies for Life* (Cambridge: MIT Press, 1986).

175 F. L. Graham and A. J. van der Eb, *Virology* 52 (1973): 456–67.

176 Interview with Angel Pellicer, New York University, New York, N.Y.

176 Michael Wigler, Saul Silverstein, Lih-Syng Lee, et al., "Transfer of Purified Herpes Virus Thymidine Kinase Gene to Cultured Mouse Cells," *Cell* 11 (May 1977): 223–32.

178 Richard Pearson, Obit: "Nicholas Lambis, Patient in NIH Anemia Research," *The Washington Post* (November 1979).

180 Interview with Richard Mulligan, Massachusetts Institute of Technology, Cambridge, Mass.

180 Interview with Paul Berg, Stanford University, Palo Alto, Calif.

182 *The Gene Doctors.*

183 Richard C. Mulligan, Bruce H. Howard, and Paul Berg, "Synthesis of Rabbit B-globin in Cultured Monkey Kidney Cells Following Infection with a SV40 B-globin Recombinant Genome," *Nature* 277 (1979) 108–14.

187 Interview with W. French Anderson, National Heart, Lung and Blood Institute.

Chapter 6

190 Interview with Charles Haskell, Veterans Administration Hospital, Los Angeles, Calif.

191 "Protecting Human Subjects: The Adequacy and Uniformity of Federal Rules and their Implementation." A report by the President's Commission for the Study of Ethical Problems in Medicine and Biomedical and Behavioral Research (December 1981). GPO Stock Number 81-6001-90. The report does not name Gale directly, but, in an interview, Atty. Barbara Mishkin, then deputy director of the commission and the person responsible for the report, identified Gale as the subject of the investigation.

194 Interview with Cesare Peschle, Istituto Superiore di Sanità, Rome, Italy.

196 Interview with Winston Salser, University of California at Los Angeles, Los Angeles, Calif.

199 Martin J. Cline et al., "Insertion of a Foreign Gene into Hematopoietic Cells of Two Patients with B-Thalassemia," unpublished.

200 Martin J. Cline et al., "Gene transfer in Intact Animals," *Nature* 284 (1980): 422–25.

200 Karen E. Mercola, Howard Stang, Jeffrey Browne, et al., "Insertion of a New Gene of Viral Origin into Bone Marrow Cells of Mice," *Science* 208 (1980): 1033–35.

202 Interview with Martin C. Cline, University of California at Los Angeles, Los Angeles, Calif.

202–5 Transcript from the UCLA Human Subjects Protection Committee meeting (June 15, 1979).

205 Transcript from the UCLA Recombinant DNA Committee meeting (July 27, 1979).

206 Letter from Martin C. Cline and Winston Salser to the UCLA Human Subjects Protection Committee and the UCLA Institutional Biosafety Committee (September 18, 1979).

207 Interview with Jeremy H. Thompson, former chairman of the UCLA Human Subjects Protection Committee, Los Angeles, Calif.

208 Letter from Martin C. Cline to Jeremy Thompson (February 29, 1980).

208 Letter from Jeremy Thompson to Martin C. Cline (April 3, 1980).

210–13 Martin J. Cline letters to Eliezer A. Rachmilewitz (March 5, 1980, and April 14, 1980) to set up collaboration.

210–13 Eliezer A. Rachmilewitz letters to Martin J. Cline about setting up the collaboration (February 27, 1980; May 20, 1980; June 2, 1980).

212 Martin J. Cline letter to Cesare Peschle (April 30, 1980), to set up collaboration.

213 Interview with Dr. Velma Gabutti.

216 National Institutes of Health Ad Hoc Committee on the UCLA Report Concerning Certain Research Activities of Dr. Martin J. Cline (May 21, 1981).

216 Interview with Leo Sacks, Weizmann Institute of Science, Rehovot, Israel.

216 Paul Jacobs, "Treatment Delayed in U.S.: Italy, Israel Quickly OKd Gene Testing on Humans," *Los Angeles Times* (Oct. 25, 1980).

217 *Los Angeles Times*, ibid.

218 Eliezer A. Rachmilewitz letter to Martin J. Cline about patient's progress after first gene treatment (July 28, 1980).

218 NIH Ad Hoc committee report, ibid.

220–23 Description of the treatment based on the NIH report, a written description by Martin J. Cline, and interviews with Martin Cline and Eliezer Rachmilewitz.

227 "Human Genetic Engineering," hearing before the U.S. House of Representatives Committee on Science and Technology, Subcommittee on Investigations and Oversight (November 16, 1982).

228 Jeremy H. Thompson, chairman of the UCLA Human Subject Protection Committee, letter to Martin Cline rejecting his application to perform gene therapy in Los Angeles (July 22, 1980).

Chapter 7

231 Interview with W. French Anderson, National Heart, Lung and Blood Institute.
233 Interview with Donald S. Fredrickson, former director of the National Institutes of Health.
233 Memo from W. French Anderson to Donald S. Fredrickson (August 25, 1980).
234 Charles R. McCarthy, director, Office for Protection from Research Risks, NIH, letter to Dr. Charles E. Young, Chancellor, UCLA (Sept. 8, 1980).
237 Henry K. Beecher, "Ethics and Clinical Research," *The New England Journal of Medicine* 274 (1966): 1354–60.
237 Saul Krugman, "The Willowbrook Hepatitis Studies Revisited: Ethical Aspects," *Reviews of Infectious Diseases* 8 (1986): 157–62.
237 The Tuskegee story, Hastings Center Report (December 21, 1978).
238 Interview with Charles R. McCarthy, former director of the NIH Office for Protection from Research Risks.
239–41 Interview with John C. Fletcher, University of Virginia Medical Center, Charlottesville, Va.
239–41 Interview with Martin C. Cline, University of California at Los Angeles, Los Angeles, Calif.
242 Martin J. Cline letter to David H. Soloman, UCLA chairman of medicine, to explain the foreign experiments (September 8, 1980).
243 Paul Jacobs, "Human Engineering: Pioneer Genetic Implants Revealed," *Los Angeles Times* (October 8, 1980).
243 Paul Jacobs, "Use of Gene for 2 Patients Stirs Protests," *Los Angeles Times* (October 12, 1980).
243 Philip J. Hilts, "Scientists Criticize Gene Engineering in People; Scientists Assail Gene Engineering in People," *The Washington Post* (October 16, 1980).
243 Gina Bari Kolata and Nicholas Wade, "Human Gene Treatment Stirs New Debate: UCLA Researchers Have Conducted Abroad a Gene Experiment They Are Not Yet Allowed to Do Here," *Science* 210 (October 24, 1980): 407.
243 Interview with Paul Jacobs, *Los Angeles Times*.
243 Interview with Winston Salser, University of California at Los Angeles, Los Angeles, Calif.

245 Martin J. Cline letter to Albert A. Barber, associate vice chancellor, research, UCLA (October 15, 1980).

247 Interview with Martin J. Cline, University of California at Los Angeles, Los Angeles, Calif.

247 Evie Cline letter to David Soloman, UCLA chairman of medicine (October 18, 1980).

247 David H. Soloman, UCLA chairman of medicine, letter asking Martin Cline to resign as chief of hematology-oncology (October 20, 1980).

248 Martin J. Cline letter to David H. Soloman (October 23, 1980).

248 Paul Jacobs, "UCLA Genetic Researcher Steps Down During Probe," *Los Angeles Times* (October 23, 1980).

248 Nicholas Wade, "Gene Therapy Caught in More Entanglements: With Five Review Committees Overseeing Him, a UCLA Researcher Did an Experiment His Own Way," *Science* 212 (1981): 24–25.

248–50 Donald S. Fredrickson diary, vol. VII (October 25, 1980).

250 Interview with Charles McCarthy, ibid.

251 W. French Anderson and John C. Fletcher, "Gene Therapy in Human Beings: When Is It Ethical to Begin?" *The New England Journal of Medicine* 303 (1980): 1293–97.

252 Karen E. Mercola and Martin J. Cline, "The Potentials of Inserting New Genetic Information," *The New England Journal of Medicine* 303 (1980): 1297–1300.

253 Charles R. McCarthy letter to Martin Cline (May 26, 1981).

253 Martin J. Cline letter to NIH Ad Hoc committee investigating him (May 15, 1981).

254 Donald S. Fredrickson, director, National Institutes of Health, statement accompanying the release of the NIH investigation into Martin Cline (May 26, 1981).

254 National Institutes of Health Ad Hoc Committee on the UCLA Report Concerning Certain Research Activities of Dr. Martin J. Cline (May 21, 1981).

254 Philip J. Hilts, "NIH to Punish Research for Genetic Work," *The Washington Post* (May 29, 1981).

254 Harold M. Schmeck, Jr., "U.S. Agency Disciplines Gene-Splicing Researcher," *Los Angeles Times* (May 29, 1981).

254 "The Crime of Scientific Zeal," *The New York Times* (June 5, 1981), editorial.

254 Martin J. Cline letter of permanent resignation to Sherman M. Mellinkoff, dean of the school of medicine, UCLA (February 26, 1981).

256 Public statement released on October 10, 1980, by 13 Italian researchers condemning the beta-thalassemia experiment in Naples, Italy.

256 Interview with Fulvio Mavilio, H.S. Raffaele Istituto di Ricovero e Cura a Carattere Scientifico, Milano, Italy.

258 Martin J. Cline, Karen Mercola, Carol LeFevre, et al., "Insertion of a

Foreign Gene into Hematopoietic Cells of Two Patients with Beta-Thalassemia." Draft paper written in mid-1983 but never accepted for publication.

262 Interview with Bernard Talbot, former special assistant to NIH director Donald S. Fredrickson.

262 Message from Three General Secretaries (of religious organizations representing Catholics, Jews, and Protestants) (June 29, 1980), reprinted in *Law, Science and Medicine,* University Casebook Series, by Judith Areen, Patricia A. King, Steven Goldberg, et al. (Westbury, N.Y.: Foundation Press, Inc., 1984).

263 Interview with Alexander C. Capron, University of Southern California, Los Angeles, Calif.

264 President's Commission for the Study of Ethical Problems in Medicine and Biomedical and Behavioral Research, *Splicing Life: The Social and Ethical Issues of Genetic Engineering with Human Beings* (Washington, D.C.: Government Printing Office, Stock no. 83-600500, 1982).

265 Donald Fredrickson in Raymond A. Zilinskas and Burke K. Zimmerman, eds., *The Gene-Splicing Wars: Reflections on the Recombinant DNA Controversy* (New York: Macmillan, 1986).

265 Hearings by House of Representatives, Committee on Science and Technology, Subcommittee on Investigations and Oversight (November 16–18, 1982). 97th Congress V.36, Record #170.

267 Theodore Friedmann, *Gene Therapy: Fact and Fiction in Biology's New Approaches to Disease* (Cold Spring Harbor, N.Y.: Cold Spring Harbor Laboratory Press, 1983).

Chapter 8

268 Interview with W. French Anderson.

268 Edward M. Scolnick, "Virogenes to Oncogenes," *NIH: An Account of Research in Its Laboratories and Clinics,* DeWitt Stetten, Jr., ed., and W. T. Carrigan, ass. ed. (San Diego: Academic Press, 1984).

274 *Gene Doctors.*

279 Richard Mann, Richard C. Mulligan, and David Baltimore, "Construction of a Retrovirus Packaging Mutant and Its Use to Produce Helper-Free Defective Retrovirus," *Cell* 33 (1983): 153–59.

279 Interview with Richard Mulligan, Massachusetts Institute of Technology, Cambridge, Mass.

281 Interview with Eli Gilboa, Memorial Sloan-Kettering Cancer Center, New York, N.Y.

282 Interview with Donna Armentano, Genzyme Corp., Boston, Mass.

285 Interview with Theodore Friedmann, University of California at San

Diego, La Jolla, Calif.

287 D. J. Jolly, A. C. Esty, H. U. Bernard, et al., "Isolation of a Genomic Clone Partially Encoding Human Hypoxanthine Phosphoribosyltransferase," *Proceedings of the National Academy of Sciences* 79 (1982): 5038–41.

288 Interview with Inder Verma, Salk Institute, La Jolla, Calif.

291 A. Dusty Miller, Douglas J. Jolly, Theodore Friedmann, et al., "A Transmissible Retrovirus Expressing Human Hypoxanthine Phosphoribosyltransferase (HPRT): Gene Transfer into Cells Obtained from Humans Deficient in HPRT," *Proceedings of the National Academy of Sciences* 80 (1983): 4709–13.

293 C. Willis Randall, et al., "Partial Phenotypic Correction of Human Lesch-Nyhan (Hypoxanthine-Guanine Phosphoribosyltransferase-deficient) Lymphoblasts with a Transmissible Retroviral Vector," *Journal of Biological Chemistry* 259 (1985): 7842–49.

293 University of California at San Diego School of Medicine press release (April 6, 1984).

293 Interview with Robertson Parkman, Children's Hospital, Los Angeles, Calif.

295 Interview with Dinko Valerio.

296 David A. Williams, Stuart H. Orkin, and Richard C. Mulligan, "Retrovirus-mediated Transfer of Human Adenosine Deaminase Gene Sequences into Cells in Culture and into Murine Hematopoietic Cells in Vivo," *Proceedings of the National Academy of Sciences* 83 (1986): 2566–70.

297 Barry Siegel, *Los Angeles Times* (December 13, 1987): 1.

298 Interview with W. French Anderson, ibid.

298 Interview with John Hutton, University of Cincinnati School of Medicine, Cincinnati, Ohio.

298 Interview with Stuart Orkin, Children's Hospital, Boston, Mass.

Chapter 9

300 W. French Anderson, "Prospects for Human Gene Therapy," *Science* 226 (1984): 401–9.

302 *Splicing Life*.

302 Interview with Philip Kantoff, Dana Farber Cancer Center, Harvard University, Boston, Mass.

304 Interview with R. Michael Blaese, National Cancer Institute.

305 Joseph Hixon, *The Patchwork Mouse: Politics and Intrigue in the Campaign to Conquer Cancer* (New York: Anchor Press, 1976).

306 Interview with Donald Kohn, Children's Hospital, Los Angeles, Calif.

307 Phillip W. Kantoff, et al., "Prospects for Gene Therapy for Immunodeficiency Diseases," *Annual Review of Immunology* 6 (1988): 581–94.

307 Phillip W. Kantoff, et al., "Correction of Adenosine Deaminase Deficiency in Human T and B Cells Using Retroviral-mediated Gene Transfer," *Proceedings of the National Academy of Sciences* 83 (1986): 6563–67.

308 D. A. Williams, I. R. Lemishka, D. G. Nathan, et al., "Introduction of New Genetic Material into Pluripotent Haematopoietic Stem Cells of the Mouse," *Nature* 310 (1984): 476–80.

311 Interview with Stuart Orkin.

312 Marilyn Chase, *The Wall Street Journal* (January 26, 1984).

312 Gina Kolata, *Science* 223 (1984): 1376.

312 Harold M. Schneck, Jr., *The New York Times* (April 10, 1984): C1.

312 Larry Thompson, *The Washington Post* (October 1984), in Health pages.

312 Interview with Richard Mulligan.

312 Institute of Medicine notes, from meeting on gene therapy on October 15, 1986.

313 Barry Siegel, "More Labs Than Patients: Desire to be First Colors Gene Studies," *Los Angeles Times* (December 14, 1987).

314 W. French Anderson, R. Michael Blaese, Arthur W. Nienhuis, et al., "Human Gene Therapy Preclinical Data Document," submitted to the Human Gene Therapy Subcommittee of the Recombinant DNA Advisory Committee (April 24, 1987).

316 Barry Siegel, *Los Angeles Times* (December 13, 1987).

316 Phil Leder, from his talk at the Institute of Medicine meeting, NAS, Washington, D.C. (October 1986).

317 Barry Siegel, "More Labs Than Patients: Desire to be First Colors Gene Studies," *Los Angeles Times* (December 14, 1987).

318 Preclinical Data Document.

318 Interview with Arthur Nienhuis, National Heart, Lung and Blood Institute.

320 Interview with Robert Cook-Deegan, Institute of Medicine, Washington, D.C., and from written comments submitted to RAC.

320 Anonymous letter (July 22, 1987) to William Gartland, Jr., on p. 21 in the RAC material for December 1987 meeting.

320 Richard Mulligan, review of Anderson's Preclinical Data Document (July 31, 1987), in the RAC material for the December 1987 meeting.

321 Dusty Miller, review of Anderson's Preclinical Data Document (July 9, 1987), in the RAC material for the December 1987 meeting. Also, interview with Dusty Miller.

321 Michael Hershfield, review of Anderson's Preclinical Data Document (July 27, 1987), and included in material for RAC December 1987 meeting.

322 French Anderson letter (November 20, 1987) to William Gartland, in

response to the critical reviews of the preclinical document, included in RAC Human Gene Therapy Subcommittee meeting supplementary material (December 1987).

323 RAC Human Gene Therapy Subcommittee Minutes of Meeting (December 7, 1987).

325 Barry Siegel, "More Labs Than Patients: Desire to be First Colors Gene Studies," *Los Angeles Times* (December 14, 1987).

326 Interview with R. Michael Blaese.

327 Interview with Kenneth Culver, National Cancer Institute.

330 Steven A. Rosenberg and John M. Barry, *The Transformed Cell: Unlocking the Mysteries of Cancer* (New York: The Putnam Publishing Group, 1992).

332 RAC Minutes of Meeting (June 3, 1988).

333 RAC Human Gene Therapy Subcommittee Minutes of Meeting (July 29, 1988).

336 RAC Human Gene Therapy Subcommittee Minutes of Telephone Conference Call (September 29, 1988).

336 French Anderson ethics lecture (March 6, 1992) at the NIH Masur auditorium.

337 RAC Minutes of Meeting (October 3, 1988).

337 Interview with James B. Wyngaarden, former NIH director.

337 Interview with Arnold Relman, former editor of *The New England Journal of Medicine.*

337 Interview with Daniel Koshland, editor of *Science* magazine.

340 Interview with Storm Whaley, former director of the NIH information office.

340 RAC Human Gene Therapy Subcommittee minutes of meeting (December 9, 1988).

341 Interview with Jeremy Rifkin, director, Foundation for Economic Trends, Washington, D.C.

342 *The Transformed Cell.*

Index

Ackerman, Barbara, 150–51
Acquired immune deficiency
   syndrome, 15, 114
Acyclovir, 28
ADA (adenosine deaminase), 16–17, 23,
   24, 28, 30, 34, 38–39, 295–301,
   303, 306–11, 314–16, 318–20,
   326–29, 342, 347, 348
   deficiency, 16–17, 21, 23–24, 27–40,
   42–50, 52, 296, 300–301, 303,
   305–8, 310, 311, 313–16, 319,
   321–22, 325, 327–29, 334, 342–45,
   347–48
Addolorata, Maria, 227–28, 258, 260,
   261
Adenosine deaminase, *see* ADA
Aeneid (Virgil), 79
African-Americans, 197, 204–5, 237
Agriculture, U.S. Department of, 156
AIDS, 15, 114
Albinism, 60
Alpha globin, 105–7, 162, 284
Alupent, 25
American Association for the
   Advancement of Science, 96
American Cancer Society, 89
American Federation of Clinical
   Research, 104
American Society of Human Genetics,
   69
Amgen, Inc., 192, 195
Amino acids, 64-66, 93, 120, 162, 286
Anderson, Daniel French, 78, 79
Anderson, Ephraim S., 144
Anderson, Kathryn Dorothy, 42, 88–89,
   94, 99, 285, 304, 349
Anderson, LaVere, 77, 78

Anderson, W. French, 13–15, 50–52,
   99–101, 126, 157, 161, 168–70,
   188, 201, 202, 207, 209, 230–35,
   239, 241, 251, 267, 268, 271, 279,
   284, 286, 294, 296, 312, 330–32,
   334, 347
   and ADA deficiency project, 15,
   17–21, 35–44, 46–52, 296–301,
   303–6, 308–11, 313–26, 328–29,
   334, 342–45
   childhood of, 77–79
   education of, 76–90
   and genetic code work, 96–99
   and iron chelation development,
   111–12, 231
   and isomerization research, 83–85
   marriage of, 88–89, 94, 99, 349
   and microinjection, 165–66, 169–72,
   174–78, 187
   and National Institutes of Health,
   14, 37, 42, 81, 90, 95, 97–99,
   101, 110, 112–13, 202, 232, 349
   and nucleic acid work, 86, 94, 96
   and protein initiation work, 101–5, 115
   and recombinant DNA Advisory
   Committee, 302, 318–20, 323–24,
   326, 328, 333–41
   and retroviruses, 271–73, 280–81
   and Roman numeral theory, 79–80
   and thalassemia research, 108–12,
   115, 160, 166, 178–79, 231, 292,
   294
Andromeda Strain, The (Crichton),
   143, 150
Anemia, 105, 218
   Cooley's, *see* Thalassemia
   sickle cell, 110, 197, 204–5, 209

About the
Author

Larry Thompson, a correspondent for Medical News Network, an interactive television news show for physicians, has been a medical journalist since 1977. He cofounded *The Washington Post*'s Health section in 1984, and served as its science editor until 1992. He also started the Science and Medicine section of *The San Jose Mercury-News* in 1982, helped launch the PBS weekly science magazine "Science Journal" in 1987, and has written for many magazines, including *Time, Discover,* and *Science*. Thompson holds a master's degree in molecular biology, was a fellow at the Yale University School of Medicine, and won the 1992 Lewis Thomas Award for Excellence in Writing about the Life Sciences, among other awards.